The series "Advances in Intelligent Systems and Computing" contains publications on theory, applications, and design methods of Intelligent Systems and Intelligent Computing. Virtually all disciplines such as engineering, natural sciences, computer and information science, ICT, economics, business, e-commerce, environment, healthcare, life science are covered. The list of topics spans all the areas of modern intelligent systems and computing such as: computational intelligence, soft computing including neural networks, fuzzy systems, evolutionary computing and the fusion of these paradigms, social intelligence, ambient intelligence, computational neuroscience, artificial life, virtual worlds and society, cognitive science and systems, Perception and Vision, DNA and immune based systems, self-organizing and adaptive systems, e-Learning and teaching, human-centered and human-centric computing, recommender systems, intelligent control, robotics and mechatronics including human-machine teaming, knowledge-based paradigms, learning paradigms, machine ethics, intelligent data analysis, knowledge management, intelligent agents, intelligent decision making and support, intelligent network security, trust management, interactive entertainment, Web intelligence and multimedia.

The publications within "Advances in Intelligent Systems and Computing" are primarily proceedings of important conferences, symposia and congresses. They cover significant recent developments in the field, both of a foundational and applicable character. An important characteristic feature of the series is the short publication time and world-wide distribution. This permits a rapid and broad dissemination of research results.

More information about this series at http://www.springer.com/series/11156

Sébastien Destercke · Thierry Denoeux
María Ángeles Gil · Przemyslaw Grzegorzewski
Olgierd Hryniewicz
Editors

Uncertainty Modelling in Data Science

 Springer

Editors
Sébastien Destercke
CNRS, Heudiasyc
Sorbonne universités, Université
   de technologie de Compiègne
Compiegne, France

Thierry Denoeux
CNRS, Heudiasyc
Sorbonne universités, Université
   de technologie de Compiègne
Compiegne, France

María Ángeles Gil
Department of Statistics and Operational
   Research and Mathematics Didactics
University of Oviedo
Oviedo, Asturias
Spain

Przemyslaw Grzegorzewski
Faculty of Mathematics and Information
   Science
Warsaw University of Technology
Warsaw, Poland

and

Systems Research Institute
Polish Academy of Sciences
Warsaw, Poland

Olgierd Hryniewicz
Department of Stochastic Methods,
   Systems Research Institute
Polish Academy of Sciences
Warsaw, Poland

ISSN 2194-5357   ISSN 2194-5365 (electronic)
Advances in Intelligent Systems and Computing
ISBN 978-3-319-97546-7   ISBN 978-3-319-97547-4 (eBook)
https://doi.org/10.1007/978-3-319-97547-4

Library of Congress Control Number: 2018950094

This Springer imprint is published by the registered company Springer Nature Switzerland AG
The registered company address is: Gewerbestrasse 11, 6330 Cham, Switzerland

Preface

This volume contains the peer-reviewed papers presented at the **9th International Conference on Soft Methods in Probability and Statistics (SMPS 2018)**, which was held in conjunction with the 5th International Conference on Belief Functions (BELIEF 2018) on 17–21 September 2018 in Compiègne, France. The series of biannual International Conference on Soft Methods in Probability and Statistics started in Warsaw in 2002. It then successfully took place in Oviedo (2004), Bristol (2006), Toulouse (2008), Oviedo/Mieres (2010), Konstanz (2012), Warsaw (2014) and Rome (2016). SMPS and BELIEF 2018 were organized by the Heudiasyc laboratory at the Université de Technologie de Compiègne.

Over the last decades, the interest for extensions and alternatives to probability and statistics has significantly grown in areas as diverse as reliability, decision-making, data mining and machine learning, optimization, etc. This interest comes from the need to enrich existing models, in order to include different facets of uncertainty such as ignorance, vagueness, randomness, conflict or imprecision. Frameworks such as rough sets, fuzzy sets, fuzzy random variables, random sets, belief functions, possibility theory, imprecise probabilities, lower previsions, desirable gambles all share this goal, but have emerged from different needs. By putting together the BELIEF and SMPS conferences, we hope to increase the interactions and discussions between the two communities and to converge towards a more unified view of uncertainty theories.

We also think that the advances, results and tools presented in this volume are important in the ubiquitous and fast-growing fields of data science, machine learning and artificial intelligence. Indeed, an important aspect of some of the learned predictive models is the trust one places in them. Modelling carefully and with principled methods, the uncertainty associated to the data and the models is one of the means to increase this trust, as the model will then be able to distinguish reliable predictions from less reliable ones. In addition, extensions such as fuzzy sets can be explicitly designed to provide interpretable predictive models, facilitating user interaction and increasing their trust.

The joint event collected 76 submissions, each reviewed by at least two reviewers. Twenty-nine of these are included in the present volume, which contains contributions of foundational, methodological and applied nature, on topics as varied as imprecise data handling, linguistic summaries, model coherence, imprecise Markov chains and robust optimization. The resulting proceedings was easily produced through the use of EasyChair.

We would like to thank all the persons that made this volume and this conference possible which include all contributing authors, organizers, programme committee members that help to build such an attractive programme. We are especially grateful to our three invited speakers, Thomas Augustin (*Ludwig-Maximilians*-Universität *München*) for his talk "Belief functions and valid statistical inference", Scott Ferson (*University of Liverpool*) for his talk "Non-Laplacian uncertainty: practical consequences of an ugly paradigm shift about how we handle not knowing" and Ryan Martin (*North Carolina State University*) for his talk "Belief functions and valid statistical inference". We would like to thank all our generous sponsors: Elsevier and the International Journal of Approximate Reasoning, the Laboratory of excellence MS2T, the Heudiasyc laboratory, the International Society of Information Fusion (ISIF), the Compiègne University of Technology, the city of Compiègne. Furthermore, we would like to thank the editor of the Springer series of Advances in Soft Computing, Prof. Janusz Kacprzyk, and Springer-Verlag for their dedication to the production of this volume.

June 2018

Sébastien Destercke
Thierry Denoeux
María Ángeles Gil
Przemyslaw Grzegorzewski
Olgierd Hryniewicz

Organization

Programme Committee

Alessandro Antonucci	IDSIA
Thomas Augustin	Department of Statistics, Univ. of Munich (LMU)
Giulianella Coletti	University of Perugia
Olivier Colot	Université Lille 1
Ana Colubi	University of Oviedo
Frank Coolen	Department of Mathematical Sciences, Durham University
Inés Couso	University of Oviedo
Fabio Cuzzolin	Oxford Brookes University
Fabio D'Andreagiovanni	Université de Technologie de Compiègne, UMR CNRS Heudiasyc
Pierpaolo D'Urso	Sapienza University of Rome
Bernard De Baets	Ghent University
Thierry Denoeux	Université de Technologie de Compiègne, UMR CNRS Heudiasyc
Sébastien Destercke	Université de Technologie de Compiègne, UMR CNRS Heudiasyc
Jean Dezert	Onera
Didier Dubois	Université de Paul Sabatier, Toulouse, UMR IRIT
Fabrizio Durante	Università del Salento, Lecce
Zied Elouedi	Institut Supérieur de Gestion de Tunis
Ramasso Emmanuel	Ecole Nationale Supérieure de Mécanique et des Microtechniques, FEMTO-ST
Maria Brigida Ferraro	Department of Statistical Sciences, Sapienza University of Rome
Maria Angeles Gil Alvarez	University of Oviedo

Lluis Godo	Artificial Intelligence Research Institute, IIIA - CSIC
Gil González-Rodríguez	University of Oviedo
Michel Grabisch	Université Paris I
Przemyslaw Grzegorzewski	Systems Research Institute Polish Academy of Sciences
Olgierd Hryniewicz	Polish Academy of Sciences, Systems Research Institute
Radim Jirousek	University of Economics
Anne-Laure Jousselme	NATO Centre for Maritime Research and Experimentation (CMRE)
Frank Klawonn	Ostfalia University of Applied Sciences
Vaclav Kratochvil	UTIA
Rudolf Kruse	University of Magdeburg
Eric Lefevre	LGI2A Université d'Artois
Liping Liu	The University of Akron
María Asunción Lubiano	Universidad de Oviedo
Arnaud Martin	Université de Rennes1/IRISA
Ronald W. J. Meester	Vrije Universiteit Amsterdam
David Mercier	Université d'Artois
Radko Mesiar	Slovak University of Technology Bratislava
Rombaut Michele	Gipsa-lab
Daniel Milan	Institute of Computer Science, The Czech Academy of Sciences
Enrique Miranda	University of Oviedo
Ignacio Montes	Carlos III University of Madrid
Susana Montes	University of Oviedo
Serafin Moral	University of Granada
Frédéric Pichon	Université d'Artois
Benjamin Quost	Université de Technologie de Compiègne, UMR CNRS Heudiasyc
Ana Belén Ramos Guajardo	University of Oviedo
Johan Schubert	Swedish Defence Research Agency
Ferson Scott	University of Liverpool, Institute for Risk and Uncertainty
Prakash P. Shenoy	University of Kansas School of Business
Beatriz Sinova	University of Oviedo
Martin Stepnicka	IRAFM, University of Ostrava
Barbara Vantaggi	Sapienza University of Rome
Jirina Vejnarova	Institute of Information Theory and Automation of the AS CR
Paolo Vicig	University of Trieste
Liu Zhunga	Northwestern Polytechnical University

Contents

Imprecise Statistical Inference for Accelerated Life Testing Data: Imprecision Related to Log-Rank Test

Abdullah A. H. Ahmadini[1,2(✉)] and Frank P. A. Coolen[1]

[1] Department of Mathematics Science, Durham University, Durham, UK
[2] Department of Mathematics, Faculty of Science, Jazan University,
Jazan, Saudi Arabia
{abdullah.ahmadini,frank.coolen}@durham.ac.uk

Abstract. In this paper we consider an imprecise predictive inference method for accelerated life testing. The method is largely nonparametric, with a basic parametric function to link different stress levels. We discuss in detail how we use the log-rank test to provide adequate imprecision for the link function parameter.

1 Introduction

To determine the reliability of a new product in a relatively short period of time, we use lifespan testing to assess a product, system or component. Accelerated life testing (ALT) is a methodology that is common in practice, where items tested under normal use (normal "level"), are not expected to fail for a very long time, far beyond the time available for testing. In ALT, units are exposed to higher stress levels (e.g. lightbulbs on at a higher than normal voltage) to induce failure more rapidly. There are several typical designs for lifespan testing, including constant-, step- and progressive-stress testing. These types of stress loading in accelerated testing are explained in detail by Nelson [5].

In this paper, we assume the Arrhenius model for the analysis of ALT with failure data under a constant level of stress. The Arrhenius model link function is a standard model for failure time data resulting from ALT. This model is used primarily in situations when the failure mechanism is driven by temperature, and has been applied to various maintenance problems in engineering [5]. If the Arrhenius model provides a realistic link between the different stress levels, then the observations transformed from the increased stress levels to the normal stress level should not be distinguishable. According to this model, an observation t^i at the stress level i, subject to stress K_i, can be transformed to an observation at the normal stress level K_0, by the equation

$$t^{i \to 0} = t^i \left(\frac{e^{\frac{\gamma}{K_0}}}{e^{\frac{\gamma}{K_i}}} \right) \tag{1}$$

© Springer Nature Switzerland AG 2019
S. Destercke et al. (Eds.): SMPS 2018, AISC 832, pp. 1–8, 2019.
https://doi.org/10.1007/978-3-319-97547-4_1

where K_i is the accelerated temperature at level i (Kelvin), K_0 is the normal temperature at level 0 (Kelvin), γ is the parameter of the Arrhenius model.

Testing equality of the survival distribution of two or more independent groups often requires a nonparametric statistical test. There are several non-parametric test procedures that can be used to test equality of the survival distributions, a popular one is the log-rank test [4,6]. We use the log-rank test to find the interval of values of the parameter γ of the Arrhenius link function for which we do not reject the null hypothesis of two or more groups of fail-ure data, possibly including right-censored observations, coming from the same underlying distribution. This can be interpreted such that, for such values of γ, the combined data at stress level K_0 are well mixed. In this paper, to illustrate our method, we assume that there are no right-censored data. In this case, the log-rank test is equal to the Wilcoxon test [3]. Note that we also assume to have failure data at the normal stress level, we comment further on this in the final section.

In this paper we propose a new log-rank test based method for predictive inference on a future unit functioning at the normal stress level. We apply the use of the log-rank test to compare the survival distributions of two groups, in combination with the Arrhenius model finding the accepted interval of γ values according to the null hypothesis. The log-rank test is used for the pairwise comparison of stress levels, leading to an interval of values for γ. This interval is used to transform the data from the increased stress levels to the normal stress level. Then, the ultimate aim is inference at the normal stress level. We consider nonparametric predictive inference at the normal stress level combined with the Arrhenius model linking observations at different stress levels.

Nonparametric Predictive Inference (NPI) is a statistical method which allows inferences about future observations to be made based on past data [2] using imprecise probability [1]. Given ordered observations $x_1 < x_2 < ... < x_n$, and defining $x_0 = 0$ and $x_{n+1} = \infty$. The NPI lower and upper survival functions for a future observation X_{n+1} are

$$\underline{S}_{X_{n+1}}(t) = \frac{n-j}{n+1}, \text{ for } t \in (x_j, x_{j+1}), \ j = 0, ..., n. \tag{2}$$

$$\overline{S}_{X_{n+1}}(t) = \frac{n+1-j}{n+1}, \text{ for } t \in (x_j, x_{j+1}), \ j = 0, ..., n. \tag{3}$$

The difference between the upper and lower survival functions, called impreci-sion, is non-zero because of the limited inferential assumptions made, and reflects the amount of information in the data.

This paper is organized as follows. Section 2 introduces the main idea of imprecise predictive inference based on ALT and log-rank test. The main novelty of our approach is that the imprecision results from a classical nonparametric test, which is the log-rank test, integrated with the Arrhenius function to link different stress levels. In Sect. 3 we explain why we do not use a single log-rank test on all stress levels. In Sect. 4 our method is illustrated in two examples. Section 5 presents some concluding remarks.

2   Predictive Inference Based on ALT Data and Pairwise Log-Rank Test

In this section we present new predictive inference based on ALT data and the log-rank test. The proposed new method consists of two steps. First, the pairwise log-rank test is used between the stress level K_i and K_0, to get the intervals $[\underline{\gamma}_i, \overline{\gamma}_i]$ of values γ_i for which we do not reject the null hypothesis that the data transformed from level i to level 0, and the original data from level 0, come from the same underlying distribution, where $i = 1, ..., m$. With these m pairs $(\underline{\gamma}_i, \overline{\gamma}_i)$, we define $\underline{\gamma} = \min \underline{\gamma}_i$ and $\overline{\gamma} = \max \overline{\gamma}_i$.

Second, we apply the data transformation using $\underline{\gamma}$ ($\overline{\gamma}$) for all levels to get transformed data at level 0 which leads to NPI lower (upper) survival function $\underline{S}$ ($\overline{S}$). Note that each observation at an increased stress level is transformed to an interval at level 0, where the interval tends to be larger if a data point was originally from a higher stress level. If the model fits really well, we expect most $\underline{\gamma}_i$ to be quite similar, and also most $\overline{\gamma}_i$. The NPI lower survival function is attained when all data observations at increased stress levels are transformed to the normal stress level using $\underline{\gamma}$, and the NPI upper survival function results from the use of $\overline{\gamma}$. If the model fits poorly, $\underline{\gamma}_i$ are likely to differ a lot, or $\overline{\gamma}_i$ differ a lot, or both. Hence, in case of poor model fit, the resulting interval $[\underline{\gamma}, \overline{\gamma}]$ tends to be wider than in the case of good model fit. A main novelty of our method is that imprecision results from pairwise comparisons via a classical test, we comment further on this in the next section.

3   Why Not to Use a Single Log-Rank Test on All Levels

In our novel method discussed in Sect. 2, we use pairwise log-rank tests between stress level K_i and K_0. An alternative would be to use one log-rank test for the data at all stress levels combined. We now explain why this would not lead to a sensible method of imprecise statistical inference. Suppose we would test the null hypothesis that data from all stress levels, transformed using parameter value γ_a, originate from the same underlying distribution. Let $[\underline{\gamma}_a, \overline{\gamma}_a]$ be the interval of such values γ_a for which this hypothesis is not rejected. If the model fits very well, we would expect $\underline{\gamma}_a$ to be close to the $\underline{\gamma}$ from Sect. 2 and also $\overline{\gamma}_a$ to be close to $\overline{\gamma}$. If however, the model fits poorly, the $[\underline{\gamma}_a, \overline{\gamma}_a]$ interval may be very small or even empty. Therefore, this leads to less imprecision if the model fits poorly, and that is the reason why we do the pairwise levels and take the minimum and the maximum of $\underline{\gamma}_i$ and $\overline{\gamma}_i$, respectively. Then, we are interested in prediction of one future observation at the normal stress level K_0. So, using the observations transformed from the increased stress levels $K_1, ..., K_m$ as well as the original data obtained at the normal stress level K_0, we apply NPI to derive lower and upper survival functions for as described in Sect. 1. The examples in Sect. 4 illustrate the proposed method of Sect. 2 as well as the problem if we would use the combined approach for all levels.

Table 1. Failure times at three temperature levels

Case	$K_0 = 283$	$K_1 = 313$	$K_2 = 353$	$K_1 = 313$ (*1.4)	$K_2 = 353$ (*0.8)
1	2692.596	241.853	74.557	338.595	59.645
2	3208.336	759.562	94.983	1063.387	75.987
3	3324.788	769.321	138.003	1077.050	110.402
4	5218.419	832.807	180.090	1165.930	144.072
5	5417.057	867.770	180.670	1214.878	144.560
6	5759.910	1066.956	187.721	1493.739	150.176
7	6973.130	1185.382	200.828	1659.535	160.662
8	7690.554	1189.763	211.913	1665.668	169.531
9	8189.063	1401.084	233.529	1961.517	186.823
10	9847.477	1445.231	298.036	2023.323	238.429

Table 2. Accepted γ for log-rank test (Ex1)

Significance level	0.99		0.95		0.90	
Stress level	Lower γ	Upper γ	Lower γ	Upper γ	Lower γ	Upper γ
$K_1 \ K_0$	3901.267	6563.545	4254.053	6251.168	4486.491	6017.435
$K_2 \ K_0$	4161.086	5555.130	4499.174	5442.667	4638.931	5353.034
$K_2 \ K_1 \ K_0$	4156.263	5652.662	4464.828	5478.451	4499.174	5419.662

4   Examples

In this section we present two examples. In example 1 we simulated data at all levels that correspond to the model for the link function we assume for the analysis. In example 2 we change these data such that the assumed link function will not provide a good fit anymore. Together, these examples illustrate our novel imprecise method, from Sect. 2, as well as the problem that could occur if we used the log-rank test on all stress levels combined, as discussed in Sect. 3.

4.1   Example 1

The method proposed in Sect. 2 is illustrated in an example, which presents the temperature-accelerated lifespan test. Data are simulated at three temperatures. The normal temperature condition was $K_0 = 283$ and the increased temperatures stress levels were $K_1 = 313$ and $K_2 = 353$ Kelvin. Ten observations were simulated from a fully specified model, using the Arrhenius link function in combination with a Weibull distribution at each temperature. The Arrhenius parameter γ was set at 5200, and the Weibull distribution at K_0 had shape parameter 3 and scale parameter 7000. This model keeps the same shape parameter at each temperature, but the scale parameter are linked by the Arrhenius relation, which led to scale parameter 1202.942 at K_1 and 183.0914 at K_2.

Ten units were tested at each temperature, for a total of 30 units used in the study. The failure times, in hours, are given in Table 1.

To illustrate the log-rank test method using these data, we assume the Arrhenius link function for the data. Note that our method does not assume a parametric distribution at each stress level. The pairwise log-rank test is used between K_1 and K_0 and between K_2 and K_0 to derive the intervals $[\underline{\gamma}_i, \overline{\gamma}_i]$ of values γ_i for which we do not reject the null hypothesis with regard to the well-mixed data transformation. The resulting intervals $[\underline{\gamma}_i, \overline{\gamma}_i]$ are giving in the first two rows of Table 2, for three test significance levels. Of course, for larger significance level the intervals become wider.

According to the accepted intervals in Table 2, we can obtain the NPI lower and upper survival functions by taking from the pairwise stress level K_1 to K_0 or K_2 to K_0 always the minimum of the $\underline{\gamma}_i$ and the maximum of the $\overline{\gamma}_i$ with levels of significance 0.99, 0.95 and 0.90 values. So, we take $\underline{\gamma} = \min \underline{\gamma}_i = 3901.267$ and the $\overline{\gamma} = \max \overline{\gamma}_i = 6563.545$ of the pairwise K_1, K_0 with 0.99 significance level, $\underline{\gamma} = \min \underline{\gamma}_i = 4254.053$ and the $\overline{\gamma} = \max \overline{\gamma}_i = 6251.168$ of the pairwise K_1, K_0 with 0.95 significance level, and $\underline{\gamma} = \min \underline{\gamma}_i = 4486.491$ and the $\overline{\gamma} = \max \overline{\gamma}_i = 6017.435$ of the pairwise K_1, K_0 with 0.90 significance level then transformed the data to the normal stress level, see Fig. 1(a). In this figure, the lower survival function $\underline{S}$ is labeled as $\underline{S}(\underline{\gamma}_i)$ and the upper survival function $\overline{S}$ is labeled as $\overline{S}(\overline{\gamma}_i)$. This figure shows that higher significance levels leads to more imprecision for the NPI lower and upper survival functions.

To illustrate the effect of using the single log-rank test for all stress levels simultaneously as discussed in Sect. 3, the final row in Table 2 provides the interval $[\underline{\gamma}_a, \overline{\gamma}_a]$ of values γ_a for all the stress levels together. From this interval we can again obtain the lower and upper survival functions using NPI, these are presented in Fig. 1(b). In this example, the data were simulated precisely with the link function as assumed in our method, so there is not much difference between the lower and upper survival functions for corresponding significance levels in Figs. 1(a) and (b). Example 2 will illustrate what happens if the model does not fit well.

4.2   Example 2

To illustrate our method in case the model does not fit the data well, and also to show what would have happened if we had used the joint log-rank test in our method instead of the pairwise tests, we use the same data as in Example 1, but we change some of these. In Scenario 1 (indicated as Ex 2-1 in Fig. 1), we multiple the data at level K_1 by 1.4. In Scenario 2 (Ex 2-2), we do the same and in addition we multiply the data at level K_2 by 0.8. The resulting data values are given in the last two columns in Table 1.

For these two scenarios, we have repeated the analysis as described in Example 1. The resulting intervals of γ values are given in Tables 3 and 4. Note that for significance level 0.90 in Scenario 2 the null hypothesis of the joint log-rank test would be rejected for all values γ_a, hence we report an empty interval, so clearly our method would not work if we had used this joint test instead of the pairwise tests.

The NPI lower and upper survival functions in Figs. 1(c) and (e), using our method as discussed in Sect. 2, have more imprecision. Note that the lower survival function is identical in both scenarios as the same γ is used, this is because the increased values at K_1 have resulted in smaller values for $\underline{\gamma}_1$ and $\overline{\gamma}_1$ and the γ in our method is equal to the $\underline{\gamma}_1$ in these cases. In Scenario 2, the observations at level K_2 have decreased, leading to larger $\underline{\gamma}_2$ and $\overline{\gamma}_2$ values, and this leads to the upper survival functions increasing in comparison to Scenario 1.

If we would have used the joint long-rank test instead of the pairwise tests, as discussed in Sect. 3, then imprecision would have decreased in these two scenarios, as can be seen from Figs. 1(d) and (f). Note that in Fig. 1(f) there are no lower and upper survival functions corresponding to the use of the joint log-rank test for significance level 0.90, as this leads to an empty interval of γ_a values. As mentioned in Sect. 3, if the model does not fit well, then we are going to sooner reject the null hypothesis for all the three levels together, see Tables 3 and 4. So we have a smaller range of values for which we do not reject the null hypothesis. But if the model fits poorly, we actually want a larger range of values, so increased imprecision. It is obvious that this is achieved by taking the minimum of the $\underline{\gamma}_i$ and the maximum of the $\overline{\gamma}_i$ of the pairwise tests, hence this is our proposed method in Sect. 2. This is illustrated by Figs. 1(a), (c) and (e).

Table 3. Accepted γ for log-rank test (Ex2-1)

Significance level	0.99		0.95		0.90	
Stress level	Lower γ	Upper γ	Lower γ	Upper γ	Lower γ	Upper γ
$K_1 * 1.4, K_0$	2907.787	5570.065	3260.574	5257.689	3493.011	5023.956
$K_2\ K_0$	4161.086	5597.978	4499.174	5442.667	4638.931	5353.034
$K_2\ K_1\ K_0$	4455.573	5568.468	4638.930	5368.780	4742.958	5257.689

Table 4. Accepted γ for log-rank test (Ex2-2)

Significance level	0.99		0.95		0.90	
Stress level	Lower γ	Upper γ	Lower γ	Upper γ	Lower γ	Upper γ
$K_1 * 1.4, K_0$	2907.787	5570.065	3260.574	5257.689	3493.011	5023.956
$K_2 * 0.8, K_0$	4479.541	5916.433	4817.629	5761.121	4957.386	5671.488
$K_2\ K_1\ K_0$	5031.547	5676.311	5220.356	5531.731	Empty set	

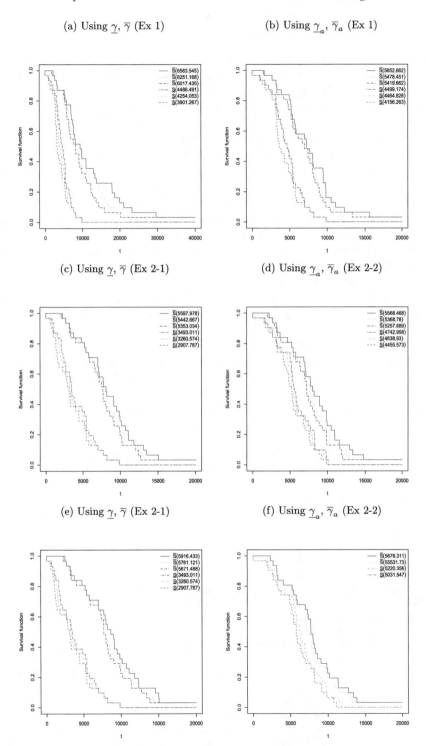

Fig. 1. The NPI lower and upper survival functions.

5   Concluding Remarks

This paper has presented an exploration of the use of a novel statistical method providing imprecise semi-parametric inference for ALT data, where the imprecision is related to the log-rank test statistics. The proposed method applies the use of the log-rank test to compare the survival distribution of pairwise stress levels, in combination with the Arrhenius model finding the accepted interval of γ values according to the null hypothesis. We explored imprecision through the use of nonparametric test for the parameter of the link function between different stress levels, which enabled us to transform the observations at increased stress levels to interval-valued observations at the normal stress level and achieve further robustness. We consider nonparametric predictive inference at the normal stress level combined with the Arrhenius model linking observations at different stress levels. We showed why, in our method, we use the imprecision from combined pairwise log-rank tests, and not from a single log-rank test on all stress levels together. The latter would lead to less imprecision if the model fits poorly, while our proposed method then leads to more imprecision. In this paper, to illustrate basic idea of our novel method, we assumed that failure data are available at all stress levels including the normal stress level. This may not be realistic. If there are no failure data at the normal stress level, or only right-censored observations, then we can apply our method using a higher stress level as the basis for the combinations, so transform data to that stress level. Then the combined data at that level could be transformed all together to the normal stress level. The log-rank test in this approach could be replaced by other comparison tests, where even the use of tests based on imprecise probability theory [7] could be explored. This is left as an interesting topic for future research.

Acknowledgements. Abdullah A.H. Ahmadini gratefully acknowledges the financial support received from Jazan University in Saudi Arabia and the Saudi Arabian Cultural Bureau (SACB) in London for pursuing his Ph.D. at Durham University. The authors thank two reviewers of this paper for supportive comments and suggestions.

References

1. Augustin, T., Coolen, F., de Cooman, G., Troffaes, M.: Introduction to Imprecise Probabilities. Wiley, Chichester (2014)
2. Coolen, F.: Nonparametric predictive inference. In: International Encyclopedia of Statistical Science, pp. 968–970. Springer, Berlin (2011)
3. Gehan, E.: A generalized Wilcoxon test for comparing arbitrarily singly-censored samples. Biometrika **52**, 203–224 (1965)
4. Mantel, N.: Evaluation of survival data and two new rank order statistics arising in its consideration. Cancer Chemother. Rep. **50**, 163–170 (1966)
5. Nelson, W.: Accelerated Testing: Statistical Models, Test Plans, and Data Analysis. Wiley, New Jersey (1990)
6. Peto, R., Peto, J.: Asymptotically efficient rank invariant test procedures. J. R. Stat. Soc. Ser. A **135**, 185–207 (1972)
7. Benavoli, A., Mangili, F., Corani, G., Zaffalon, M., Ruggeri, F.: A Bayesian Wilcoxon signed-rank test based on the Dirichlet process. In: Proceedings of the 30th International Conference on Machine Learning (ICML 2014), pp. 1–9 (2014)

Descriptive Comparison of the Rating Scales Through Different Scale Estimates: Simulation-Based Analysis

Irene Arellano[1], Beatriz Sinova[1], Sara de la Rosa de Sáa[2], María Asunción Lubiano[1], and María Ángeles Gil[1(✉)]

[1] Departamento de Estadística, I.O. y D.M., Universidad de Oviedo, C/ Federico García Lorca 18, 33007 Oviedo, Spain
{uo239511,sinovabeatriz,lubiano,magil}@uniovi.es
[2] Oficina de Evaluación de Tecnologías Sanitarias, Servicio de Salud del Principado de Asturias, Asturias, Spain
sara.delarosa@sespa.es

Abstract. In dealing with intrinsically imprecise-valued magnitudes, a common rating scale type is the natural language-based Likert. Along the last decades, fuzzy scales (more concretely, fuzzy linguistic scales/variables and fuzzy ratig scales) have also been considered for rating values of these magnitudes. A comparative descriptive analysis focussed on the variability/dispersion associated with the magnitude depending on the considered rating scale is performed in this study. Fuzzy rating responses are simulated and associated with Likert responses by means of a 'Likertization' criterion. Then, each 'Likertized' datum is encoded by means of a fuzzy linguistic scale. In this way, with the responses available in the three scales, the value of the different dispersion estimators is calculated and compared among the scales.

Keywords: Fuzzy linguistic scale · Fuzzy rating scale · Likert scale
Scale estimates

1 Introduction

The Likert-type scales are frequently used in designing questionnaires to rate characteristics or attributes that cannot be numerically measured (like satisfaction, perceived quality, perception...). Although they are easy to answer and they do not require a special training to use them, respondents often do not find accurate answers to items and the available statistical methodology to analyze the data from these questionnaires is rather limited. This is mainly due to the fact that Likert scales are discrete with a very small number of responses to choose for each item (often 4 to 7). To overcome this concern, Hesketh *et al.* [5] proposed the so-called fuzzy rating scale to allow a complete freedom and expressiveness in responding, without respondents being constrained to choose among a few pre-specified responses.

© Springer Nature Switzerland AG 2019
S. Destercke et al. (Eds.): SMPS 2018, AISC 832, pp. 9–16, 2019.
https://doi.org/10.1007/978-3-319-97547-4_2

By drawing the fuzzy number that best represents the respondent's valuation, the fuzzy rating scale captures the logical imprecision associated with such variables. Moreover, this fuzzy rating scale allows us to have a rich continuous scale of measurement, unlike the case of a posterior numerical or fuzzy encoding (the latter encoding Likert points with fuzzy numbers from a linguistic scale, and usually made by trained experts).

In previous studies (see Gil *et al.* [3], Lubiano *et al.* [6–8]) we have confirmed that the results when fuzzy rating scales are considered sometimes differ importantly from the conclusions drawn from numerically or fuzzy linguistically encoded Likert values.

As differences can often be even clearer from the dispersion than for the location perspective, this paper aims to examine, by means of simulation developments, how location-based 'scale' estimates are affected by the considered scale of measurement.

2 Preliminaries

A (bounded) **fuzzy number** is a mapping $\widetilde{U} : \mathbb{R} \rightarrow [0,1]$ such that for all $\alpha \in [0,1]$, the α-level set $\widetilde{U}_\alpha = \{x \in \mathbb{R} : \widetilde{U}(x) \geq \alpha\}$ if $\alpha \in (0,1]$, and $\widetilde{U}_0 = \mathrm{cl}\{x \in \mathbb{R} : \widetilde{U}(x) > 0\}$ (with 'cl' denoting the closure of the set) is a nonempty compact interval.

In dealing with fuzzy number-valued data, distances will be computed by considering two different metrics introduced by Diamond and Kloeden [1]: the 2-norm metric ρ_2 and the 1-norm metric ρ_1, which for fuzzy numbers $\widetilde{U}$ and $\widetilde{V}$ are given by

$$\rho_2(\widetilde{U}, \widetilde{V}) = \sqrt{\frac{1}{2} \int_{[0,1]} \left[(\inf \widetilde{U}_\alpha - \inf \widetilde{V}_\alpha)^2 + (\sup \widetilde{U}_\alpha - \sup \widetilde{V}_\alpha)^2 \right] d\alpha},$$

$$\rho_1(\widetilde{U}, \widetilde{V}) = \frac{1}{2} \int_{[0,1]} \left[|\inf \widetilde{U}_\alpha - \inf \widetilde{V}_\alpha| + |\sup \widetilde{U}_\alpha - \sup \widetilde{V}_\alpha| \right] d\alpha.$$

3 Scales Measures for Fuzzy Data

In developing statistics with fuzzy data coming from intrinsically imprecise-valued attributes, random fuzzy numbers constitute a well-formalized model within the probabilistic setting for the random mechanisms generating such data.

Let $\mathcal{X}$ be a random fuzzy number (as defined by Puri and Ralescu [9]) associated with a probability space, i.e., a fuzzy number-valued mapping $\mathcal{X}$ associated with a probability space and such that, for each α, the α-level interval-valued mapping is a random interval associated with the probability space.

Let $\widetilde{\mathbf{x}}_n = (\widetilde{x}_1, \ldots, \widetilde{x}_n)$ be a sample of observations from $\mathcal{X}$. The **sample Aumann-type mean** is the fuzzy number such that for each α

$$(\overline{\widetilde{\mathbf{x}}}_n)_\alpha = \left[\sum_{i=1}^{n} \inf(\widetilde{x}_i)_\alpha / n, \sum_{i=1}^{n} \sup(\widetilde{x}_i)_\alpha / n \right],$$

and the **sample 1-norm median** is the fuzzy number such that for each α

$$(\widehat{\text{Me}}(\widetilde{\mathbf{x}}_n))_\alpha = [\text{Me}_i \inf(\widetilde{x}_i)_\alpha, \text{Me}_i \sup(\widetilde{x}_i)_\alpha].$$

In De la Rosa de Sáa et al. [2] one can find together the most commonly used location-based scale estimates, namely: the **sample Fréchet-type ρ_2-Standard Deviation** and, for $D \in \{\rho_1, \rho_2\}$ and $\widetilde{M} \in \{\overline{\widetilde{\mathbf{x}}}_n, \widehat{\text{Me}}(\widetilde{\mathbf{x}}_n)\}$, the **sample D-Average Distance Deviation** and the **sample D-Median Distance Deviation**, which are respectively given by

$$\rho_2\text{-SD}(\widetilde{\mathbf{x}}_n) = \sqrt{\frac{1}{n} \sum_{i=1}^n \left[\rho_2(\widetilde{x}_i, \overline{\widetilde{\mathbf{x}}}_n)\right]^2},$$

$$\widehat{D\text{-ADD}}(\widetilde{\mathbf{x}}_n, \widetilde{M}) = \frac{1}{n} \sum_{i=1}^n D(\widetilde{x}_i, \widetilde{M}), \ \ \widehat{D\text{-MDD}}(\widetilde{\mathbf{x}}_n, \widetilde{M}) = \text{Me}_i \left\{D(\widetilde{x}_i, \widetilde{M})\right\}.$$

4   Generating Fuzzy Data for Simulation Studies

In this work, simulations have been inspired by real-life datasets in connection with fuzzy rating scale-based experiments.

To generate fuzzy data from a trapezoidal-valued random fuzzy number $\mathcal{X} = \text{Tra}(\inf \mathcal{X}_0, \inf \mathcal{X}_1, \sup \mathcal{X}_1, \sup \mathcal{X}_0)$, Sinova *et al.* [10] suggest to use an alternative characterization, $\mathcal{X} = \text{Tra}\langle X_1, X_2, X_3, X_4 \rangle$, where (see Fig. 1)

$$X_1 = \text{mid}\mathcal{X}_1 = (\inf \mathcal{X}_1 + \sup \mathcal{X}_1)/2, \quad X_2 = \text{spr}\mathcal{X}_1 = (\sup \mathcal{X}_1 - \inf \mathcal{X}_1)/2,$$

$$X_3 = \text{lspr}\mathcal{X}_0 = \inf \mathcal{X}_1 - \inf \mathcal{X}_0, \quad X_4 = \text{uspr}\mathcal{X}_0 = \sup \mathcal{X}_0 - \sup \mathcal{X}_1,$$

(i.e., X_1 = core mid-point, X_2 = core radius, X_3 = 'left distance' between core and support, X_4 = 'right distance' between core and support) whence

$$\mathcal{X} = \text{Tra}\langle X_1, X_2, X_3, X_4 \rangle = \text{Tra}(X_1 - X_2 - X_3, X_1 - X_2, X_1 + X_2, X_1 + X_2 + X_4).$$

In fact, fuzzy data will be generated by simulating the four real-valued random variables X_1, X_2, X_3 and X_4, so that the $\mathbb{R} \times [0, \infty) \times [0, \infty) \times [0, \infty)$-valued random vector (X_1, X_2, X_3, X_4) will provide us with the 4-tuples (x_1, x_2, x_3, x_4) with x_1 = center and x_2 = radius of the core, and x_3 = lower and x_4 = upper spread of the fuzzy number. To each generated 4-tuple (x_1, x_2, x_3, x_4) we associate the fuzzy number $\text{Tra}\langle x_1, x_2, x_3, x_4 \rangle$.

According to the simulation procedure, data have been generated from random fuzzy numbers with a bounded reference set and abstracting and mimicking what we have observed in real-life examples employing the fuzzy rating scale (FRS). More concretely, fuzzy data have been generated such that

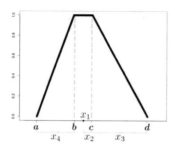

Fig. 1. 4-Tuples to be generated for the simulation procedures

- $100 \cdot \omega_1 \%$ of the data have been obtained by first considering a simulation from a simple random sample of size 4 from a beta $\beta(p, q)$ distribution, ordering the corresponding 4-tuple, and finally computing the values x_i. The values of p and q vary in most cases to cover different distributions (namely, symmetrical weighting central values, symmetrical weighting extreme values, and asymmetric ones). In most of the comparative studies involving simulations, the values from the beta distribution are re-scaled and translated to an interval $[l_0, u_0]$ different from $[0, 1]$.
- $100 \cdot \omega_2 \%$ of the data have been obtained considering a simulation of four random variables $X_i = (u_0 - l_0) \cdot Y_i + l_0$ as follows:
    $Y_1 \sim \beta(p, q)$,
    $Y_2 \sim \text{Uniform}\big[0, \min\{1/10, Y_1, 1 - Y_1\}\big]$,
    $Y_3 \sim \text{Uniform}\big[0, \min\{1/5, Y_1 - Y_2\}\big]$,
    $Y_4 \sim \text{Uniform}\big[0, \min\{1/5, 1 - Y_1 - Y_2\}\big]$.
- $100 \cdot \omega_3 \%$ of the data have been obtained considering a simulation of four random variables $X_i = (u_0 - l_0) \cdot Y_i + l_0$ as follows:

$Y_1 \sim \beta(p, q)$,

$$Y_2 \sim \begin{cases} \text{Exp}(200) & \text{if } Y_1 \in [0.25, 0.75] \\ \text{Exp}(100 + 4\,Y_1) & \text{if } Y_1 < 0.25 \\ \text{Exp}(500 - 4\,Y_1) & \text{otherwise} \end{cases}$$

$$Y_3 \sim \begin{cases} \gamma(4, 100) & \text{if } Y_1 - Y_2 \geq 0.25 \\ \gamma(4, 100 + 4\,Y_1) & \text{otherwise} \end{cases}$$

$$Y_4 \sim \begin{cases} \gamma(4, 100) & \text{if } Y_1 + Y_2 \geq 0.25 \\ \gamma(4, 500 - 4\,Y_1) & \text{otherwise.} \end{cases}$$

5   Results

First, FRS data will be simulated in accordance with the above described realistic simulation procedure. Later, fuzzy data based on a fuzzy rating scale can fairly be associated/classified in accordance with labels in a Likert scale (more concretely, with their numerical encoding). This process is to be called "Likertization". Furthermore, the associated Likert values could also be later encoded by means of values from a fuzzy linguistic scale.

For carrying out the Likertization, the "minimum distance Likertization criterion" will be employed (see Fig. 2):

FRS-based response

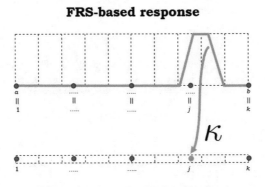

k-point κ-associated response

Fig. 2. Minimum distance criterion scheme when the reference interval equals $[1, k]$

In this way, if the considered Likert scale is a k-point one, given a metric D between fuzzy data and $\widetilde{U}$ the free fuzzy response to be classified, then $\widetilde{U}$ is associated with the integer $\kappa(\widetilde{U})$ such that

$$\kappa(\widetilde{U}) = \arg \min_{j \in \{1,\ldots,k\}} D(\widetilde{U}, \mathbb{1}_{\{j\}}).$$

Each FRS-based datum will be first Likertized by means of the minimum distance criterion, and it will later be encoded by means of a fuzzy linguistic scale. We have chosen the most usual (see, for instance, Herrera *et al.* [4]) balanced semantic representations of the linguistic hierarchies of $k = 4$ and $k = 5$ levels (Fig. 3).

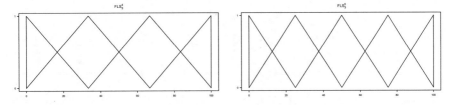

Fig. 3. Usual balanced semantic representation of the linguistic hierarchies of $k = 4$ and $k = 5$ levels

Simulations-based tables (Tables 1 and 2) collect the percentages of Euclidean distances between the sample scale estimates $\widehat{D}$ for the FRS-simulated data and for their numerically (NEL) and fuzzy linguistically (FLS) encoded ρ_1 Likertization that are over $\varepsilon \in \{1, 5, 10, 15\}$.

Table 1. % of simulated samples of size n for which the Euclidean distance between the sample scale estimate $\widehat{\text{D}}$ associated with the FRS and the one associated with either the NEL (numerically encoded Likert) or the FLS (fuzzy linguistic scale) with $k = 4$ different values is greater than $\varepsilon \in \{1, 5, 10, 15\}$ and (from top to bottom) $\beta(p,q) \equiv \beta(1,1), \beta(.75,.75), \beta(4,2),$ and $\beta(6,1)$

$\%\ \left|\widehat{\text{D}}(\text{FRS}) - \widehat{\text{D}}(\text{S})\right| > \varepsilon \qquad (k = 4,\ \beta(p,q) \equiv \beta(1,1))$

$\widehat{\text{D}}$	n	$\varepsilon = 1$ S = NEL	S = FLS	$\varepsilon = 5$ S = NEL	S = FLS	$\varepsilon = 10$ S = NEL	S = FLS	$\varepsilon = 15$ S = NEL	S = FLS
$\widehat{\rho_2\text{-SD}}(\tilde{\mathbf{x}}_n)$	10	82.2	64.3	16.8	4.1	0.2	0	0	0
	30	85.7	51.9	4.8	0.1	0	0	0	0
	100	94.9	44.2	0	0	0	0	0	0
$\widehat{\rho_2\text{-ADD}}(\tilde{\mathbf{x}}_n, \overline{\tilde{\mathbf{x}}}_n)$	10	78.2	70.1	11.3	6.1	0.2	0	0	0
	30	69	52.2	1.6	0.1	0	0	0	0
	100	79.1	25.7	0	0	0	0	0	0
$\widehat{\rho_1\text{-ADD}}(\tilde{\mathbf{x}}_n, \widehat{\text{Me}}(\tilde{\mathbf{x}}_n))$	10	79.2	76.2	14.2	16.7	0.2	0.1	0	0
	30	63.5	72.3	1.9	3.1	0	0	0	0
	100	64.9	58.9	0	0	0	0	0	0
$\widehat{\rho_2\text{-MDD}}(\tilde{\mathbf{x}}_n, \overline{\tilde{\mathbf{x}}}_n)$	10	91	86.3	47.9	41.1	14.3	9.4	2.3	0.7
	30	88.7	89.1	42.8	42.8	8.7	7.7	0.4	0
	100	94.9	95.7	55.3	55.8	4.3	3.7	0	0
$\widehat{\rho_1\text{-MDD}}(\tilde{\mathbf{x}}_n, \widehat{\text{Me}}(\tilde{\mathbf{x}}_n))$	10	93	91.2	66.2	57.5	40.7	25	15.7	8.3
	30	97.1	86.6	83.5	37.8	47.8	9	11.2	3.2
	100	100	74.3	97.5	12.5	44.7	0.7	1.2	0.1

$\%\ \left|\widehat{\text{D}}(\text{FRS}) - \widehat{\text{D}}(\text{S})\right| > \varepsilon \qquad (k = 4,\ \beta(p,q) \equiv \beta(0.75, 0.75))$

$\widehat{\text{D}}$	n	$\varepsilon = 1$ S = NEL	S = FLS	$\varepsilon = 5$ S = NEL	S = FLS	$\varepsilon = 10$ S = NEL	S = FLS	$\varepsilon = 15$ S = NEL	S = FLS
$\widehat{\rho_2\text{-SD}}(\tilde{\mathbf{x}}_n)$	10	83	67.9	19.9	3.8	0	0	0	0
	30	90	59.7	6.2	0	0	0	0	0
	100	98.8	70	0.1	0	0	0	0	0
$\widehat{\rho_2\text{-ADD}}(\tilde{\mathbf{x}}_n, \overline{\tilde{\mathbf{x}}}_n)$	10	76.4	71.4	12.1	5.3	0	0	0	0
	30	70.7	55.4	2.2	0	0	0	0	0
	100	82.1	40	0	0	0	0	0	0
$\widehat{\rho_1\text{-ADD}}(\tilde{\mathbf{x}}_n, \widehat{\text{Me}}(\tilde{\mathbf{x}}_n))$	10	78	78.4	14.1	14	0	0	0	0
	30	65.9	77.8	1.3	2.7	0	0	0	0
	100	65.3	77.6	0	0	0	0	0	0
$\widehat{\rho_2\text{-MDD}}(\tilde{\mathbf{x}}_n, \overline{\tilde{\mathbf{x}}}_n)$	10	87.9	87.8	53.2	41.9	17.4	9.9	1.3	0.2
	30	91.5	92	59.5	56.1	20.4	15.3	0.7	0.3
	100	98.5	99.2	84.8	86.8	31.1	28.6	1.2	0.2
$\widehat{\rho_1\text{-MDD}}(\tilde{\mathbf{x}}_n, \widehat{\text{Me}}(\tilde{\mathbf{x}}_n))$	10	94	91.1	68.5	57.5	39.7	22.6	15.5	6.1
	30	96.5	83.9	74.7	34.6	36.3	9.5	5.6	2.1
	100	98.8	78.1	74.6	18.6	13.5	4.8	0.3	0.3

$\%\ \left|\widehat{\text{D}}(\text{FRS}) - \widehat{\text{D}}(\text{S})\right| > \varepsilon \qquad (k = 4,\ \beta(p,q) \equiv \beta(4, 2))$

$\widehat{\text{D}}$	n	$\varepsilon = 1$ S = NEL	S = FLS	$\varepsilon = 5$ S = NEL	S = FLS	$\varepsilon = 10$ S = NEL	S = FLS	$\varepsilon = 15$ S = NEL	S = FLS
$\widehat{\rho_2\text{-SD}}(\tilde{\mathbf{x}}_n)$	10	81	71.7	18.9	4.9	0.6	0.6	0	0
	30	81	50.6	3.5	0	0	0	0	0
	100	93.1	22.3	0	0	0	0	0	0
$\widehat{\rho_2\text{-ADD}}(\tilde{\mathbf{x}}_n, \overline{\tilde{\mathbf{x}}}_n)$	10	77.6	73.1	10.3	0.3	0.3	0	0	0
	30	70	75.5	5	5.7	0	0	0	0
	100	85.8	95.9	0.6	0.4	0	0	0	0
$\widehat{\rho_1\text{-ADD}}(\tilde{\mathbf{x}}_n, \widehat{\text{Me}}(\tilde{\mathbf{x}}_n))$	10	79.2	84.2	16	20.3	0.1	0	0	0
	30	76.5	95.7	5.6	16.4	0	0	0	0
	100	92.8	100	0.2	5.8	0	0	0	0
$\widehat{\rho_2\text{-MDD}}(\tilde{\mathbf{x}}_n, \overline{\tilde{\mathbf{x}}}_n)$	10	92.5	92.8	62.4	61.9	21.7	13	2.6	0
	30	97.1	97.4	90.7	91.4	51.9	37	3.2	0.2
	100	100	100	100	100	86.3	61.7	1.3	0
$\widehat{\rho_1\text{-MDD}}(\tilde{\mathbf{x}}_n, \widehat{\text{Me}}(\tilde{\mathbf{x}}_n))$	10	96.5	96.1	84.9	80.6	42	36.5	9.2	4
	30	98.1	99.3	95.4	95.4	73.7	72.4	7.7	5.7
	100	100	100	100	100	96.1	96.1	0.7	0.7

$\%\ \left|\widehat{\text{D}}(\text{FRS}) - \widehat{\text{D}}(\text{S})\right| > \varepsilon \qquad (k = 4,\ \beta(p,q) \equiv \beta(6, 1))$

$\widehat{\text{D}}$	n	$\varepsilon = 1$ S = NEL	S = FLS	$\varepsilon = 5$ S = NEL	S = FLS	$\varepsilon = 10$ S = NEL	S = FLS	$\varepsilon = 15$ S = NEL	S = FLS
$\widehat{\rho_2\text{-SD}}(\tilde{\mathbf{x}}_n)$	10	95.4	78	44.2	8.6	0.3	0	0	0
	30	99.5	76.7	44.9	0.4	0	0	0	0
	100	100	90.3	37.3	0	0	0	0	0
$\widehat{\rho_2\text{-ADD}}(\tilde{\mathbf{x}}_n, \overline{\tilde{\mathbf{x}}}_n)$	10	92.8	80.6	46.9	11.1	1.1	0	0	0
	30	99.2	91.1	63.8	2.5	0	0	0	0
	100	100	99.4	78.4	0	0	0	0	0
$\widehat{\rho_1\text{-ADD}}(\tilde{\mathbf{x}}_n, \widehat{\text{Me}}(\tilde{\mathbf{x}}_n))$	10	81.1	68.7	23.4	3.9	1.5	0	0	0
	30	86.7	52.9	20.6	0	0	0	0	0
	100	95.2	25.9	6.4	0	0	0	0	0
$\widehat{\rho_2\text{-MDD}}(\tilde{\mathbf{x}}_n, \overline{\tilde{\mathbf{x}}}_n)$	10	79.4	71.9	25.2	8.4	2.6	0.3	0.2	0
	30	86.6	62.2	24.6	1.5	0.1	0	0	0
	100	98.3	60.9	12.1	0.2	0.1	0	0	0
$\widehat{\rho_1\text{-MDD}}(\tilde{\mathbf{x}}_n, \widehat{\text{Me}}(\tilde{\mathbf{x}}_n))$	10	100	98.7	66.2	57.7	7.1	3.7	0.2	0
	30	100	100	88.6	85.2	2.4	2.3	0.3	0.2
	100	100	100	99.1	99	0.1	0.1	0.1	0.1

Table 2. % of simulated samples of size n for which the Euclidean distance between the sample scale estimate $\widehat{D}$ associated with the FRS and the one associated with either the NEL (numerically encoded Likert) or the FLS (fuzzy linguistic scale) with $k = 5$ different values is greater than $\varepsilon \in \{1, 5, 10, 15\}$ and (from top to bottom) $\beta(p,q) \equiv \beta(1,1), \beta(.75, .75), \beta(4,2),$ and $\beta(6,1)$

% $\left|\widehat{D}(\mathrm{FRS}) - \widehat{D}(\mathrm{S})\right| > \varepsilon$ $(k = 5, \beta(p,q) \equiv \beta(1,1))$

$\widehat{D}$	n	$\varepsilon = 1$ S = NEL	S = FLS	$\varepsilon = 5$ S = NEL	S = FLS	$\varepsilon = 10$ S = NEL	S = FLS	$\varepsilon = 15$ S = NEL	S = FLS
$\widehat{\rho_2\text{-SD}}(\tilde{x}_n)$	10	71.6	58.6	4.4	1.1	0	0	0	0
	30	66.2	40.5	0	0	0	0	0	0
	100	75.2	25.1	0	0	0	0	0	0
$\widehat{\rho_2\text{-ADD}}(\tilde{x}_n, \overline{\tilde{x}}_n)$	10	65.8	64.9	2.4	1.9	0	0	0	0
	30	48.6	60.2	0.2	0.2	0	0	0	0
	100	25.1	72.7	0	0	0	0	0	0
$\widehat{\rho_1\text{-ADD}}(\tilde{x}_n, \widehat{Me}(\tilde{x}_n))$	10	68.3	70.4	3.7	5.3	0	0	0	0
	30	50.2	60.8	0.2	0.3	0	0	0	0
	100	20.7	75.1	0	0	0	0	0	0
$\widehat{\rho_2\text{-MDD}}(\tilde{x}_n, \overline{\tilde{x}}_n)$	10	83.9	83.2	34.3	28.4	3.6	2.3	0.1	0.1
	30	79.3	79.8	24.3	21.2	1.9	0.9	0.1	0.1
	100	71.2	70.6	7.2	6.2	0	0	0	0
$\widehat{\rho_1\text{-MDD}}(\tilde{x}_n, \widehat{Me}(\tilde{x}_n))$	10	89.7	87.3	53	46.1	16.7	13.1	2.4	1.9
	30	85.3	85.4	33.1	29.7	5.5	4.6	0.3	0.2
	100	73.3	73.4	7.3	7.2	0.5	0.4	0	0

% $\left|\widehat{D}(\mathrm{FRS}) - \widehat{D}(\mathrm{S})\right| > \varepsilon$ $(k = 5, \beta(p,q) \equiv \beta(0.75, 0.75))$

$\widehat{D}$	n	$\varepsilon = 1$ S = NEL	S = FLS	$\varepsilon = 5$ S = NEL	S = FLS	$\varepsilon = 10$ S = NEL	S = FLS	$\varepsilon = 15$ S = NEL	S = FLS
$\widehat{\rho_2\text{-SD}}(\tilde{x}_n)$	10	72.8	64.5	3.8	0.8	0	0	0	0
	30	72.4	49.1	0	0	0	0	0	0
	100	83.8	46.1	0	0	0	0	0	0
$\widehat{\rho_2\text{-ADD}}(\tilde{x}_n, \overline{\tilde{x}}_n)$	10	68.6	69.4	2.2	2.2	0	0	0	0
	30	49.3	61.6	0	0	0	0	0	0
	100	21	78.3	0	0	0	0	0	0
$\widehat{\rho_1\text{-ADD}}(\tilde{x}_n, \widehat{Me}(\tilde{x}_n))$	10	67.3	74.2	4	5.3	0	0.1	0	0
	30	50.8	64.6	0	0.1	0	0	0	0
	100	24.4	79.7	0	0	0	0	0	0
$\widehat{\rho_2\text{-MDD}}(\tilde{x}_n, \overline{\tilde{x}}_n)$	10	86.4	83.5	36.7	29.8	4.3	2.6	0	0
	30	86.3	86.8	33.5	28.8	2.5	1	0	0
	100	79.4	78.9	13.7	12.4	0.2	0.1	0	0
$\widehat{\rho_1\text{-MDD}}(\tilde{x}_n, \widehat{Me}(\tilde{x}_n))$	10	90	85.1	52.3	43.8	14.1	10.8	2	1.4
	30	85.7	86.4	33.9	32.6	6.1	4.8	0.2	0
	100	78.3	79.2	12.1	13.4	0.7	0.6	0	0

% $\left|\widehat{D}(\mathrm{FRS}) - \widehat{D}(\mathrm{S})\right| > \varepsilon$ $(k = 5, \beta(p,q) \equiv \beta(4,2))$

$\widehat{D}$	n	$\varepsilon = 1$ S = NEL	S = FLS	$\varepsilon = 5$ S = NEL	S = FLS	$\varepsilon = 10$ S = NEL	S = FLS	$\varepsilon = 15$ S = NEL	S = FLS
$\widehat{\rho_2\text{-SD}}(\tilde{x}_n)$	10	70.7	60.7	4.4	1.2	0	0	0	0
	30	57.5	39.1	0	0	0	0	0	0
	100	53.5	12.6	0	0	0	0	0	0
$\widehat{\rho_2\text{-ADD}}(\tilde{x}_n, \overline{\tilde{x}}_n)$	10	67.9	64.4	3.3	2.4	0	0	0	0
	30	48.4	43.7	0.2	0	0	0	0	0
	100	33.1	19.5	0	0	0	0	0	0
$\widehat{\rho_1\text{-ADD}}(\tilde{x}_n, \widehat{Me}(\tilde{x}_n))$	10	72.2	74.7	6.7	7.3	0	0	0	0
	30	58	66.5	0.2	0.5	0	0	0	0
	100	42.8	61.4	0	0	0	0	0	0
$\widehat{\rho_2\text{-MDD}}(\tilde{x}_n, \overline{\tilde{x}}_n)$	10	85.1	84.7	36.8	31.1	4.7	2.1	0.1	0
	30	82.3	80	32.1	26.2	2.3	0.7	0	0
	100	86.5	83.3	29.4	19.7	0.3	0.1	0	0
$\widehat{\rho_1\text{-MDD}}(\tilde{x}_n, \widehat{Me}(\tilde{x}_n))$	10	89.7	90.4	66.8	62.9	23.9	18.8	2.2	1.4
	30	92.8	96.8	79.2	70.5	53.1	35.6	2.2	1.4
	100	95.3	100	92.8	84.1	84.5	56.3	0.2	0.1

% $\left|\widehat{D}(\mathrm{FRS}) - \widehat{D}(\mathrm{S})\right| > \varepsilon$ $(k = 5, \beta(p,q) \equiv \beta(6,1))$

$\widehat{D}$	n	$\varepsilon = 1$ S = NEL	S = FLS	$\varepsilon = 5$ S = NEL	S = FLS	$\varepsilon = 10$ S = NEL	S = FLS	$\varepsilon = 15$ S = NEL	S = FLS
$\widehat{\rho_2\text{-SD}}(\tilde{x}_n)$	10	88.5	63.3	13.6	0.5	0	0	0	0
	30	96.7	48.2	2.3	0	0	0	0	0
	100	100	39.4	0	0	0	0	0	0
$\widehat{\rho_2\text{-ADD}}(\tilde{x}_n, \overline{\tilde{x}}_n)$	10	91.1	66.3	16.8	0.7	0	0	0	0
	30	99.1	69.5	9.5	0	0	0	0	0
	100	100	86.7	1.5	0	0	0	0	0
$\widehat{\rho_1\text{-ADD}}(\tilde{x}_n, \widehat{Me}(\tilde{x}_n))$	10	81.2	59.2	15.4	0.7	0	0	0	0
	30	92.6	44.3	9.6	0	0	0	0	0
	100	99.8	48.1	5.9	0	0	0	0	0
$\widehat{\rho_2\text{-MDD}}(\tilde{x}_n, \overline{\tilde{x}}_n)$	10	83.4	69.7	22.2	9.2	1	0	0	0
	30	90.3	72.5	24.5	3.8	0.5	0	0	0
	100	98.7	89.5	34	0.5	0	0	0	0
$\widehat{\rho_1\text{-MDD}}(\tilde{x}_n, \widehat{Me}(\tilde{x}_n))$	10	99.1	93.4	57.4	40	5.4	2.9	1.5	0
	30	100	96.5	79.2	66.7	11.6	8.2	9.7	0
	100	100	99.5	96	89.7	21	19.9	21	0

The percentages have been quantified over 1000 samples of $n \in \{10, 30, 100\}$ FRS simulated (with different betas) data with reference interval $[0, 100]$ (this last fact being irrelevant for the study). On the basis of Tables 1 and 2 we cannot get very general conclusions, but we can definitely assert that scale measures mostly vary more from the FRS-based data to the encoded Likert ones.

Furthermore, one can state some approximate behaviour patterns, such as

- for almost all situations, the robust scale estimate (the last one) provides us with much higher percentages than non-robust ones; more concretely, $\rho_1\text{-}\widehat{\text{MDD}}(\widetilde{\mathbf{x}}_n, \widehat{\text{Me}}(\widetilde{\mathbf{x}}_n))$ is almost generally more sensitive to the change in the rating scale type; this is especially clear for small samples;
- distances are uniformly lower for $k = 5$ than for $k = 4$ when the midpoint of the 1-level is beta distributed with $(p, q) \in \{(1, 1), (0.75, 0.75), (4, 2)\}$; when $(p, q) = (6, 1)$ such a conclusion is appropriate for robust estimates and $\varepsilon \in \{1, 5\}$, but there is no clear conclusion for non-robust estimates or greater values of ε.

Acknowledgements. The research is this paper has been partially supported by the Spanish Ministry of Economy, Industry and Competitiveness Grant MTM2015-63971-P. Its support is gratefully acknowledged.

References

1. Diamond, P., Kloeden, P.: Metric spaces of fuzzy sets. Fuzzy Sets Syst. **35**, 241–249 (1990)
2. De la Rosa de Sáa, S., Lubiano, S., Sinova, S., Filzmoser, P.: Robust scale estimators for fuzzy data. Adv. Data Anal. Classif. **11**, 731–758 (2017)
3. Gil, M.A., Lubiano, M.A., De la Rosa de Sáa, S., Sinova, B.: Analyzing data from a fuzzy rating scale-based questionnaire: a case study. Psicothema **27**, 182–191 (2015)
4. Herrera, F., Herrera-Viedma, E., Martínez, L.: A fuzzy linguistic methodology to deal with unbalanced linguistic term sets. IEEE Trans. Fuzzy Syst. **16**(2), 354–370 (2008)
5. Hesketh, T., Pryor, R., Hesketh, B.: An application of a computerized fuzzy graphic rating scale to the psychological measurement of individual differences. Int. J. Man-Mach. Stud. **29**, 21–35 (1988)
6. Lubiano, M.A., De la Rosa de Sáa, S., Montenegro, M., Sinova, B., Gil, M.A.: Descriptive analysis of responses to items in questionnaires. Why not using a fuzzy rating scale? Inf. Sci. **360**, 131–148 (2016)
7. Lubiano, M.A., Montenegro, M., Sinova, B., De la Rosa de Sáa, S., Gil, M.A.: Hypothesis testing for means in connection with fuzzy rating scale-based data: algorithms and applications. Eur. J. Oper. Res. **251**, 918–929 (2016)
8. Lubiano, M.A., Salas, A., Gil, M.A.: A hypothesis testing-based discussion on the sensitivity of means of fuzzy data with respect to data shape. Fuzzy Sets Syst. **328**, 54–69 (2017)
9. Puri, M.L., Ralescu, D.A.: Fuzzy random variables. J. Math. Anal. Appl. **114**, 409–422 (1986)
10. Sinova, B., Gil, M.A., Colubi, A., Van Aelst, S.: The median of a random fuzzy number. The 1-norm distance approach. Fuzzy Sets Syst. **200**, 99–115 (2012)

Central Moments of a Fuzzy Random Variable Using the Signed Distance: A Look Towards the Variance

Rédina Berkachy[✉] and Laurent Donzé

Applied Statistics and Modelling, Department of Informatics,
Faculty of Economics and Social Sciences, University of Fribourg,
Fribourg, Switzerland
{Redina.Berkachy,Laurent.Donze}@unifr.ch
http://diuf.unifr.ch/asam

Abstract. The central moments of a random variable are extensively used to understand the characteristics of distributions in classical statistics. It is well known that the second central moment of a given random variable is simply its variance. When fuzziness in data occurs, the situation becomes much more complicated. The central moments of a fuzzy random variable are often very difficult to be calculated because of the analytical complexity associated with the product of two fuzzy numbers. An estimation is needed. Our research showed that the so-called signed distance is a great tool for this task. The main contribution of this paper is to present the central moments of a fuzzy random variable using this distance. Furthermore, since we are interested in the statistical measures of the distribution, particularly the variance, we put an attention on its estimation using the signed distance. Using this distance in approximating the square of a fuzzy difference, we can get an unbiased estimator of the variance. Finally, we prove that in some conditions our methodology related to the signed distance returns an exact crisp variance.

Keywords: Central moments · Estimation of the variance
Fuzzy variance · Fuzzy statistics · Signed distance · Unbiased estimator

1 Introduction and Motivation

The central moments of a given random variable are always needed for specifying the distribution or in calculating different statistical measures. The variance is known as the second central moment. The classical approaches are very clear on the ways of computing these measures. When the data are exposed to fuzziness, fuzzy methods are well suited for such situations. Thus, calculating central moments of a fuzzy random variable can be of good use. But, computing them is not totally understood, especially regarding the multiplication of two fuzzy sets. Computational difficulties arise and are not evident to overcome. Since we know that the product of two fuzzy numbers given by the extension principle of

© Springer Nature Switzerland AG 2019
S. Destercke et al. (Eds.): SMPS 2018, AISC 832, pp. 17–24, 2019.
https://doi.org/10.1007/978-3-319-97547-4_3

Zadeh seen in Zadeh [1] and Zimmermann [2] is not necessarily a closed bounded fuzzy set and that the space of fuzzy sets is not linear but semilinear, an estimation of this product is needed. Therefore, in this case only approximation of the moments can be computed. Many researchers in the field have presented ways to estimate the variance for example, such as Akbari et al. [3], Colubi et al. [4] etc.

The measure of the distance between two fuzzy numbers is in this matter necessary. We propose to use a particular measure called the signed distance. This distance was first described by Yao and Wu [5]. It has been shown that this signed distance has nice properties in the process of defuzzification. For example, Berkachy and Donzé [6] computed using simulations, the statistical measures of distributions obtained by this method in order to reveal their characteristics. A principal objective of this paper is to give the central moments of a fuzzy random variable using this relatively new distance. This latter will be applied when it comes to calculating the difference between two fuzzy numbers. Furthermore, we provide an unbiased estimator of the variance of a fuzzy random variable. One of our main findings is that in some specific conditions this estimator returns the crisp variance.

To sum up, Sect. 2 is devoted to some basic definitions of fuzzy sets. We expose in Sect. 3 definitions regarding a fuzzy random variable, its expectation and variance. In Sect. 4, we display a definition of the signed distance. We close the paper in the Sect. 5 by the central moments and give the estimator of the variance using the signed distance.

2   Definitions on Fuzzy Sets

Let us recall some basic definitions.

Definition 1 (Fuzzy set).
We denote by A a collection of objects denoted generically by x. A fuzzy set $\tilde{X}$ in A is a set of ordered pairs:

$$\tilde{X} = \{(x, \mu_{\tilde{X}}(x)) : x \in A\}, \tag{1}$$

where $\mu_{\tilde{X}}(x)$ is the membership function of x in $\tilde{X}$ which maps A to the closed interval [0,1]. This interval characterizes the degree of membership of x in $\tilde{X}$.

Definition 2 (Fuzzy number).
A fuzzy number $\tilde{X}$ is a convex and normalized fuzzy set on $\mathbb{R}$, such that its membership function $\mu_{\tilde{X}} : \mathbb{R} \mapsto [0,1]$ is continuous and its support is bounded.

Definition 3 (α-cut of a fuzzy number).
We denote by the non-fuzzy set $\tilde{X}_\alpha$, the α-cut of a fuzzy number $\tilde{X}$ defined as:

$$\tilde{X}_\alpha = \{x \in \mathbb{R} : \mu_{\tilde{X}}(x) \geqslant \alpha\}. \tag{2}$$

We know that a given fuzzy number $\tilde{X}$ can be represented by the family set $\{\tilde{X}_\alpha : \alpha \in [0,1]\}$ of his α-cuts. Its α-cuts can be seen as the closed intervals of the left and right α-cuts $[\tilde{X}_\alpha^L, \tilde{X}_\alpha^R]$ as follows

$$\tilde{X}_\alpha^L = \inf\{x \in \mathbb{R} : \mu_{\tilde{X}}(x) \geqslant \alpha)\} \quad \text{and} \quad \tilde{X}_\alpha^R = \sup\{x \in \mathbb{R} : \mu_{\tilde{X}}(x) \geqslant \alpha)\}.$$

3 Fuzzy Random Variable, Its Expectation and Variance

First, let us define as seen in Puri and Ralescu [7] and Viertl [8] the definition of a fuzzy random variable. The expectation and variance are given as well in order to be able to introduce the central moments in next sections.

Definition 4 (Fuzzy random variable).
Let $(\Omega, \mathscr{A}, \mathbb{P})$ be a probability space, and $\mathbb{F}$ be the family of fuzzy numbers on $\mathbb{R}$. A fuzzy random variable on this space is a Borel-measurable function $\tilde{X} : \Omega \to \mathbb{F}(\mathbb{R})$ such that

$$\{(\omega, x) : \omega \in \Omega, x \in \tilde{X}_\alpha(\omega)\} \in \mathscr{A} \times \mathscr{B}, \qquad \forall \alpha \in [0,1], \tag{3}$$

where $\mathscr{B}$ denotes the Borel subsets of $\mathbb{R}$.

Let us now define the expectation of a fuzzy random variable given by its α-cuts in the following manner:

Definition 5 (Expectation of a fuzzy random variable).
Let $\tilde{X}$ be a fuzzy random variable. The fuzzy expectation $\tilde{\mathbb{E}}(\tilde{X})$ of $\tilde{X}$ is a closed fuzzy number given by the following family set of α-cuts:

$$\left(\tilde{\mathbb{E}}(\tilde{X})\right)_\alpha = \{\mathbb{E}(X) \mid X : \Omega \to \mathbb{R}^n, X(\omega) = \tilde{X}_\alpha(\omega)\}, \tag{4}$$

for which $(\tilde{\mathbb{E}}(\tilde{X}))_\alpha^L$ and $(\tilde{\mathbb{E}}(\tilde{X}))_\alpha^R$ are its respective left and right continuous α-cuts with respect to α.

Given $\{\tilde{X}_i\}_{i=1}^n$ a fuzzy random sample, we are able to define the fuzzy sample mean $\overline{\tilde{X}}$. It is seen as an estimator of the expectation $\tilde{\mathbb{E}}(\tilde{X})$ of the fuzzy random variable $\tilde{X}$, i.e. $\tilde{\mathbb{E}}(\tilde{X}) = \overline{\tilde{X}}$. Its α-cuts can be written in the following manner:

Definition 6 (α-cuts of the fuzzy sample mean).
Let $\tilde{X}_i$, $i = 1, \ldots, n$ be a fuzzy sample. We denote by $\overline{\tilde{X}} = \frac{1}{n} \sum_{i=1}^n \tilde{X}_i$ the fuzzy sample mean of the fuzzy sample $\tilde{X}_i$, $i = 1, \ldots, n$. Its α-cuts are as follows:

$$\overline{\tilde{X}}_\alpha = \left[\frac{1}{n} \sum_{i=1}^n \tilde{X}_{i\alpha}^L, \; \frac{1}{n} \sum_{i=1}^n \tilde{X}_{i\alpha}^R\right], \tag{5}$$

where $\tilde{X}_{i\alpha}^L$ and $\tilde{X}_{i\alpha}^R$ are respectively the left and right α-cuts of the fuzzy number $\tilde{X}_i$.

We are now able to define the variance of a fuzzy random variable using the concept of the fuzzy expectations as follows:

Definition 7 (Fuzzy variance).
Let $\tilde{X}$ be a fuzzy random variable. We denote by $\tilde{V}(\tilde{X})$ its fuzzy variance as follows:

$$\tilde{V}(\tilde{X}) = \tilde{\mathbb{E}}(< \tilde{X}, \tilde{X} >) \ominus < \tilde{\mathbb{E}}(\tilde{X}), \tilde{\mathbb{E}}(\tilde{X}) >, \tag{6}$$

where $< \tilde{X}, \tilde{X} >$ is a scalar multiplication between $\tilde{X}$ and $\tilde{X}$, and "$\ominus$" is the fuzzy difference operator.

Remark 1. Some researchers used to calculate the variance of a fuzzy number in specific conditions. For sake of clarification, the variance described in this paper is the one of the fuzzy random variable and not of a fuzzy number.

4   The Signed Distance

This section is devoted to briefly describe the signed distance, particularly defended by Yao and Wu [5]. In addition, this method has been used in different contexts such as the evaluation of linguistic questionnaires or the defuzzification of a fuzzy decision in fuzzy hypotheses testing (Berkachy and Donzé [9], [10] and others).

We start by defining this signed distance denoted by d as a function d: $\mathbb{F} \times \mathbb{F} \to \mathbb{R}$, where $\mathbb{F}$ is the space of fuzzy numbers.

Definition 8 (The signed distance of a real number a).
The signed distance measured from zero $d(a,0)$ for $a \in \mathbb{R}$ is a, i.e. $d(a,0) = a$.

Consequently, if $a < 0$, $-d(a,0) = -a$. The signed distance between two values a and $b \in \mathbb{R}$ is $d(a,b) = a - b$.

Definition 9 (The signed distance between two fuzzy sets).
We consider two fuzzy sets $\tilde{X}$ and $\tilde{Y}$ of the family $\mathbb{F}$ of the fuzzy numbers on $\mathbb{R}$ such that their α-cuts exist and are integrable for $\alpha \in [0,1]$. The signed distance between $\tilde{X}$ and $\tilde{Y}$ is

$$d(\tilde{X}, \tilde{Y}) = \frac{1}{2} \int_0^1 [\tilde{X}_L(\alpha) + \tilde{X}_R(\alpha) - \tilde{Y}_L(\alpha) - \tilde{Y}_R(\alpha)]d\alpha. \tag{7}$$

The signed distance of a fuzzy set $\tilde{X} \in \mathbb{F}$ measured from the origin can be calculated from Eq. 7. Some properties of this distance are very useful for further discussions.

Property 1. The signed distance d: $\mathbb{F} \times \mathbb{F} \to \mathbb{R}$ is continuous and differentiable.

Property 2. The signed distance is commutative, associative, transitive and distributive.

These properties can be easily proven, see Yao and Wu [5].

Lemma 1.
$$\overline{d(\tilde{X},\tilde{0})} = d(\overline{\tilde{X}},\tilde{0}), \tag{8}$$

where $\overline{d(\tilde{X},\tilde{0})}$ is the mean of the signed distances of the fuzzy random sample measured from the fuzzy origin.

Proof. Using the Property 2, the proof of this lemma is direct.

$$d(\overline{\tilde{X}},\tilde{0}) = d\left(\frac{1}{n}\sum_{i=1}^{n}\tilde{X}_i,\tilde{0}\right) = \frac{1}{n}d\left(\sum_{i=1}^{n}\tilde{X}_i,\tilde{0}\right) = \frac{1}{n}\sum_{i=1}^{n}d\left(\tilde{X}_i,\tilde{0}\right) = \overline{d(\tilde{X},\tilde{0})}.$$

5 Central Moments of a Fuzzy Random Variable

First, the moments of a fuzzy random variable are defined. As stated previously, the signed distance will be used as an approximation of the difference between two fuzzy sets. In addition, since we are particularly interested in the variance of a fuzzy random variable, we display an unbiased estimator of it using the signed distance.

If we reason similarly to the classical statistical approaches, we can write the moments of a fuzzy random variable as follows:

Definition 10 (Central moments of a fuzzy random variable).
Let $\tilde{X}$ be a fuzzy random variable. We denote by $\tilde{\nu}_k$ its central moments given by:

$$\tilde{\nu}_k = \frac{1}{n}\sum_{i=1}^{n}\underbrace{(\tilde{X}_i \ominus \overline{\tilde{X}}) \otimes \ldots \otimes (\tilde{X}_i \ominus \overline{\tilde{X}})}_{k\ times}, \tag{9}$$

where n is the number of observations, $\overline{\tilde{X}}$ is the fuzzy sample mean, "$\otimes$" and "$\ominus$" are respectively the fuzzy multiplication and difference operators.

Regarding our interest in the signed distance and the complexity associated with calculating a fuzzy product, we re-write these moments using this distance. We get the following:

Definition 11 (Central moments of a fuzzy random variable using the signed distance).
Let $\tilde{X}$ be a fuzzy random variable. We denote by ν_k^S the central moments using the signed distance given by:

$$\nu_k^S = \frac{1}{n}\sum_{i=1}^{n}\underbrace{d\left(\tilde{X}_i,\overline{\tilde{X}}\right) \times \ldots \times d\left(\tilde{X}_i,\overline{\tilde{X}}\right)}_{k\ times} = \frac{1}{n}\sum_{i=1}^{n}d^k\left(\tilde{X}_i,\overline{\tilde{X}}\right), \tag{10}$$

where n is the number of observations, $\overline{\tilde{X}}$ is the fuzzy sample mean, d is the signed distance and "$\times$" is the crisp multiplication operator.

In particular, the first central moment is given by:

$$\nu_1^S = \frac{1}{n}\sum_{i=1}^{n} d(\tilde{X}_i, \overline{\tilde{X}}).$$

(11)

The second one is written as follows:

$$\nu_2^S = \frac{1}{n}\sum_{i=1}^{n} d(\tilde{X}_i, \overline{\tilde{X}}) \times d(\tilde{X}_i, \overline{\tilde{X}}).$$

(12)

Note: Analogously, the moments ν'^S_k can be written as $\nu'^S_k = \frac{1}{n}\sum_{i=1}^{n} d^k(\tilde{X}_i, \tilde{0})$.

Lemma 2 (First central moment of a fuzzy random variable).
The first central moment using the signed distance of a fuzzy random variable $\tilde{X}$ is null.

Proof. Using the Definition 11, the Property 2 and the Lemma 1, we have

$$\nu_1^S = \frac{1}{n}\sum_{i=1}^{n} d(\tilde{X}_i, \overline{\tilde{X}}) = \frac{1}{n}\sum_{i=1}^{n}\left(d(\tilde{X}_i, \tilde{0}) - d(\overline{\tilde{X}}, \tilde{0})\right) = \overline{d(\tilde{X}, \tilde{0})} - d(\overline{\tilde{X}}, \tilde{0}) = 0.$$

We can easily see that the second moment using the signed distance is nothing but the crisp version of the variance of the fuzzy random variable $\tilde{X}$. Consequently, one can simultaneously construct the crisp version of the asymmetry and kurtosis measures using the second, third and fourth central moments using the signed distance.

We are now able to introduce an unbiased estimator of the variance of a fuzzy random variable. Since we know that the linearity of the difference between two fuzzy sets is not always assured, the signed distance is a way to approximate this difference, between $\tilde{X}_i$ and the mean $\overline{\tilde{X}}$ in our case.

Proposition 1. *Let $\tilde{X} = (\tilde{X}_1, \ldots, \tilde{X}_n)$ be a fuzzy random sample with $\tilde{X}_i$ independent and identically distributed. The estimator S_n^2 of the fuzzy variance using the signed distance given by*

$$S_n^2 = \frac{1}{n-1}\sum_{i=1}^{n} d^2(\tilde{X}_i, \overline{\tilde{X}}),$$

(13)

is an unbiased crisp estimator of the variance $\tilde{V}(\tilde{X}_i)$ of the distribution of $\tilde{X}_i$, where $\overline{\tilde{X}}$ is the fuzzy sample mean shown in Definition 6.

Proof. Let $\tilde{X}$ be the fuzzy random sample defined on $\mathbb{F}^n$. Consider consequently that $d(\tilde{X}, \tilde{0})$ is a random variable defined on $\mathbb{R}^n$. An unbiased estimator called V_n^2 of the variance of the random variable $d(\tilde{X}, \tilde{0})$ can be written as:

$$V_n^2 = \frac{1}{n-1}\sum_{i=1}^{n}\left(d(\tilde{X}_i, \tilde{0}) - \overline{d(\tilde{X}, \tilde{0})}\right)^2.$$

(14)

To prove that S_n^2 is an unbiased estimator of the variance of $\tilde{X}$, we have to be sure that its expectation is nothing but the variance of $\tilde{X}$, and indeed we have:

$$\mathbb{E}[S_n^2] = \mathbb{E}\left[\frac{1}{n-1}\sum_{i=1}^{n}d^2(\tilde{X}_i,\overline{\tilde{X}})\right] = \mathbb{E}\left[\frac{1}{n-1}\sum_{i=1}^{n}\left(d(\tilde{X}_i,\tilde{0}) - d(\overline{\tilde{X}},\tilde{0})\right)^2\right],$$

$$= \mathbb{E}\left[\frac{1}{n-1}\sum_{i=1}^{n}\left(d(\tilde{X}_i,\tilde{0}) - \overline{d(\tilde{X},\tilde{0})}\right)^2\right],$$

$$= \mathbb{E}[V_n^2] = \tilde{V}\left(d(\tilde{X},\tilde{0})\right) = d\left(\tilde{V}(\tilde{X}),\tilde{0}\right),$$

which obviously is the distance between the fuzzy variance and the fuzzy origin i.e. the crisp version of the fuzzy variance.

Proposition 2. *Consider a fuzzy random sample $\tilde{X} = (\tilde{X}_1,\ldots,\tilde{X}_n)$ with $\tilde{X}_i$ independent and identically distributed and modelled by triangular isosceles fuzzy numbers or trapezoidal isosceles fuzzy numbers. The unbiased estimator of the fuzzy variance using the signed distance returns an exact crisp variance. For the fuzzy random sample $\tilde{X}$ with $\tilde{X}_i$ modelled by triangular isosceles fuzzy numbers, $\tilde{X}_i = (p_i, q_i, r_i)$, and for which the fuzzy sample mean $\overline{\tilde{X}}$ is given by the tuple $\overline{\tilde{X}} = (p_m, q_m, r_m)$, the crisp variance is given by:*

$$S_n^2 = \frac{1}{n-1}\sum_{i=1}^{n}\left(q_i - q_m\right)^2. \tag{15}$$

In the case of trapezoidal isosceles fuzzy numbers given by $\tilde{X}_i = (p_i, q_i, r_i, s_i)$ with the fuzzy sample mean $\overline{\tilde{X}} = (p_m, q_m, r_m, s_m)$, the crisp variance is as follows:

$$S_n^2 = \frac{1}{n-1}\sum_{i=1}^{n}\left(\frac{q_i + r_i}{2} - \frac{q_m + r_m}{2}\right)^2. \tag{16}$$

Proof. We know that the signed distance of a triangular fuzzy number $\tilde{X}_i = (p_i, q_i, r_i)$ is $d(\tilde{X}_i, \tilde{0}) = \frac{1}{4}(p_i + 2q_i + r_i)$. In addition, a triangular isosceles fuzzy number has the following property: $p_i + r_i = 2q_i$. The fuzzy sample mean is a triangular isosceles fuzzy number as well and the previous property holds $p_m + r_m = 2q_m$. The unbiased estimator of the fuzzy variance S_n^2 will be written as follows:

$$S_n^2 = \frac{1}{n-1}\sum_{i=1}^{n}d^2(\tilde{X}_i,\overline{\tilde{X}}) = \frac{1}{n-1}\sum_{i=1}^{n}\left(d(\tilde{X}_i,\tilde{0}) - d(\overline{\tilde{X}},\tilde{0})\right)^2,$$

$$= \frac{1}{n-1}\sum_{i=1}^{n}\left(\frac{1}{4}(p_i + 2q_i + r_i) - \frac{1}{4}(p_m + 2q_m + r_m)\right)^2,$$

$$= \frac{1}{n-1}\sum_{i=1}^{n}\frac{1}{16}\left((4q_i)^2 + (4q_m)^2 - 2(4q_i)(4q_m)\right) = \frac{1}{n-1}\sum_{i=1}^{n}\left(q_i - q_m\right)^2.$$

For the case of trapezoidal isosceles fuzzy numbers $\tilde{X}_i = (p_i, q_i, r_i, s_i)$, $i = 1, \ldots, n$ with the property $p_i + s_i = q_i + r_i$, the proof is similar.

6 Conclusion

The purpose of this paper is to present the central moments of a fuzzy random variable. We estimated the difference between two fuzzy numbers by the signed distance method. We particularly put our attention on the first and second central moments because of their direct relation with the variance. Nevertheless, we remind that other statistical characteristics such as skewness, kurtosis etc. can be consequently calculated. Using the signed distance, we gave an unbiased estimator of the fuzzy variance. Finally, we note that even though the defended version of the variance is unbiased and it returns in some cases an exact crisp expression of it, this version is nothing but an approximation. Therefore, one has to differentiate between the true variance modelled by fuzzy sets and its approximations given by crisp sets. These estimations can be of good use in several statistical domains, testing hypotheses as instance. A future direction would be to investigate the differences between the approximation by the signed distance from one side, and some other approximations often used in literature as mentioned above.

References

1. Zadeh, L.: Fuzzy sets. Inf. Control **8**(3), 338–353 (1965)
2. Zimmermann, H.-J.: The Extension Principle and Applications. Springer, Dordrecht (2001)
3. Akbari, M.G., Rezaei, A.H., Waghei, Y.: Statistical inference about the variance of fuzzy random variables. Sankhyā Indian J. Stat. **71–B**, 206–221 (2009)
4. Colubi, A., Coppi, R., D'urso, P., Gil, M.A.: Statistics with fuzzy random variables. Metron Int. J. Stat. **LXV**(3), 277–303 (2007)
5. Yao, J., Wu, K.: Ranking fuzzy numbers based on decomposition principle and signed distance. Fuzzy Sets Syst. **116**(2), 275–288 (2000)
6. Berkachy, R., Donzé, L.: Statistical characteristics of distributions obtained using the signed distance defuzzification method compared to other methods. In: Proceedings of the International Conference on Fuzzy Management Methods ICFM-Square, Fribourg, Switzerland, pp. 48–58, September 2016
7. Puri, M.L., Ralescu, D.A.: Fuzzy random variables. J. Math. Anal. Appl. **114**(2), 409–422 (1986)
8. Viertl, R.: Statistical Methods for Fuzzy Data. Wiley, Hoboken (2011)
9. Berkachy, R., Donzé, L.: Individual and global assessments with signed distance defuzzification, and characteristics of the output distributions based on an empirical analysis. In: Proceedings of the 8th International Joint Conference on Computational Intelligence - Volume 1: FCTA, pp. 75–82 (2016)
10. Berkachy, R., Donzé, L.: A new approach of testing fuzzy hypotheses by confidence intervals and defuzzification of the fuzzy decision by the signed distance (2018, Under Review)

On Missing Membership Degrees: Modelling Non-existence, Ignorance and Inconsistency

Michal Burda$^{(\boxtimes)}$, Petra Murinová, and Viktor Pavliska

Institute for Research and Applications of Fuzzy Modeling,
Centre of Excellence IT4Innovations, Division University of Ostrava,
30. dubna 22, 701 03 Ostrava, Czech Republic
{michal.burda,petra.murinova,viktor.pavliska}@osu.cz

Abstract. In real-world applications, mathematical models must often deal with values that are missing or undefined. The aim of this paper is to provide a survey on types and reasons for such non-availability. It motivates the need to handle different reasons for missingness in a different, but appropriate way. In particular, non-existence, ignorance, and inconsistency are studied. The paper also presents a novel way of how to compute with different types of missing values at the same time.

Keywords: Missing membership degrees · Non-existence · Ignorance
Inconsistency · Fuzzy set

1 Introduction

Real-world data processing often faces the challenges of handling data that is imprecise, vague, inconsistent, or missing. The objective of handling such data is not new. Fundamental grounds in mathematical logic were established by Kleene, Bochvar, Sobociński and others, who studied the properties of three-valued logics $0/1/*$, which were also studied by Łukasiewicz in [1] in 1920. These authors showed that the third value $*$ may represent an unknown, undefined or indeterminate truth value. An overview of the main contributions can be found, e.g., in [2].

The logic called *Bochvar's internal* [3] (also known as Kleene's weak) defines the third truth value $*$ as an annihilator. *Sobociński's* variant handles $*$ as an ignorable non-sense, so that $*$ is treated like 1 (resp. 0) in conjunction (resp. disjunction) with a defined (i.e. non-$*$) truth value. *Kleene's* (strong) logic's indeterminate value $*$ preserves absorbing elements 0 (in conjunction) and 1 (in disjunction), while annihilating in other cases. See Table 1.

Zadeh's fuzzy sets [4] provide a powerful tool for modelling partial membership. A fuzzy set is a mapping $F : \mathcal{U} \to [0,1]$ that assigns to each member u of the universe of discourse $\mathcal{U}$ a membership degree $F(u)$ such that $F(u) = 0$

© Springer Nature Switzerland AG 2019
S. Destercke et al. (Eds.): SMPS 2018, AISC 832, pp. 25–32, 2019.
https://doi.org/10.1007/978-3-319-97547-4_4

Table 1. Sobociński's $\wedge_S$, Bochvar's internal $\wedge_B$ and Kleene's (strong) $\wedge_K$ variants for handling of the third truth value $*$ in conjunction

$\wedge_S$	0	$*$	1
0	0	0	0
$*$	0	$*$	1
1	0	1	1

$\wedge_B$	0	$*$	1
0	0	$*$	0
$*$	$*$	$*$	$*$
1	0	$*$	1

$\wedge_K$	0	$*$	1
0	0	0	0
$*$	0	$*$	$*$
1	0	$*$	1

means that u is not a member of the fuzzy set F, $F(u) = 1$ represents full membership and other values indicate various intensities of the membership. Fuzzy sets and fuzzy logic allow, e.g., to mitigate the problem of hard borders or crisp thresholds between categories (e.g. categorization of numeric values into intervals always faces the problem of where to put borderline subjects). The operation of intersection of fuzzy sets is based on t-norms, which represent the general class of multiplications. These are binary operations $\otimes : [0,1]^2 \rightarrow [0,1]$ which were mainly studied by Klement, Mesiar and Pap in [5] and later elaborated by many others. A concept associated with a t-norm is the triangular conorm (t-conorm) $\oplus : [0,1]^2 \rightarrow [0,1]$, which is the base for disjunction of fuzzy sets.

The above-mentioned work on the third truth value $*$ logics was brought to fuzzy logic by Novák, Běhounek and Daňková in [6,7], where they propose a fuzzy propositional partial logic and later a predicate partial fuzzy logic based on expansions of a well-known MTL_Δ fuzzy logic of left-continuous t-norms (see [8]) that handles $*$ with several types of fuzzy logical connectives, each of which treat $*$ in a different way. [9] presented a study of the fuzzy type theory with partial functions, which are used for characterization of undefined values. In contrast to [7], the authors of [10] introduce several $*$-like truth values together with a single set of fuzzy logical connectives.

In practical data analytical tasks, missing data is preferably *imputed* if the missingness is not dependent on the missing value itself [11]. Imputation is done by substituting missing values with an average or a value obtained from regression or application of some of the machine learning techniques [12]. However, as discussed later, not all missing values can be imputed. Besides non-random missingness, whose imputations are hard to estimate, some values are missing intentionally, simply because they refer to a non-existent object.

Other authors try to adapt data analysis techniques to handle missing values directly. For instance, Hájek et al. proposed, within their association analysis (GUHA) method [13], an extension capable of searching for association rules on data with missing values. In [14], Dubois' approach [15] for fuzzy association rules was generalized to work with undefined values.

The main objective of this paper is to provide a survey on the types and reasons for the non-availability of data. We identify the need to handle different reasons for missingness in a different, but appropriate way. The paper also presents a novel way of how to compute with different types of missing values at the same time by proposing a mathematical structure for representation of missing values (unknown or concealed, non-existent or inconsistent).

In the subsequent sections, we discuss various sources of missingness: ignorance (i.e. unknown data) in Sect. 2, non-existence (i.e. data values that can not be imputed because they refer to a non-existent object) in Sect. 3, and inconsistency (i.e. truth and falsity hold at the same time) in Sect. 4. In Sect. 5, we develop a framework that is capable of representing all of the previously mentioned types of missingness together. Section 6 concludes the paper and draws directions for possible future research.

This preliminary paper is focused on discussing variants and alternatives for the representation of missing values in the fuzzy set theory. As the authors stem the research from practice, a lot of examples are provided from real-world situations in order to motivate and establish reader's intuition.

2   Ignorance

Ignorance is perhaps the most common reason for missingness. Data may be missing because it was not measured, forgotten, denied to obtain or lost. That is, the correct value is virtually detectable, i.e. existing, but unknown. Example: respondent u of a sociological questionnaire refuses to tell his or her age. That is, the value surely exists (every living human has some age), but it is unknown. Hence a fuzzy set of "young" respondents has an unknown membership degree for respondent u.

In classical logic, the *unknown* truth value $*$ may be handled accordingly to Kleene: an operation with unknown $(*)$ results in unknown if different substitutions of $*$ give different results. For instance, $1 \wedge * = 0 \vee * = *$. On the other hand, sometimes the result does not depend on $*$: $0 \wedge * = 0$, $1 \vee * = 1$, etc.

In the fuzzy set theory, *Interval-Valued Fuzzy Sets* [16,17] generalize *unknown* by replacing a membership degree with an interval of membership degrees. An interval-valued fuzzy set is a mapping $F : u \to [l(u), h(u)]$, where $u \in \mathcal{U}$ is an element of the universe and $0 \leq l(u) \leq h(u) \leq 1$. The interval $F(u) = [l(u), h(u)]$ then may be interpreted as an interval of values that are all possible, without knowing, which one is correct. The interval $[0, 1]$ represents a complete ignorance.

Let F, G be interval-valued fuzzy sets such that $F(u) = [l_F(u), h_F(u)]$ and $G(u) = [l_G(u), h_G(u)]$. The set operations are defined as follows [18]:

$$(F \cap G)(u) = [\min\{l_F(u), l_G(u)\}, \min\{h_F(u), h_G(u)\}],$$
$$(F \cup G)(u) = [\max\{l_F(u), l_G(u)\}, \max\{h_F(u), h_G(u)\}].$$

As noted in [18], interval-valued fuzzy sets are mathematically equivalent to "Intuitionistic Fuzzy Sets" by Atanassov [19].

Interval-valued fuzzy sets are a special case of set-valued fuzzy sets, which are a special case of L-fuzzy sets [20]. We can e.g. consider a fuzzy set H over an universe $\mathcal{U}$ as a mapping $H : \mathcal{U} \to \mathscr{P}([0, 1])$ that assigns to each element $u \in \mathcal{U}$ an arbitrary subset of $[0, 1]$. If it happens that $H(u)$ results always to a non-empty closed interval, these set-valued fuzzy sets collapse to interval-valued fuzzy sets.

Note that set-valued fuzzy sets (and thus also interval-valued fuzzy sets) may be viewed as representations of a limited knowledge. $H(u)$ simply enumerates possible membership degrees. Results of set operations on set-valued fuzzy sets capture all possibilities obtainable from arbitrary selections, hence, the set operations may be constructed as follows:

$$(G \bowtie H)(u) = \{g \otimes h : g \in G(u), h \in H(u)\}, \tag{1}$$
$$(G \uplus H)(u) = \{g \oplus h : g \in G(u), h \in H(u)\}, \tag{2}$$

where $\otimes$ is a t-norm and $\oplus$ a corresponding (dual) t-conorm.

Note that ignorance is generally not a truth-functional notion. The above-presented definitions of $\bowtie$ and $\uplus$ work purely in a truth-functional way, which may cause unnecessary loss of accuracy in complex expressions. See e.g. [21] for more details.

3 Non-existence

Sometimes, a fuzzy set F is defined on the universe $\mathcal{U}$, $F : \mathcal{U} \to [0,1]$, but there is needed to work in a universe $\mathcal{U}_0 \supset \mathcal{U}$. That is, for some $u \in \mathcal{U}_0 - \mathcal{U}$, $F(u)$ is undefined. Example: respondents are asked to tell their spouse's age. A respondent u is single, i.e. asking for the age of his or her spouse makes no sense, hence, a fuzzy set of respondents, whose spouse is "young", has undefined membership degree of u. Other examples comprise membership degrees that are derived from mathematical operations which are not defined everywhere, such as logarithm or square root of a negative number, division by zero, etc.

Whereas missing values caused by ignorance may be resolved by imputation, i.e. replacement with sensible defaults (such as average), non-existent values cannot be treated that way. Non-existence is a new value carrying special information, incomparable with others.

In classical logic, non-existence may be treated accordingly to Bochvar or Sobociński [2] by introducing the third truth-value $*$. Bochvar defined missing value $*$ as an annihilator, i.e. $0 \wedge * = 1 \wedge * = 0 \vee * = 1 \vee * = *$. Sobociński style treated $*$ as an ignorable non-sense, i.e. $0 \wedge * = 0$, $1 \wedge * = 1$ etc.

Non-existence of a membership degree can be modelled in set-valued fuzzy sets with an empty set, i.e. by putting $H(u) = \emptyset$.

4 Inconsistency

Inconsistency arises if both membership and non-membership is true at the same time. It is typically a result of contradictory information such that different sources say different things. On the other hand, inconsistent information may appear naturally e.g. in expert systems. For instance, consider a relation $R \subseteq A \times F$ that assigns each animal $a \in A$ a set of features from F. Then birds have assigned a feature "has wings", while dogs don't, i.e. $\langle \text{bird}, \text{has wings} \rangle \in R$ and $\langle \text{dog}, \text{has wings} \rangle \notin R$. A difficulty may arise if considering whether dogs swim.

Some dogs, such as labrador, swim very well, while others, such as those with short legs, may have serious difficulties in water. So virtually, both $\langle dog, swims \rangle \in R$ and $\langle dog, swims \rangle \notin R$ seem to be valid. Another examples are whether people have hair, whether cars have four wheels, etc; some of them do, some do not. Generally, these examples follow a scheme where an object of the universe, in fact, represents a set of individuals, which may have different properties, so that asking if the given set of individuals has a given property results in true and false at the same time.

Inconsistent membership degrees may be handled with the same apparatus as unknown (ignored) memberships. We can apply Kleene's approach. Let $*$ denote inconsistency (0 and 1 at the same time). Then $* \wedge 1 = *$, but $* \wedge 0 = 0$ etc. If we are working in fuzzy, one may model partial inconsistency with interval-valued fuzzy sets or with set-valued fuzzy sets similarly as discussed in Sect. 2. That is, interval-valued fuzzy sets as well as set-valued fuzzy sets can represent either ignorance, or inconsistency, depending on their interpretation only.

5 Putting It All Together

Sections 2 and 4 provide examples of missing values caused by ignorance and inconsistency. As discussed above, the same mathematical tools may be used for both of these types. If we adopt set-valued fuzzy sets framework, we also enable non-existence, for free. However, a theory may be helpful that allows to use both ignorance and inconsistency distinguishably and at the same time. This can be achieved by nesting a set-valued fuzzy sets definition, as discussed in this section.

Before going further, let us provide some motivational examples of combinations of reasons for missing values. A questionnaire may ask for the age of respondent's spouse. So a set of people that have "young spouse" may utilize the following types of membership degrees:

1. *fuzziness:* a value from the interval $[0, 1]$ that represents a degree of "youngness";
2. *non-existence:* a non-existent membership degree for single person without any husband or wife;
3. *inconsistency:* in a hypothetical case of polygamy, it may be at the same time true that the spouse is young (1st spouse) and old (2nd spouse);
4. *ignorance:* a respondent may reveal his or her marriage status but refuse to answer any question regarding spouse's age, so that the membership may be represented with a set of all possibilities, i.e. the whole interval $[0, 1]$.

Additionally to that, an even more complex situation may arise:

– the respondent refuses to answer anything about his or her marriage: then the set of all possible degrees of membership to the set of respondents with young spouses should contain any value from $[0, 1]$ and also non-existence (if the respondent is single, which is unknown too), or virtually, all finite subsets of $[0, 1]$ (in case of polygamy, if it cannot be rejected).

5.1   Set-Valued Fuzzy Sets over Power Sets

To represent all types of missing values discussed above, the domain of membership degrees may be built as follows. Let $\mathcal{U}$ be a set of respondents. If we put $L = [0, 1]$ then $F : \mathcal{U} \to L$ is a fuzzy set capable of capturing various intensities of membership, perhaps a degree of "youngness" of one's spouse. If we do not insist on capturing fuzziness, e.g. a degree to which someone's spouse is "young", we can put $L = \{0, 1\}$ or use some other set of membership degrees as needed (closed to the operations).

Let $M = \mathscr{P}(L)$ (a power set of L) then $H : \mathcal{U} \to M$ is a set-valued fuzzy set, whose membership degrees from M can be interpreted as an inconsistency, if the cardinality of the membership degree is greater than 1, i.e. $|H(u)| > 1$, for some $u \in \mathcal{U}$. Note that $H(u) = \emptyset$ may be interpreted as a non-existent membership degree – see Sect. 3 for more details. Alternatively, we may consider a subset of $\mathscr{P}(L)$ as M, but care has to be taken for M to remain closed with respect to all the required operations.

We can also continue further and define $P : \mathcal{U} \to \mathscr{P}(M)$, i.e. a set-valued fuzzy set with membership degrees from $\mathscr{P}(M)$. That is, the membership degrees of P are subsets of M, which can be now interpreted as *possible alternatives*, i.e. in terms of *ignorance*, as discussed in Sect. 2.

Example. Membership degrees of P are sets of sets of elements from $L = [0, 1]$. The "inner" sets capture inconsistency while the "outer" sets represent alternatives caused by ignorance. Several important types of membership degrees are:

- $P(u) = \{\{a\}\}$, for $a \in L$, denotes the membership of an element $u \in \mathcal{U}$ to a set-valued fuzzy set P on power-set M to be of degree a, equivalently as $F(u) = a$. (i.e. $P(u) = \{\{1\}\}$ means that u's spouse is young.)
- $P(u) = \{\{0, 1\}\}$ represents an inconsistent membership degree, i.e. u is a member of P in degree 0 and 1 at the same time. Similarly, $P(u) = \{[0, 1]\}$ represents a complete inconsistency so that u is a member of P in all degrees from the interval $[0, 1]$.
- $P(u) = \{\{\}\} = \{\emptyset\}$ denotes a non-existent membership degree. (i.e. spouse does not exist.)
- $P(u) = \{\{0\}, \ldots, \{1\}\}$, i.e. a set of all singleton subsets of L is interpreted as complete ignorance, i.e. the concrete membership degree is not known. (i.e. a spouse exists, but his or her age is revealed.)
- $P(u) = \{\emptyset, \{0\}, \ldots, \{1\}\}$, i.e. as in previous bullet, but also with $\emptyset$ as another possibility, which can be interpreted as not knowing the age of the spouse and not knowing even if the spouse even exists. (Here we reject polygamy. Complete ignorance with polygamy allowed would be expressed with $P(u) = \mathscr{P}(M)$.)

Another example illustrating the richness of the proposed framework is:

- $P(u) = \{\{1\}\} \cup \{\{x, 1\} : x \in [0, 1)\}$ represents a constrained inconsistency indicating a knowledge that u is a member of P in degree 1 and inconsistently at the same time at unknown degree from interval $[0, 1)$. (This models a hypothetical polygamy marriage with two spouses, one of which is known to be young in degree 1, while the other spouse's age is unknown.)

The set operations $\boxtimes$ and $\uplus$, which were constructed for set-valued fuzzy sets in (1) and (2), can be introduced by Zadeh's extension principle for set-valued fuzzy sets on $\mathscr{P}(M)$ as follows. Let $G, H : \mathcal{U} \to \mathscr{P}(M)$. Then

$$(G \boxtimes H)(u) = \{T(g, h) : g \in G(u), h \in H(u)\},$$
$$(G \uplus H)(u) = \{S(g, h) : g \in G(u), h \in H(u)\},$$

where

$$T(g, h) = \{g' \otimes h' : g' \in g, h' \in h\},$$
$$S(g, h) = \{g' \oplus h' : g' \in g, h' \in h\},$$

where $\otimes$ is a t-norm and $\oplus$ is a corresponding (dual) t-conorm.

6   Discussion and Conclusion

This paper summarized some common reasons for missing values in data. We have surveyed several different existing approaches for handling missing values in fuzzy logic or fuzzy set theory.

Set-valued fuzzy sets over power sets, as introduced in the previous section, provide a fine-grained framework for representing inconsistency, ignorance, and non-existent membership degrees. Its advantage is the capability of capturing detailed information of possible alternatives and combinations of inconsistent knowledge that appear at the same time. However, it is honest to admit that the detailness is also its greatest drawback limiting the framework from a direct use in computer-based applications. The intention of the proposed framework is mainly in supporting the research on missing values from the theoretical perspective. For practical applications, simplification would be more appropriate.

In future research, we would like to define a subset of the complicated membership degree domain $\mathscr{P}(M)$ and define a limited number of special $*$ values that represent non-existence, ignorance, inconsistency, and perhaps their combinations (such as ignored inconsistency or ignored non-existence, etc.).

Acknowledgements. Authors acknowledge support by project "LQ1602 IT4Innovations excellence in science" and by GAČR 16-19170S.

References

1. Łukasiewicz, J.: O logice trojwartosciowej. Ruch Filozoficzny **5**, 170–171 (1920)
2. Malinowski, G.: The Many Valued and Nonmonotonic Turn in Logic. North-Holand, Amsterdam (2007)

3. Bergmann, M.: An Introduction to Many-Valued and Fuzzy Logic: Semantics, Algebras, and Derivation Systems. Cambridge University Press, Cambridge/New York (2008)
4. Zadeh, L.A.: Fuzzy sets. Inf. Control **8**, 338–353 (1965)
5. Klement, E.P., Mesiar, R., Pap, E.: Triangular Norms. Kluwer, Dordrecht (2000)
6. Běhounek, L., Novák, V.: Towards fuzzy partial logic. In: In Proceedings of the IEEE 45th International Symposium on Multiple-Valued Logics (ISMVL 2015), pp. 139–144 (2015)
7. Běhounek, L., Daňková, M.: Towards fuzzy partial set theory. In: Information Processing and Management of Uncertainty in Knowledge-Based Systems (IPMU 2016), pp. 482–494 (2016)
8. Hájek, P.: Metamathematics of Fuzzy Logic. Kluwer, Dordrecht (1998)
9. Novák, V.: Towards fuzzy type theory with partial functions. In: Advances in Fuzzy Logic and Technology, pp. 25–37 (2017)
10. Burda, M., Murinová, P., Pavliska, V.: Undefined values in fuzzy logic. In: Advances in Fuzzy Logic and Technology, pp. 604–610 (2017)
11. Gelman, A., Hill, J.: Data Analysis Using Regression and Multilevel/Hierarchical Models. Analytical Methods for Social Research. Cambridge University Press, New York (2007)
12. Liu, Y., Gopalakrishnan, V.: An overview and evaluation of recent machine learning imputation methods using cardiac imaging data. Data **2**(1), 8 (2017)
13. Hájek, P., Havránek, T.: Mechanizing Hypothesis Formation (Mathematical Foundations for a General Theory). Springer, Berlin (1978)
14. Burda, M., Murinová, P., Pavliska, V.: Fuzzy association rules on data with undefined values. In: Information Processing and Management of Uncertainty in Knowledge-Based Systems (IPMU 2018) (2018, page accepted)
15. Dubois, D., Hüllermeier, E., Prade, H.: A systematic approach to the assessment of fuzzy association rules. Data Min. Knowl. Disc. **13**(2), 167–192 (2006)
16. Chen, Q., Kawase, S.: An approach towards consistency degrees of fuzzy theories. Fuzzy Sets Syst. **113**, 237–251 (2000)
17. Cornelis, C., Deschrijver, G., Kerre, E.E.: Implication in intuitionistic fuzzy and interval-valued fuzzy set theroy: construction, classification, application. Int. J. Approx. Reason. **35**, 55–95 (2004)
18. Dubois, D., Prade, H.: Interval-valued fuzzy sets, possibility theory and imprecise probability. In: EUSFLAT Conference (2005)
19. Atanassov, K.T.: Intuitionistic fuzzy sets. Fuzzy Sets Syst. **20**(1), 87–96 (1986)
20. Goguen, J.A.: L-fuzzy sets. J. Math. Anal. Appl. **18**(1), 145–174 (1967)
21. Dubois, D.: On ignorance and contradiction considered as truth-values*. Logic J. IGPL **16**(2), 195–216 (2008)

Characterization of Conditional Submodular Capacities: Coherence and Extension

Giulianella Coletti[1], Davide Petturiti[2(✉)], and Barbara Vantaggi[3]

[1] Dip. Matematica e Informatica, University of Perugia, Perugia, Italy
giulianella.coletti@unipg.it
[2] Dip. Economia, University of Perugia, Perugia, Italy
davide.petturiti@unipg.it
[3] Dip. S.B.A.I., "La Sapienza" University of Rome, Rome, Italy
barbara.vantaggi@sbai.uniroma1.it

Abstract. We provide a representation in terms of a linearly ordered class of (unconditional) submodular capacities of an axiomatically defined conditional submodular capacity. This allows to provide a notion of coherence for a partial assessment and a related notion of coherent extension.

Keywords: Conditional submodular capacity · Coherence · Extension

1 Introduction

Inferential processes play a central role in decisions under uncertainty and in knowledge acquisition. In particular, the adopted notion of conditioning is crucial in drawing inferences as it impacts on the final result of inferential procedures: prototypically, conditional probability can be seen as a derived notion, as in the usual Kolmogorovian setting [27] or, in the more general context due to de Finetti, Rényi and Dubins [15,20,30], as a primitive concept defined axiomatically.

The probabilistic framework is the most common choice for dealing with uncertainty, nevertheless when the decision maker has only a partial, imprecise probabilistic assessment it is convenient to accept a more general view than the probabilistic one (see, e.g., [9,16,22,33]). In particular, if the probability assessment is known only on a part of the events of interest or even on a different set of events, it is unavoidable to act under ambiguity, since uncertainty must be evaluated by means of imprecise probabilities or classes of probabilities and their envelopes (see, e.g., [33]).

It is well-known that the upper envelope of the coherent extensions of a coherent probability is, in general, a not necessarily submodular capacity (see, for instance, Example 1 in [4]). Nevertheless, starting from a probability on a Boolean algebra, the upper envelope of the extensions to another Boolean

© Springer Nature Switzerland AG 2019
S. Destercke et al. (Eds.): SMPS 2018, AISC 832, pp. 33–41, 2019.
https://doi.org/10.1007/978-3-319-97547-4_5

algebra is a plausibility function (see [8,17,18,25,33]). Moreover, under specific (logical and numerical) constraints one can even obtain a possibility measure (see [9,14,21,28]).

The quoted capacities can be taken themselves as a framework for modelling uncertainty, nevertheless for them a commonly accepted notion of conditioning is still lacking, even if many proposals have been addressed (see, e.g., [5,8,11, 16,18,25,26,29,31].

Conditioning for capacities is usually faced emulating what is done for conditional probability. Hence, the existing proposals can be essentially divided into two classes: those introducing a conditional measure as a derived notion obtained, through a suitable conditioning rule, from an unconditional measure (e.g., [13,16,18,21,23,31]), and those presenting a conditional measure as a primitive concept, that is a function of two variables satisfying a set of axioms (e.g., [1,7,10–12,34]).

Focusing on submodular capacities, the paper adopts a "focusing" conditioning rule expressed by an axiomatic definition of conditional submodular capacity allowing for conditioning to "null" events. This definition generalizes that given in [16] for plausibility functions and permits the introduction of a notion of coherence for a partial assessment and a related notion of coherent extension. In turn, these notions seem of particular interest since conditional submodular capacities can be used to define conditional coherent risk measures [6].

2   Preliminaries

Let $\mathcal{A}$ be a Boolean algebra of *events* E's, and denote with $(\cdot)^c$, $\vee$ and $\wedge$ the usual Boolean operations of negation, disjunction and conjunction, respectively, and with $\subseteq$ the partial order of implication. The *sure event* Ω and the *impossible event* $\emptyset$ coincide, respectively, with the top and bottom elements of $\mathcal{A}$. If $\mathcal{A}$ is finite, as will always be in this paper, we denote with $\mathcal{C_A} = \{C_1, \ldots, C_m\}$ the subset of its *atoms* which form the finer partition of Ω contained in $\mathcal{A}$. Denoting $\mathcal{A}^0 = \mathcal{A} \setminus \{\emptyset\}$, $\mathcal{H} \subseteq \mathcal{A}^0$ is an *additive class* if it is closed under finite disjunctions. For an arbitrary index set I, $\mathbf{alg}(\{E_i\}_{i \in I})$ indicates the minimal Boolean algebra containing $\{E_i\}_{i \in I}$, while $\mathbf{add}(\{E_i\}_{i \in I})$ stands for the minimal additive class containing $\{E_i\}_{i \in I}$.

We recall that a *(normalized) capacity* (see [19,24]) on $\mathcal{A}$ is a function $\psi :$ $\mathcal{A} \to [0,1]$ such that $\psi(\emptyset) = 0$, $\psi(\Omega) = 1$ and $\psi(A) \leq \psi(B)$ whenever $A \subseteq B$, for $A, B \in \mathcal{A}$. A *submodular (or 2-alternating) capacity* ψ on $\mathcal{A}$ further satisfies, for every $A_1, A_2 \in \mathcal{A}$,

$$\psi(A_1 \vee A_2) + \psi(A_1 \wedge A_2) \leq \psi(A_1) + \psi(A_2).$$

The *dual* function φ defined, for every $A \in \mathcal{A}$, as $\varphi(A) = 1 - \psi(A^c)$, is a *supermodular (or 2-monotone) capacity* and satisfies the above inequality in the opposite direction.

The functions ψ and φ on $\mathcal{A}$ are completely singled out by the *Möbius inverse* of φ [3,24], defined, for every $A \in \mathcal{A}$, as

$$m(A) = \sum_{B \subseteq A} (-1)^{|\{C_r \in \mathcal{C}_{\mathcal{A}} : C_r \subseteq A \wedge B^c\}|} \varphi(B).$$

The function $m : \mathcal{A} \to \mathbb{R}$ is such that $m(\emptyset) = 0$, $\sum_{A \in \mathcal{A}} m(A) = 1$, and, for every $A \in \mathcal{A}$,

$$\varphi(A) = \sum_{B \subseteq A} m(B) \quad \text{and} \quad \psi(A) = \sum_{B \wedge A \neq \emptyset} m(B).$$

In particular, if m is non-negative then ψ and φ are a *plausibility* and a *belief function* [3,16,24,31]. Recall that submodular and supermodular capacities are distinguished subclasses of *(coherent) upper* and *lower probabilities* [32,34].

3   Conditional Submodular Capacities

A *conditional event* $E|H$ is an ordered pair of events (E, H) with $H \neq \emptyset$. In particular, any event E can be identified with the conditional event $E|\Omega$. In this section we consider a finite Boolean algebra $\mathcal{A}$.

Definition 1. *Let $\mathcal{H} \subseteq \mathcal{A}^0$ be an additive class. A function $\psi : \mathcal{A} \times \mathcal{H} \to [0,1]$ is a* **conditional submodular capacity** *if it satisfies the following conditions:*

(i) $\psi(E|H) = \psi(E \wedge H|H)$, for every $E \in \mathcal{A}$ and $H \in \mathcal{H}$;
(ii) $\psi(\cdot|H)$ is a submodular capacity on $\mathcal{A}$, for every $H \in \mathcal{H}$;
(iii) $\psi(E \wedge F|H) = \psi(E|H) \cdot \psi(F|E \wedge H)$, for every $E \wedge H, H \in \mathcal{H}$ and $E, F \in \mathcal{A}$.

Moreover, given a conditional submodular capacity $\psi(\cdot|\cdot)$, the dual **conditional supermodular capacity** *$\varphi(\cdot|\cdot)$ is defined for every event $E|H \in \mathcal{A} \times \mathcal{H}$ as*

$$\varphi(E|H) = 1 - \psi(E^c|H).$$

Following the terminology of [20], a conditional submodular capacity is said to be *full on $\mathcal{A}$* if its domain is $\mathcal{A} \times \mathcal{A}^0$.

A conditional submodular capacity is a complex object that cannot be obtained, in general, by means of a single (unconditional) submodular capacity, but a class is needed instead.

Definition 2. *For fixed $\mathcal{A}$ and $\mathcal{H}$, a linearly ordered class of (unconditional) submodular capacities $\{\psi_0, \ldots, \psi_k\}$ on $\mathcal{A}$ is said a* **minimal agreeing class** *if there exists a decreasing chain $H_0^0 \supseteq \ldots \supseteq H_0^k$ of distinct elements of $\mathcal{H}$, whose order agrees with the indices of capacities, such that:*

(i) $\psi_\alpha(A) = 0$ for every $A \wedge H_0^\alpha = \emptyset$, for $\alpha = 0, \ldots, k$;
(ii) for every $H \in \mathcal{H}$ there exists $\alpha \in \{0, \ldots, k\}$ such that $\psi_\alpha(H) > 0$;
(iii) if $\psi_\alpha(A) > 0$ then $A \wedge H_0^\beta = \emptyset$, for $\beta > \alpha$.

Notice that conditions *(i)–(iii)* above imply $H_0^0 = \bigvee_{H \in \mathcal{H}} H$. The following result holds.

Theorem 1. *Every conditional submodular capacity $\psi(\cdot|\cdot)$ on $\mathcal{A} \times \mathcal{H}$ is in bijection with a minimal agreeing class $\{\psi_0, \ldots, \psi_k\}$ on $\mathcal{A}$.*

Proof. Given a conditional submodular capacity $\psi(\cdot|\cdot)$ set:

- $\psi_0(\cdot) = \psi(\cdot|H_0^0)$ with $H_0^0 = \bigvee_{H \in \mathcal{H}} H$;
- for $\alpha > 0$, let $H_0^\alpha = \bigvee \{H \in \mathcal{H} : \psi_\beta(H) = 0, \beta = 0, \ldots, \alpha-1\}$, if $H_0^\alpha \neq \emptyset$, then $\psi_\alpha(\cdot) = \psi(\cdot|H_0^\alpha)$, and the construction stops at index k such that $H_0^{k+1} = \emptyset$;

that is easily verified to be a minimal agreeing class. Vice versa, given a minimal agreeing class $\{\psi_0, \ldots, \psi_k\}$, for every $E|H \in \mathcal{A} \times \mathcal{H}$, denoting with α_H the minimum index in $\{0, \ldots, k\}$ such that $\psi_{\alpha_H}(H) > 0$, it holds that

$$\psi(E|H) = \frac{\psi_{\alpha_H}(E \wedge H)}{\psi_{\alpha_H}(H)},$$

that is easily verified to be a conditional submodular capacity.

If condition *(ii)* in Definition 1 is reinforced by requiring that $\psi(\cdot|H)$ is a plausibility function on $\mathcal{A}$, for every $H \in \mathcal{H}$, the resulting conditional measure is a *conditional plausibility function* according to [11], for which a minimal agreeing class representation has been given in [2].

Example 1. Let $\mathcal{A}$ be the finite Boolean algebra with atoms $\mathcal{C}_\mathcal{A} = \{C_1, C_2, C_3\}$, and take the additive class $\mathcal{H} = \{C_{12}, C_{23}, \Omega\}$, where $C_{ij} = C_i \vee C_j$. Let $\psi(\cdot|\cdot)$ be the conditional submodular capacity defined on $\mathcal{A} \times \mathcal{H}$ as

$\mathcal{A}$	$\emptyset$	C_1	C_2	C_3	C_{12}	C_{13}	C_{23}	Ω
$\psi(\cdot|\Omega)$	0	1	0	0	1	1	0	1
$\psi(\cdot|C_{12})$	0	1	0	0	1	1	0	1
$\psi(\cdot|C_{23})$	0	0	$\frac{1}{3}$	1	$\frac{1}{3}$	1	1	1

The minimal agreeing class representing $\psi(\cdot|\cdot)$ is $\{\psi_0, \psi_1\}$ with $H_0^0 = \Omega$ and $H_0^1 = C_{23}$, where

$\mathcal{A}$	$\emptyset$	C_1	C_2	C_3	C_{12}	C_{13}	C_{23}	Ω
ψ_0	0	1	0	0	1	1	0	1
ψ_1	0	0	$\frac{1}{3}$	1	$\frac{1}{3}$	1	1	1

4   Coherence and Extension

In this section we first show that a submodular capacity on a finite domain can always be extended to a full conditional submodular capacity defined on a finite Boolean super-algebra.

Theorem 2. *Let $\mathcal{A}, \mathcal{B}$ be finite Boolean algebras with $\mathcal{A} \subseteq \mathcal{B}$ and $\mathcal{H} \subseteq \mathcal{A}^0$ an additive class. If $\psi(\cdot|\cdot)$ is a conditional submodular capacity on $\mathcal{A} \times \mathcal{H}$, then there exists a full conditional submodular capacity $\psi'(\cdot|\cdot)$ on $\mathcal{B} \times \mathcal{B}^0$ extending it.*

Proof. By Theorem 1 the conditional submodular capacity $\psi(\cdot|\cdot)$ is in bijection with a minimal agreeing class $\{\psi_0, \ldots, \psi_k\}$ on $\mathcal{A}$ that, in turn, is in bijection with the corresponding linearly ordered class $\{m_0, \ldots, m_k\}$ of Möbius inverses on $\mathcal{A}$. For $\alpha = 0, \ldots, k$, define a Möbius inverse m'_α on $\mathcal{B}$ setting $m'_\alpha(A) = m_\alpha(A)$ if $A \in \mathcal{A}$ and $m'_\alpha(A) = 0$ if $A \in \mathcal{B} \setminus \mathcal{A}$. It is easily proven that every such m'_α is the Möbius inverse of a submodular capacity ψ'_α on $\mathcal{B}$, in particular, the class $\{\psi'_0, \ldots, \psi'_k\}$ on $\mathcal{B}$ determines the chain $K_0^0 \supseteq \ldots \supseteq K_0^k$ of elements of $\mathcal{B}^0$, with $K_0^0 = \Omega$ and $K_0^\alpha = \bigvee \{H \in \mathcal{B}^0 : \psi'_\beta(H) = 0, \beta = 0, \ldots, \alpha - 1\} \neq \emptyset$, for $\alpha > 0$. Define $K_0^{k+1} = \bigvee \{H \in \mathcal{B}^0 : \psi'_\beta(H) = 0, \beta = 0, \ldots, \alpha - 1\}$.

If $K_0^{k+1} = \emptyset$ then $\{\psi'_0, \ldots, \psi'_k\}$ is a minimal agreeing class on $\mathcal{B}$ in bijection with a full conditional submodular capacity on $\mathcal{B}$ extending $\psi(\cdot|\cdot)$ by construction.

Otherwise, if $K_0^{k+1} \neq \emptyset$ then, taking the submodular capacity ψ'_{k+1} on $\mathcal{B}$ whose Möbius inverse is such that $m'_{k+1}(K_0^{k+1}) = 1$, the class $\{\psi'_0, \ldots, \psi'_k, \psi'_{k+1}\}$ is a minimal agreeing class on $\mathcal{B}$ in bijection with a full conditional submodular capacity on $\mathcal{B}$ extending $\psi(\cdot|\cdot)$ by construction.

The above theorem shows just one of the possible ways for extending $\psi(\cdot|\cdot)$ on the entire $\mathcal{B} \times \mathcal{B}^0$: in general there are infinitely many of such extensions.

Now we consider an assessment defined on an arbitrary finite set of conditional events $\mathcal{G} = \{E_i|H_i\}_{i \in I}$. Recall that $\mathcal{G}$ can always be embedded into a minimal set $\mathcal{A} \times \mathcal{H}$, where $\mathcal{A} = \mathbf{alg}(\{E_i, H_i\}_{i \in I})$ and $\mathcal{H} = \mathbf{add}(\{H_i\}_{i \in I})$.

Definition 3. *Let $\mathcal{G} = \{E_i|H_i\}_{i \in I}$ be an arbitrary finite set of conditional events. An assessment $\psi : \mathcal{G} \to [0,1]$ is a* **coherent conditional submodular capacity** *if there is a conditional submodular capacity $\psi'(\cdot|\cdot)$ on $\mathcal{A} \times \mathcal{H}$ such that $\psi'_{|\mathcal{G}} = \psi$, where $\mathcal{A} = \mathbf{alg}(\{E_i, H_i\}_{i \in I})$ and $\mathcal{H} = \mathbf{add}(\{H_i\}_{i \in I})$.*

The following theorem provides a characterization of coherence.

Theorem 3. *Let $\mathcal{G} = \{E_i|H_i\}_{i \in I}$ be an arbitrary finite set of conditional events. The following statements are equivalent:*

(i) the assessment $\psi : \mathcal{G} \to [0,1]$ is a coherent conditional submodular capacity;

(ii) there exists a minimal agreeing class $\{\psi_0, \ldots, \psi_k\}$ of submodular capacities on $\mathcal{A}$ whose Möbius inverses $\{m_0, \ldots, m_k\}$ solve the sequence of linear systems $\mathcal{S}_0, \ldots, \mathcal{S}_k$ with

$$\mathcal{S}_\alpha : \begin{cases} \displaystyle\sum_{B \wedge E_i \wedge H_i \neq \emptyset} m_\alpha(B) - \psi(E_i|H_i) \cdot \sum_{B \wedge H_i \neq \emptyset} m_\alpha(B) = 0, \ \forall i \in I_\alpha, \\[2ex] m_\alpha(C_r) \geq 0, \ \forall C_r \in \mathcal{C}_\mathcal{A}, \\[2ex] \displaystyle\sum_{C_r \vee C_s \subseteq B \subseteq A} m_\alpha(B) \geq 0, \ \forall A \wedge H_0^\alpha \neq \emptyset \text{ and } \forall C_r, C_s \subseteq A, \ C_r \neq C_s, \\[2ex] \displaystyle\sum_{A \in \mathcal{A}^0} m_\alpha(A) = 1, \\[2ex] \displaystyle\sum_{B \wedge A \neq \emptyset} m_\alpha(B) = 0, \ \forall A \wedge H_0^\alpha = \emptyset, \end{cases}$$

with $I_0 = I$, $I_\alpha = \{i \in I : \psi_\beta(H_i) = 0, \beta = 0, \ldots, \alpha - 1\}$, for $\alpha > 0$, and $H_0^\alpha = \bigvee_{h \in I_\alpha} H_h$.

Proof. Let $\mathcal{H} = \mathbf{add}(\{H_i\}_{i \in I})$. By Theorem 1, the assessment $\psi(\cdot|\cdot)$ on $\mathcal{G}$ is coherent with a conditional submodular capacity if and only if we can solve the following sequence of systems $\mathcal{S}_0^*, \ldots, \mathcal{S}_k^*$ with

$$\mathcal{S}_\alpha^* : \begin{cases} \psi_\alpha(E_i \wedge H_i) - \psi(E_i|H_i) \cdot \psi_\alpha(H_i) = 0, \ \forall i \in I_\alpha, \\[1ex] \psi_\alpha \text{ is a submodular capacity on } \mathcal{A}, \\[1ex] \psi_\alpha(A) = 0, \ \forall A \wedge H_0^\alpha = \emptyset, \end{cases}$$

where I_α and H_0^α are defined as in statement *(ii)*. Finally, by Proposition 1 and Corollary 2 in [3], every system $\mathcal{S}_\alpha^*$ has solution if and only if the corresponding system $\mathcal{S}_\alpha$ has solution.

The following theorem shows that coherence is a necessary and sufficient condition to have a *coherent extension* (i.e., a coherent submodular capacity extending the initial assessment) on a finite super-set of conditional events.

Theorem 4. *Let $\mathcal{G} = \{E_i|H_i\}_{i \in I}$ and $\mathcal{G}' = \{E_j|H_j\}_{j \in J}$ be arbitrary finite sets of conditional events with $\mathcal{G} \subseteq \mathcal{G}'$, and $\psi : \mathcal{G} \to [0,1]$ an assessment. There exists a coherent conditional submodular capacity $\psi' : \mathcal{G}' \to [0,1]$ extending ψ if and only if ψ is a coherent conditional submodular capacity.*

Proof. If ψ' is a coherent conditional submodular capacity, then its restriction $\psi = \psi'_{|\mathcal{G}}$ is trivially coherent. Vice versa, if ψ is coherent, then there exists a conditional submodular capacity $\tilde{\psi}(\cdot|\cdot)$ on $\mathcal{A} \times \mathcal{H}$ extending it, with $\mathcal{A} = \mathbf{alg}(\{E_i, H_i\}_{i \in I})$ and $\mathcal{H} = \mathbf{alg}(\{H_i\}_{i \in I})$. By Theorem 2 such $\tilde{\psi}(\cdot|\cdot)$ can be extended to a full conditional submodular capacity $\tilde{\psi}'(\cdot|\cdot)$ on $\mathcal{B} = \mathbf{alg}(\{E_j, H_j\}_{j \in J})$. Hence, it is sufficient to take $\psi' = \tilde{\psi}'_{|\mathcal{G}'}$.

Example 2. Consider the events A, B with $A \subseteq B$ and take $\mathcal{A} = \mathbf{alg}(\{A, B\})$ with $\mathcal{C}_\mathcal{A} = \{C_1, C_2, C_3\}$, where $C_1 = A \wedge B$, $C_2 = A^c \wedge B$ and $C_3 = A^c \wedge B^c$.

    Take the assessment $\psi(A|B) = \frac{3}{8}$ and $\psi(B|A^c) = \frac{5}{9}$. So, we have $\mathcal{H} = \mathbf{add}(\{B, A^c\}) = \{B, A^c, \Omega\}$. To prove its coherence we first set $H_0^0 = B \vee A^c =$

Ω and search for a submodular capacity ψ_0 on $\mathcal{A}$ having Möbius inverse m_0. To avoid cumbersome notation we set $x_i^0 = m_0(C_i)$, $x_{ij}^0 = m_0(C_i \vee C_j)$ and $x_{123}^0 = m_0(\Omega)$, so, the system $\mathcal{S}_0$ in Theorem 3 becomes

$$\mathcal{S}_0 : \begin{cases} (x_1^0 + x_{12}^0 + x_{13}^0 + x_{123}^0) - \frac{3}{8} \cdot (x_1^0 + x_2^0 + x_{12}^0 + x_{13}^0 + x_{23}^0 + x_{123}^0) = 0, \\ (x_2^0 + x_{12}^0 + x_{23}^0 + x_{123}^0) - \frac{5}{9} \cdot (x_2^0 + x_3^0 + x_{12}^0 + x_{13}^0 + x_{23}^0 + x_{123}^0) = 0, \\ x_i^0 \geq 0, \text{ for } i = 1, 2, 3, \\ x_{ij}^0 \geq 0, \text{ for } i, j = 1, 2, 3, \ i \neq j, \\ x_{123}^0 + x_{ij}^0 \geq 0, \text{ for } i, j = 1, 2, 3, \ i \neq j, \\ x_1^0 + x_2^0 + x_3^0 + x_{12}^0 + x_{13}^0 + x_{23}^0 + x_{123}^0 = 1, \end{cases}$$

for which a solution is $x_1^0 = \frac{1}{10}$, $x_2^0 = \frac{3}{10}$, $x_3^0 = x_{12}^0 = x_{13}^0 = x_{23}^0 = \frac{2}{10}$ and $x_{123}^0 = -\frac{2}{10}$, determining the Möbius inverse m_0 and the corresponding submodular capacity ψ_0. Notice that $H_0^1 = \{H \in \mathcal{H} : \psi_0(H) = 0\} = \emptyset$, so, $\{\psi_0\}$ is a minimal agreeing class representing a conditional submodular capacity $\psi'(\cdot|\cdot)$ on $\mathcal{A} \times \mathcal{H}$ extending the assessment ψ.

Consider now another event C such that $A \subseteq B \subseteq C$ and $\mathcal{B} = \mathbf{alg}(\{A, B, C\})$ with $\mathcal{C}_\mathcal{B} = \{D_1, D_2, D_3, D_4\}$, where $D_1 = A \wedge B \wedge C = C_1$, $D_2 = A^c \wedge B \wedge C = C_2$, $D_3 = A^c \wedge B^c \wedge C$, $D_4 = A^c \wedge B^c \wedge C^c$, and $C_3 = D_3 \vee D_4$.

Following the procedure in the proof of Theorem 2, the conditional submodular capacity ψ_0 on $\mathcal{A}$ is extended to the submodular capacity ψ_0' on $\mathcal{B}$ below, where $D_{ij} = D_i \vee D_j$ and $D_{ijk} = D_i \vee D_j \vee D_k$:

	$\emptyset$	D_1	D_2	D_3	D_4	D_{12}	D_{13}	D_{14}	D_{23}	D_{24}	D_{34}	D_{123}	D_{124}	D_{134}	D_{234}	Ω
ψ_0'	0	$\frac{3}{10}$	$\frac{5}{10}$	$\frac{4}{10}$	$\frac{4}{10}$	$\frac{8}{10}$	$\frac{7}{10}$	$\frac{7}{10}$	$\frac{9}{10}$	$\frac{9}{10}$	$\frac{4}{10}$	1	1	$\frac{7}{10}$	$\frac{9}{10}$	1

Also in this case we have $K_0^0 = \Omega$ and $K_0^1 = \{H \in \mathcal{B}^0 : \psi_0'(H) = 0\} = \emptyset$, so, $\{\psi_0'\}$ is a minimal agreeing class representing a full conditional submodular capacity $\psi''(\cdot|\cdot)$ on $\mathcal{B} \times \mathcal{B}^0$ extending ψ' and, so, the assessment ψ.

References

1. Bouchon-Meunier, B., Coletti, G., Marsala, C.: Independence and possibilistic conditioning. Ann. Math. Artif. Intell. **35**(1–4), 107–123 (2002)
2. Capotorti, A., Coletti, G., Vantaggi, B.: Standard and nonstandard representability of positive uncertainty orderings. Kybernetika **50**(2), 189–215 (2014)
3. Chateauneuf, A., Jaffray, J.-Y.: Some characterizations of lower probabilities and other monotone capacities through the use of Möbous inversion. Math. Soc. Sci. **17**, 263–283 (1989)
4. Coletti, G., Petturiti, D., Vantaggi, B.: Bayesian inference: the role of coherence to deal with a prior belief function. Stat. Methods Appl. **23**, 519–545 (2014)
5. Coletti, G., Petturiti, D., Vantaggi, B.: Conditional belief functions as lower envelopes of conditional probabilities in a finite setting. Inf. Sci. **339**, 64–84 (2016)

6. Coletti, G., Petturiti, D., Vantaggi, B.: Conditional submodular coherent risk measures. In: Medina, J. et al. (eds.) Information Processing and Management of Uncertainty in Knowledge-Based Systems. Theory and Foundations. IPMU 2018. CCIS, vol. 854, pp. 239–250. Springer, Cham (2018)

7. Coletti, G., Scozzafava, R.: From conditional events to conditional measures: a new axiomatic approach. Ann. Math. Artif. Intell. **32**(1–4), 373–392 (2001)

8. Coletti, G., Scozzafava, R.: Toward a general theory of conditional beliefs. Int. J. Intell. Syst. **21**(3), 229–259 (2006)

9. Coletti, G., Scozzafava, R., Vantaggi, B.: Inferential processes leading to possibility and necessity. Inf. Sci. **245**, 132–145 (2013)

10. Coletti, G., Vantaggi, B.: Possibility theory: conditional independence. Fuzzy Sets Syst. **157**(11), 1491–1513 (2006)

11. Coletti, G., Vantaggi, B.: A view on conditional measures through local representability of binary relations. Int. J. Approx. Reason. **47**, 268–283 (2008)

12. Coletti, G., Vantaggi, B.: T-conditional possibilities: coherence and inference. Fuzzy Sets Syst. **160**(3), 306–324 (2009)

13. de Cooman, G.: Possibility theory II: conditional possibility. Int. J. Gen. Syst. **25**, 25–325 (1997)

14. de Cooman, G., Aeyels, D.: Supremum preserving upper probabilities. Inf. Sci. **118**, 173–212 (1999)

15. de Finetti, B.: Probability, Induction and Statistics: The Art of Guessing. Wiley, London (1972)

16. Dempster, A.P.: Upper and lower probabilities induced by a multivalued mapping. Ann. Math. Stat. **2**, 325–339 (1967)

17. Dempster, A.P.: A generalization of Bayesian inference. J. Roy. Stat. Soc. B **30**(2), 205–247 (1968)

18. Denneberg, D.: Conditioning (updating) non-additive measures. Ann. Oper. Res. **52**(1), 21–42 (1994)

19. Denneberg, D.: Non-additive Measure and Integral. Kluwer Academic Publishers, Dordrecht (1994)

20. Dubins, L.E.: Finitely additive conditional probabilities, conglomerability and disintegrations. Ann. Probab. **3**, 89–99 (1975)

21. Dubois, D., Prade, H.: When upper probabilities are possibility measures. Fuzzy Sets Syst. **49**(1), 65–74 (1992)

22. Dubois, D., Prade, H., Smets, P.: Representing partial ignorance. IEEE Trans. Syst. Man Cybern. A **26**(3), 361–377 (1996)

23. Gilboa, I., Schmeidler, D.: Updating ambiguous beliefs. J. Econ. Theor. **59**(1), 33–49 (1993)

24. Grabisch, M.: Set Functions, Games and Capacities in Decision Making. Springer, Cham (2016)

25. Fagin, R., Halpern, J.Y.: A New Approach to Updating Beliefs, pp. 347–374. Elsevier Science Publishers (1991)

26. Jaffray, J.-Y.: Bayesian updating and belief functions. IEEE Trans. Man Cybern. **22**, 1144–1152 (1992)

27. Kolmogorov, A.N.: Foundations of the Theory of Probability. Chelsea, New York (1950)

28. Miranda, E., Couso, I., Gil, P.: Relationships between possibility measures and nested random sets. Int. J. Uncertain. Fuzziness Knowl. Based Syst. **10**(1), 1–15 (2002)

29. Miranda, E., Montes, I.: Coherent updating of non-additive measures. Int. J. Approx. Reason. **56**, 159–177 (2015)

30. Rényi, A.: On conditional probability spaces generated by a dimensionally ordered set of measures. Theor. Probab. Appl. **1**(1), 55–64 (1956)
31. Shafer, G.: A Mathematical Theory of Evidence. Princeton University Press, Princeton (1976)
32. Troffaes, M.C.M., de Cooman, G.: Lower Previsions. Wiley, Hoboken (2014)
33. Walley, P.: Coherent lower (and upper) probabilities. Technical report, Department of Statistics, University of Warwick (1981)
34. Walley, P.: Statistical Reasoning with Imprecise Probabilities. Chapman and Hall, London (1991)

Some Partial Order Relations on a Set of Random Variables

Bernard De Baets[1]([✉]) and Hans De Meyer[2]

[1] Department of Data Analysis and Mathematical Modelling, Ghent University,
Coupure links 653, 9000 Gent, Belgium
`bernard.debaets@ugent.be`
[2] Department of Applied Mathematics, Computer Science and Statistics,
Ghent University, Krijgslaan 281 S9, 9000 Gent, Belgium
`hans.demeyer@ugent.be`

1 Introduction

Reciprocal relations, i.e. $[0, 1]$-valued relations Q satisfying $Q(a, b) + Q(b, a) = 1$, provide a convenient tool for expressing the result of the pairwise comparison of a set of alternatives. Some time ago, we have studied reciprocal relations that are generated from the pairwise comparison of the components of a random vector [2]. The comparison strategy was based on the concept of winning probability between two random variables (RVs): to a given a random vector $(X_1, X_2, \ldots, X_m)$ we associate a reciprocal relation Q that is defined as follows:

$$Q(X_i, X_j) = \mathrm{Prob}\{X_i > X_j\} + \frac{1}{2}\mathrm{Prob}\{X_i = X_j\}, \quad i, j = 1, \ldots, m. \quad (1)$$

This reciprocal relation Q can be visualized by means of a weighted directed graph with m nodes, where each node is associated with one component of the random vector and the weights correspond to the winning probabilities.

Clearly, to compute these winning probabilities it is not required to know the joint cumulative distribution function $F_{X_1,\ldots,X_n}$ of the random vector; instead, it suffices to know all bivariate marginal distribution functions F_{X_i, X_j}. Henceforth, we have explored any possible bivariate couplings between the RVs $X_1, \ldots, X_m$ rather than focusing on the m-dimensional dependence structure that characterizes the random vector $(X_1, \ldots, X_m)$. Note that if we say that X and Y are coupled by means of a copula C, this means that $F_{X,Y}(x, y) = C(F_X^{-1}(x), F_Y^{-1}(y))$.

A standard method to turn a $[0, 1]$-valued relation into a crisp relation is inspired by the concept of strict α-cut of a fuzzy relation. More specifically, given a cutting level α, all elements strictly greater than α are set equal to 1 and all other elements are set equal to 0. In case of a reciprocal relation, we consider cutting levels $\alpha \in [1/2, 1]$ and denote by Q^α the result of strictly cutting a given reciprocal relation Q at α.

The following question has gained our interest: given a method for associating with a given random vector $(X_1, \ldots, X_m)$ a reciprocal relation Q, for which values of α is the relation Q^α free of cycles for any random vector $(X_1, \ldots, X_m)$?

© Springer Nature Switzerland AG 2019
S. Destercke et al. (Eds.): SMPS 2018, AISC 832, pp. 42–45, 2019.
https://doi.org/10.1007/978-3-319-97547-4_6

We were able to answer this question for the situation where the reciprocal relation Q is constructed as described in Eq. (1) and under the assumption that all bivariate couplings are realized with a same Frank copula [4]. Note that the latter is an artificial device, in the sense that these couplings do not have to agree with the true bivariate dependencies induced by the joint cumulative distribution function of the given random vector. In general, cutting Q at a level α such that Q^α is cycle-free does not imply that Q^α is transitive. To turn Q^α into a transitive and cycle-free relation, or, equivalently, into a strict partial order relation, it suffices to compute the transitive closure $\bar{Q}^\alpha$ of Q^α. The covering graph of $\bar{Q}^\alpha$ is a so-called Hasse-diagram which visualizes how the method of constructing Q generates a strict partial order relation on the components of the given random vector at cutting level α.

Obviously, the set of cutting levels α such that Q^α is cycle-free (i.e. such that $\bar{Q}^\alpha$ is a strict partial order relation), is an interval of the type $[\alpha_0, 1]$, with α_0 the smallest cutting level at which a strict partial order relation is generated. If α is gradually increased from α_0 to 1, at certain stages one or more edges gradually disappear from the corresponding Hasse diagram, whence the power to distinguish between the RVs diminishes and more RVs become pairwisely incomparable. On the other hand, the Hasse diagram obtained from $\bar{Q}^{\alpha_0}$ is the one that expresses the highest sensitivity for small differences between the RVs. For practical purposes, often a Hasse diagram in between these two extremes will be preferred. Anyhow, it is advantageous to assure that the spectrum of the Hasse diagrams is as broad as possible, hence we will seek for methods for generating a reciprocal relation Q for which it holds that $\alpha_0 = 1/2$. This problem is the scope of the present contribution.

2 Transitivity of Reciprocal Relations

To simplify the problem we will shortcut the step of computing the transitive closure of Q^α in the sense that we impose the stronger condition that Q^{α_0} should be a transitive relation. Since $\alpha_0 \geq 1/2$, this relation is then also cycle-free, and therefore it is a strict partial order relation itself.

It is not surprising that the transitivity of the relation $Q^{1/2}$ is in some way related to a transitivity property of the reciprocal relation Q. There is, however, such a huge diversity of types of transitivity that can be defined for reciprocal relations that we felt the need for establishing what is now known as the cycle-transitivity framework [3], to cover most of these types. Here, we only make use of the concepts of T-transitivity and of stochastic transitivity, which we briefly recall.

Definition 1. *Let T be a t-norm. A reciprocal relation Q on A is called T-transitive if for any $(a, b, c) \in A^3$ it holds that*

$$T(Q(a,b), Q(b,c)) \leq Q(a,c). \tag{2}$$

The three main continuous t-norms are the minimum operator $T_\mathbf{M}$, the algebraic product $T_\mathbf{P}$ and the Lukasiewicz t-norm $T_\mathbf{L}$ (defined by $T_\mathbf{L}(x,y) = \max(x + y - 1, 0)$), and they respectively coincide with the copulas M, Π and W. It is well known that if Q is $T_\mathbf{M}$-transitive, then Q^α is transitive for any $\alpha \in [0,1]$. In particular, $Q^{1/2}$ is a strict partial order relation.

Definition 2. *Let g be an increasing $[1/2, 1]^2 \rightarrow [0,1]$ mapping such that $g(1/2, 1/2) \leq 1/2$. A reciprocal relation Q on A is called g-stochastic transitive if for any $(a, b, c) \in A^3$ it holds that*

$$(Q(a,b) \geq 1/2 \ \wedge \ Q(b,c) \geq 1/2) \ \Rightarrow \ Q(a,c) \geq g(Q(a,b), Q(b,c)). \quad (3)$$

Our special attention goes to the situation where g is again the minimum operator $T_\mathbf{M}$ (in which case the type of transitivity is called moderate stochastic transitivity), since it holds that if Q is moderately stochastic transitive, then Q^α is transitive for any $\alpha \in [1/2, 1]$, whence $Q^{1/2}$ is a strict partial order relation.

We have thus at hand two types of transitivity of reciprocal relations each of which ensures that the relation obtained by cutting at level $1/2$ is transitive.

3   A Partial Order Relation Generated by Winning Probabilities

We first investigate the reciprocal relations Q that are obtained by means of winning probabilities, as defined in Eq. (1), assuming that the bivariate couplings are expressed with a same Frank copula.

Proposition 1. [5] *Consider a random vector $(X_1, \ldots, X_m)$. Then the reciprocal relation Q_W, defined as in Eq. (1), with all bivariate couplings given by W, is $T_\mathbf{M}$-transitive.*

Once more it should be emphasized that the coupling with W is artificial in the sense that it does not correspond to any of the actual pairwise couplings between the components of the given random vector. In fact, there exists no random vector with more than two components that has all its bivariate couplings with W. The winning probability between two (continuous) random variables that are coupled by W has the following interpretation [1]. Given two random variables X, Y, let u be any point such that $F_X(u) + F_Y(u) = 1$, then $Q_W(X,Y) = F_Y(u)$. This interpretation of $Q_W(X,Y)$ shows that its use as a measure of the difference between the marginal cumulative distribution functions is not obvious, as it involves only the distance between the distribution functions F_X and F_Y at one point u. This motivates us to search for other methods to compare random variables than Eq. (1) relying on winning probabilities.

4 Partial Order Relations Generated by Proportional Expected Differences

As an alternative to the method of Eq. (1) to associate to a random vector a reciprocal relation that in some sense expresses a way of pairwisely comparing its components, we propose the following construction of a reciprocal relation Q^{PED} using the concept of proportional expected difference:

$$Q^{\mathrm{PED}}(X,Y) = \frac{\mathbf{E}[(X-Y)_+]}{\mathbf{E}[|X-Y|]}, \quad i,j = 1,\dots,m. \tag{4}$$

Herein $(X-Y)_+$ equals $(X-Y)$ if $X \geq Y$, and 0 otherwise. Again we assume that the components are pairwisely coupled with a same Frank copula C. We have investigated the extreme situations of all components either being coupled in a comonotone $(C = M)$ or countermonotone manner $(C = W)$, as well as the situation of independent components $(C = \Pi)$. Surprisingly, as far as the transitivity of the generated reciprocal relation is concerned, respectively denoted by Q_M^{PED}, Q_W^{PED} and Q_Π^{PED}, the three types of (artificial) couplings lead to the same result.

Proposition 2. *Consider a random vector* $(X_1,\dots,X_m)$. *Then the reciprocal relations* Q_M^{PED}, Q_W^{PED} *and* Q_Π^{PED} *are moderately stochastic transitive.*

It follows that in these three cases cutting at any level $\alpha \in [1/2, 1]$ yields a strict partial order relation.

It turns out that especially the method with the assumption that the random variables are comonotone $(C = M)$, admits an interesting geometric interpretation. Indeed, in the case of continuous random variables X_i, X_j, Eq. (4) can be expressed as

$$Q_M^{\mathrm{PED}}(X,Y) = \frac{\int (F_Y(x) - F_X(x))_+ \, \mathrm{d}x}{\int |F_Y(x) - F_X(x)| \, \mathrm{d}x}, \quad i,j = 1,\dots,m. \tag{5}$$

References

1. De Baets, B., De Meyer, H.: Towards graded and non-graded variants of stochastic dominance. In: Perception-based Data Mining and Decision Making in Economics and Finance, pp. 261–274. Springer (2007)
2. De Baets, B., De Meyer, H.: On the cycle-transitive comparison of artificially coupled random variables. Int. J. Approx. Reason. **47**, 306–322 (2008)
3. De Baets, B., De Meyer, H., De Schuymer, B., Jenei, S.: Cyclic evaluation of transitivity of reciprocal relations. Soc. Choice Welf. **26**, 217–238 (2006)
4. De Meyer, H., De Baets, B.: Cycle-free cuts of the reciprocal relation generated by random variables that are pairwisely coupled by a Frank copula. In: Proceedings of the SMPS 2018/BELIEF 2018 Conference (Université de Technologie de Compiègne, 17–21 September, 2018) (2018, submitted)
5. De Meyer, H., De Baets, B., De Schuymer, B.: On the transitivity of the comonotonic and countermonotonic comparison of random variables. J. Multivar. Anal. **98**, 177–193 (2007)

A Desirability-Based Axiomatisation
for Coherent Choice Functions

Jasper De Bock$^{(\boxtimes)}$ and Gert de Cooman

SYSTeMS, ELIS, Ghent University, Ghent, Belgium
{jasper.debock,gert.decooman}@ugent.be

Abstract. Choice functions constitute a simple, direct and very general mathematical framework for modelling choice under uncertainty. In particular, they are able to represent the set-valued choices that typically arise from applying decision rules to imprecise-probabilistic uncertainty models. We provide them with a clear interpretation in terms of attitudes towards gambling, borrowing ideas from the theory of sets of desirable gambles, and we use this interpretation to derive a set of basic axioms. We show that these axioms lead to a full-fledged theory of coherent choice functions, which includes a representation in terms of sets of desirable gambles, and a conservative inference method.

1 Introduction

When uncertainty is described by probabilities, decision making is usually done by maximising expected utility. Except in degenerate cases, this leads to a unique optimal decision. If, however, the probability measure is only partially specified—for example by lower and upper bounds on the probabilities of specific events—this method no longer works. Essentially, the problem is that two different probability measures that are both compatible with the given bounds may lead to different optimal decisions. In this context, several generalisations of maximising expected utility have been proposed; see [7] for an nice overview.

A common feature of many such generalisations is that they yield *set-valued choices*: when presented with a set of options, they generally return a subset of them. If this turns out to be a singleton, then we have a unique optimal decision, as before. If, however, it contains multiple options, this means that they are incomparable and that our uncertainty model does not allow us to choose between them. Obtaining a single decision then requires a more informative uncertainty model, or perhaps a secondary decision criterion, as the information present in the uncertainty model does not allow us to single out an optimal option. Set-valued choice is also a typical feature of decision criteria based on other uncertainty models that generalise the probabilistic ones to allow for imprecision and indecision, such as lower previsions and sets of desirable gambles.

Choice functions provide an elegant unifying mathematical framework for studying such set-valued choice. They map option sets to option sets: for any given set of options, they return the corresponding set-valued choice. Hence,

S. Destercke et al. (Eds.): SMPS 2018, AISC 832, pp. 46–53, 2019.
https://doi.org/10.1007/978-3-319-97547-4_7

when working with choice functions, it is immaterial whether there is some underlying decision criterion. The primitive objects of this framework are simply the set-valued choices themselves, and the choice function that represents all these choices, serves as an uncertainty model in and by itself.

A major advantage of working with choice functions is that they allow us to impose axioms on choices, aimed at characterising what it means for choices to be rational and internally consistent; see for example the seminal work by Seidenfeld et al. [6]. Here, we undertake a similar mission, yet approach it from a different angle. Rather than think of choice in an intuitive manner, we provide it with a concrete interpretation in terms of attitudes towards gambling, borrowing ideas from the theory of sets of desirable gambles [1,2,4,9]. From this interpretation alone, and nothing more, we develop a theory of coherent choice that includes a full set of axioms, a representation in terms of sets of desirable gambles, and a natural extension theorem.

Due to length constraints, proofs have been relegated to the appendix of an extended arXiv version of this contribution [3].

2   Choice Functions

A choice function C is a set-valued operator on sets of options. In particular, for any set of options A, the corresponding value of C is a subset $C(A)$ of A. The options themselves are typically actions amongst which a subject wishes to choose. As is customary in decision theory, every action has a corresponding reward that depends on the state of a variable X, about which the subject is typically uncertain. Hence, the reward is uncertain too. The purpose of a choice function is to represent our subject's choices between such uncertain rewards.

Let us make this more concrete. First of all, the variable X takes values x in some set of states $\mathcal{X}$. The reward that corresponds to a given option is then a function u on $\mathcal{X}$. We will assume that this reward can be expressed in terms of a real-valued linear utility scale, allowing us to identify every option with a real-valued function on $\mathcal{X}$.[1] We take these functions to be bounded and call them *gambles*. We use $\mathcal{L}$ to denote the set of all such gambles and also let

$$\mathcal{L}_{>0} := \{u \in \mathcal{L} \colon u \geq 0 \text{ and } u \neq 0\} \text{ and } \mathcal{L}_{\leq 0} := \{u \in \mathcal{L} \colon u \leq 0\}.$$

Option sets can now be identified with subsets of $\mathcal{L}$, which we call *gamble sets*. We restrict our attention here to *finite* gamble sets and will use $\mathcal{Q}$ to denote the set of all such finite subsets of $\mathcal{L}$, including the empty set.

Definition 1 (Choice function). *A choice function C is a map from $\mathcal{Q}$ to $\mathcal{Q}$ such that $C(A) \subseteq A$ for every $A \in \mathcal{Q}$.*

Gambles in A that do not belong to $C(A)$ are said to be *rejected*. This leads to an alternative representation in terms of so-called rejection functions.

---

[1] A more general approach, which takes options to be elements of an arbitrary vector space, encompasses the horse lottery approach, and was explored by Van Camp [8]. Our results here can be easily extended to this more general framework.

Definition 2 (Rejection function). *The* rejection function R_C *corresponding to a choice function C is a map from $\mathcal{Q}$ to $\mathcal{Q}$, defined by $R_C(A) := A \setminus C(A)$ for all $A \in \mathcal{Q}$.*

Since a choice function is completely determined by its rejection function, any interpretation for rejection functions automatically implies an interpretation for choice functions. This allows us to focus on the former.

Our interpretation for rejection functions now goes as follows. Consider a subject whose uncertainty about X is represented by a rejection function R_C, or equivalently, by a choice function C. Then for a given gamble set $A \in \mathcal{Q}$, the statement that a gamble $u \in A$ is rejected from A—that is, that $u \in R_C(A)$—is taken to mean that *there is at least one gamble v in A that our subject strictly prefers over u.*

This interpretation is of course still meaningless, because we have not yet explained the meaning of strict preference. Fortunately, that problem has already been solved elsewhere: strict preference between elements of $\mathcal{L}$ has an elegant interpretation in terms of desirability [5,9], and it is this interpretation that we intend to borrow here. To allow us to do so, we first provide a brief introduction to the theory of sets of desirable gambles.

3   Sets of Desirable Gambles

A gamble $u \in \mathcal{L}$ is said to be *desirable* if our subject strictly prefers it over the zero gamble, meaning that rather than not gamble at all, she strictly prefers to commit to the gamble where, after the true value x of the uncertain variable X has been determined, she will receive the (possibly negative) reward $u(x)$.

A *set of desirable gambles D* is then a subset of $\mathcal{L}$, whose interpretation will be that it consists of gambles $u \in \mathcal{L}$ that our subject considers desirable. The set of all sets of desirable gambles is denoted by $\mathbf{D}$. In order for a set of desirable gambles to represent a rational subject's beliefs, it should satisfy a number of rationality, or *coherence*, criteria.

Definition 3. *A set of desirable gambles $D \in \mathbf{D}$ is called* coherent *if it satisfies the following axioms [1,2,4,5]:*

D_1. $0 \notin D$;
D_2. $\mathcal{L}_{>0} \subseteq D$;
D_3. *if $u, v \in D$, $\lambda, \mu \geq 0$ and $\lambda + \mu > 0$, then $\lambda u + \mu v \in D$.*

We denote the set of all coherent sets of desirable gambles by $\overline{\mathbf{D}}$.

Axioms D_1 and D_2 follow immediately from the meaning of desirability: zero cannot be strictly preferred to itself, and any gamble that is never negative but sometimes positive should be strictly preferred to the zero gamble. Axiom D_3 is implied by the assumed linearity of our utility scale.

Every coherent set of desirable gambles $D \in \overline{\mathbf{D}}$ induces a binary preference order $>_D$—a strict vector ordering—on $\mathcal{L}$, defined by $u >_D v \Leftrightarrow u - v \in D$, for all $u, v \in \mathcal{L}$. The intuition behind this definition is that a subject strictly prefers the uncertain reward u over v if she strictly prefers trading v for u over not trading at all, or equivalently, if she strictly prefers the net uncertain reward $u - v$ over the zero gamble. The preference order $>_D$ fully characterises D: one can easily see that $u \in D$ if and only if $u >_D 0$. Hence, sets of desirable gambles are completely determined by binary strict preferences between gambles.

4 Sets of Desirable Gamble Sets

Let us now go back to our interpretation for choice functions, which is that a gamble u in A is rejected from A if and only if there is some gamble v in A that our subject strictly prefers over u. We will from now on interpret this preference in terms of desirability: we take it to mean that $v - u$ is desirable. In this way, we arrive at the following interpretation for a choice function C. Consider any $A \in \mathcal{Q}$ and $u \in A$, then

$$u \notin C(A) \Leftrightarrow u \in R_C(A) \Leftrightarrow (\exists v \in A)\, v - u \text{ is desirable.} \qquad (1)$$

In other words, if we let $A - \{u\} := \{v - u \colon v \in A\}$, then according to our interpretation, the statement that u is rejected from A is taken to mean that $A - \{u\}$ contains at least one desirable gamble.

A crucial observation here is that this interpretation does not require our subject to specify a set of desirable gambles. Instead, all that is needed is for her to specify those gamble sets $A \in \mathcal{Q}$ that to her contain at least one desirable gamble. We call such gamble sets *desirable gamble sets* and collect them in a *set of desirable gamble sets* $K \subseteq \mathcal{Q}$. As can be seen from Eq. (1), such a set of desirable gamble sets K completely determines a choice function C and its rejection function R_C:

$$u \notin C(A) \Leftrightarrow u \in R_C(A) \Leftrightarrow A - \{u\} \in K, \text{ for all } A \in \mathcal{Q} \text{ and } u \in A.$$

The study of choice functions can therefore be reduced to the study of sets of desirable gamble sets. We will from now on work directly with the latter. We will use the collective term *choice models* for choice functions, rejection functions, and sets of desirable gamble sets.

Let $\mathbf{K}$ denote the set of all sets of desirable gamble sets $K \subseteq \mathcal{Q}$, and consider any such K. The first question to address is when to call K *coherent*: which properties should we impose on a set of desirable gamble sets in order for it to reflect a rational subject's beliefs? We propose the following axiomatisation, using $(\lambda, \mu) > 0$ as a shorthand notation for '$\lambda \geq 0$, $\mu \geq 0$ and $\lambda + \mu > 0$'.

Definition 4 (Coherence). *A set of desirable gamble sets $K \subseteq \mathcal{Q}$ is called coherent if it satisfies the following axioms:*

K_0. $\emptyset \notin K$;
K_1. $A \in K \Rightarrow A \setminus \{0\} \in K$, for all $A \in \mathcal{Q}$;
K_2. $\{u\} \in K$, for all $u \in \mathcal{L}_{>0}$;

K_3. *if* $A_1, A_2 \in K$ *and if, for all* $u \in A_1$ *and* $v \in A_2$, $(\lambda_{u,v}, \mu_{u,v}) > 0$, *then*

$$\{\lambda_{u,v} u + \mu_{u,v} v : u \in A_1, v \in A_2\} \in K;$$

K_4. $A_1 \in K$ *and* $A_1 \subseteq A_2 \Rightarrow A_2 \in K$, *for all* $A_1, A_2 \in \mathcal{Q}$.

We denote the set of all coherent sets of desirable gamble sets by $\overline{\mathbf{K}}$.

Since a desirable gamble set is by definition a set of gambles that contains at least one desirable gamble, Axioms K_0 and K_4 are immediate. The other three axioms follow from the principles of desirability that also lie at the basis of Axioms D_1–D_3: the zero gamble is not desirable, the elements of $\mathcal{L}_{>0}$ are all desirable, and any finite positive linear combination of desirable gambles is again desirable. Axioms K_1 and K_2 follow naturally from the first two of these principles. The argument for Axiom K_3 is more subtle; it goes as follows. Since A_1 and A_2 are two desirable gamble sets, there must be at least one desirable gamble $u \in A_1$ and one desirable gamble $v \in A_2$. Since for these two gambles, the positive linear combination $\lambda_{u,v} u + \mu_{u,v} v$ is again desirable, we know that at least one of the elements of $\{\lambda_{u,v} u + \mu_{u,v} v : u \in A_1, v \in A_2\}$ is a desirable gamble. Hence, it must be a desirable gamble set.

5   The Binary Case

Because we interpret them in terms of desirability, one might be inclined to think that sets of desirable gamble sets are simply an alternative representation for sets of desirable gambles. However, this is not the case: we will see that sets of desirable gamble sets constitute a much more general uncertainty framework than sets of desirable gambles. What lies behind this added generality is that it need not be known which gambles are actually desirable. For example, within the framework of sets of desirable gamble sets, it is possible to express the belief that at least one of the gambles u or v is desirable while remaining undecided about which of them actually is; in order to express this belief, it suffices to state that $\{u, v\} \in K$. This is impossible within the framework of sets of desirable gambles.

Any set of desirable gamble sets $K \in \mathbf{K}$ determines a unique set of desirable gambles based on its binary choices only, given by

$$D_K := \{u \in \mathcal{L} : \{u\} \in K\}.$$

That choice models typically represent more than just binary choice is reflected in the fact that different K can have the same D_K. Nevertheless, there are sets of desirable gamble sets $K \in \mathbf{K}$ that *are* completely characterised by a set of desirable gambles, in the sense that there is a (necessarily unique) set of desirable gambles $D \in \mathbf{D}$ such that $K = K_D$, with

$$K_D := \{A \in \mathcal{Q} : A \cap D \neq \emptyset\}.$$

It follows from the discussion at the end of Sect. 3 that such sets of desirable gamble sets are completely determined by binary preferences between gambles.

We therefore call them, and their corresponding choice functions, *binary*. For any such binary set of desirable gamble sets K, the unique set of desirable gambles $D \in \mathbf{D}$ such that $K = K_D$ is given by D_K.

Proposition 5. *Consider any set of desirable gamble sets $K \in \mathbf{K}$. Then K is binary if and only if $K_{D_K} = K$.*

The coherence of a binary set of desirable gamble sets is completely determined by the coherence of its corresponding set of desirable gambles.

Proposition 6. *Consider any binary set of desirable gamble sets $K \in \mathbf{K}$ and let $D_K \in \mathbf{D}$ be its corresponding set of desirable gambles. Then K is coherent if and only if D_K is.*

6   Representation in Terms of Sets of Desirable Gambles

That there are sets of desirable gamble sets that are completely determined by a set of desirable gambles is nice, but such binary choice models are typically *not* what we are interested in here, because then we could just as well use sets of desirable gambles to represent choice. It is the non-binary coherent choice models that we have in our sights here. But it turns out that our axioms lead to a representation result that allows us to still use sets of desirable gambles, or rather, sets of them, to completely characterise *any* coherent choice model.

Theorem 7 (Representation). *Every coherent set of desirable gamble sets $K \in \overline{\mathbf{K}}$ is dominated by at least one binary set of desirable gamble sets: $\overline{\mathbf{D}}(K) := \{D \in \overline{\mathbf{D}} : K \subseteq K_D\} \neq \emptyset$. Moreover, $K = \bigcap\{K_D : D \in \overline{\mathbf{D}}(K)\}$.*

This powerful representation result allows us to incorporate a number of other axiomatisations [8] as special cases in a straightforward manner, because the binary models satisfy the required axioms, and these axioms are preserved under taking arbitrary non-empty intersections.

7   Natural Extension

In many practical situations, a subject will typically not specify a full-fledged coherent set of desirable gamble sets, but will only provide some partial *assessment* $\mathcal{A} \subseteq \mathcal{Q}$, consisting of a number of gamble sets for which she is comfortable about assessing that they contain at least one desirable gamble. We now want to extend this assessment $\mathcal{A}$ to a coherent set of desirable gamble sets in a manner that is as conservative—or uninformative—as possible. This is the essence of *conservative inference*.

   We say that a set of desirable gamble sets K_1 is less informative than (or rather, at most as informative as) a set of desirable gamble sets K_2, when $K_1 \subseteq K_2$: a subject whose beliefs are represented by K_2 has more (or rather, at least as many) desirable gamble sets—sets of gambles that definitely contain a desirable gamble—than a subject with beliefs represented by K_1. The resulting partially ordered set $(\mathbf{K}, \subseteq)$ is a complete lattice with intersection as infimum and union as supremum. The following theorem, whose proof is trivial, identifies an interesting substructure.

Theorem 8. *Let $\{K_i\}_{i \in I}$ be an arbitrary non-empty family of sets of desirable gamble sets, with intersection $K := \bigcap_{i \in I} K_i$. If K_i is coherent for all $i \in I$, then so is K. This implies that $(\overline{\mathbf{K}}, \subseteq)$ is a complete meet-semilattice.*

This result is important, as it allows us to a extend a partially specified set of desirable gamble sets to the most conservative coherent one that includes it. This leads to the conservative inference procedure we will call natural extension.

Definition 9 (Consistency and natural extension). *For any assessment $\mathcal{A} \subseteq \mathcal{Q}$, let $\overline{\mathbf{K}}(\mathcal{A}) := \{K \in \overline{\mathbf{K}} : \mathcal{A} \subseteq K\}$. We call the assessment $\mathcal{A}$ consistent if $\overline{\mathbf{K}}(\mathcal{A}) \neq \emptyset$, and we then call $\mathrm{Ex}(\mathcal{A}) := \bigcap \overline{\mathbf{K}}(\mathcal{A})$ the natural extension of $\mathcal{A}$.*

In other words: an assessment $\mathcal{A}$ is consistent if it can be extended to some coherent rejection function, and then its natural extension $\mathrm{Ex}(\mathcal{A})$ is the least informative such coherent rejection function.

Our final result provides a more 'constructive' expression for this natural extension and a simpler criterion for consistency. In order to state it, we need to introduce the set $\mathcal{L}^{\mathrm{s}}_{>0} := \{\{u\} : u \in \mathcal{L}_{>0}\}$ and two operators on—transformations of—$\mathbf{K}$. The first is denoted by Rs, and defined by

$$\mathrm{Rs}(K) := \{A \in \mathcal{Q} : (\exists B \in K) B \setminus \mathcal{L}_{\leq 0} \subseteq A\} \text{ for all } K \in \mathbf{K},$$

so $\mathrm{Rs}(K)$ contains all gamble sets A in K, all versions of A with some of their non-positive options removed, and all supersets of such sets. The second is denoted by Posi, and defined for all $K \in \mathbf{K}$ by

$$\mathrm{Posi}(K) := \left\{ \left\{ \sum_{k=1}^{n} \lambda_k^{u_{1:n}} u_k : u_{1:n} \in \times_{k=1}^{n} A_k \right\} : n \in \mathbb{N}, (A_1, \ldots, A_n) \in K^n, \right.$$
$$\left. \left(\forall u_{1:n} \in \times_{k=1}^{n} A_k \right) \lambda_{1:n}^{u_{1:n}} > 0 \right\},$$

where we used the notations $u_{1:n}$ and $\lambda_{1:n}^{u_{1:n}}$ for n-tuples of options u_k and real numbers $\lambda_k^{u_{1:n}}$, $k \in \{1, \ldots, n\}$, so $u_{1:n} \in \mathcal{L}^n$ and $\lambda_{1:n}^{u_{1:n}} \in \mathbb{R}^n$. We also used $\lambda_{1:n}^{u_{1:n}} > 0$ as a shorthand for '$\lambda_k^{u_{1:n}} \geq 0$ for all $k \in \{1, \ldots, n\}$ and $\sum_{k=1}^{n} \lambda_k^{u_{1:n}} > 0$'.

Theorem 10 (Natural extension). *Consider any assessment $\mathcal{A} \subseteq \mathcal{Q}$. Then $\mathcal{A}$ is consistent if and only if $\emptyset \notin \mathcal{A}$ and $\{0\} \notin \mathrm{Posi}(\mathcal{L}^{\mathrm{s}}_{>0} \cup \mathcal{A})$. Moreover, if $\mathcal{A}$ is consistent, then $\mathrm{Ex}(\mathcal{A}) = \mathrm{Rs}(\mathrm{Posi}(\mathcal{L}^{\mathrm{s}}_{>0} \cup \mathcal{A}))$.*

8   Conclusion

Our representation result shows that binary choice *is* capable of representing general coherent choice functions, provided we extend its language with a 'disjunction' of desirability statements—as is implicit in our interpretation—, next to the 'conjunction' and 'negation' that are already implicit in the language of sets of desirable gambles—see [5] for a clear exposition of the latter claim.

In addition, we have found recently that by adding a convexity axiom, and working with more general vector spaces of options to allow for the incorporation of horse lotteries, our interpretation and corresponding axiomatisation allows for a representation in terms of lexicographic sets of desirable gambles [8], and therefore encompasses the one by Seidenfeld et al. [6] (without archimedeanity). We will report on these findings in more detail elsewhere.

Future work will address (i) dealing with the consequences of merging our accept-reject statement framework [5] with the choice function approach to decision making; (ii) discussing the implications of our axiomatisation and representation for conditioning, independence, and indifference (exchangeability); and (iii) expanding our natural extension results to deal with the computational and algorithmic aspects of conservative inference with coherent choice functions.

Acknowledgements. This work owes a large intellectual debt to Teddy Seidenfeld, who introduced us to the topic of choice functions. His insistence that we ought to pay more attention to non-binary choice if we wanted to take imprecise probabilities seriously, is what eventually led to this work.

The discussion in Arthur Van Camp's PhD thesis [8] was the direct inspiration for our work here, and we would like to thank Arthur for providing a pair of strong shoulders to stand on.

As with most of our joint work, there is no telling, after a while, which of us two had what idea, or did what, exactly. We have both contributed equally to this paper. But since a paper must have a first author, we decided it should be the one who took the first significant steps: Jasper, in this case.

References

1. De Cooman, G., Quaeghebeur, E.: Exchangeability and sets of desirable gambles. Int. J. Approx. Reason. **53**(3), 363–395 (2012). special issue in honour of Henry E. Kyburg, Jr
2. Couso, I., Moral, S.: Sets of desirable gambles: conditioning, representation, and precise probabilities. Int. J. Approx. Reason. **52**(7), 1034–1055 (2011)
3. De Bock, J., De Cooman, G.: A desirability-based axiomatisation for coherent choice functions (2018). https://arxiv.org/abs/1806.01044
4. De Cooman, G., Miranda, E.: Irrelevance and independence for sets of desirable gambles. J. Artif. Intell. Res. **45**, 601–640 (2012)
5. Quaeghebeur, E., De Cooman, G., Hermans, F.: Accept & reject statement-based uncertainty models. Int. J. Approx. Reason. **57**, 69–102 (2015)
6. Seidenfeld, T., Schervish, M.J., Kadane, J.B.: Coherent choice functions under uncertainty. Synthese **172**(1), 157–176 (2010)
7. Troffaes, M.C.M.: Decision making under uncertainty using imprecise probabilities. Int. J. Approx. Reason. **45**(1), 17–29 (2007)
8. Van Camp, A.: Choice Functions as a Tool to Model Uncertainty. Ph.D. thesis, Ghent University, Faculty of Engineering and Architecture, January 2018
9. Walley, P.: Towards a unified theory of imprecise probability. Int. J. Approx. Reason. **24**, 125–148 (2000)

Cycle-Free Cuts of the Reciprocal Relation Generated by Random Variables that are Pairwisely Coupled by a Frank Copula

Hans De Meyer[1]([⊠]) and Bernard De Baets[2]

[1] Department of Applied Mathematics, Computer Science and Statistics,
Ghent University, Krijgslaan 281 S9, 9000 Gent, Belgium
hans.demeyer@ugent.be
[2] Department of Data Analysis and Mathematical Modelling, Ghent University,
Coupure links 653, 9000 Gent, Belgium
bernard.debaets@ugent.be

1 Introduction

Some years ago, we have investigated the transitivity properties of the reciprocal relation generated from the pairwise comparison of the components of a random vector [2]. Using the notion of winning probability to compare two random variables (RVs), the reciprocal relation Q associated with a random vector $(X_1, X_2, \ldots, X_m)$ is given by

$$ Q(X_i, X_j) = \mathrm{Prob}\{X_i > X_j\} + \frac{1}{2}\mathrm{Prob}\{X_i = X_j\}, \quad i, j = 1, \ldots, m. $$

Clearly, the bivariate marginal distribution function F_{X_i, X_j} contains all the information necessary for computing the winning probability $Q(X_i, X_j)$. In general, the m-copula underlying a random vector can be quite intricate and contain different marginal 2-copulas. However, the picture is still far from clear, which is commonly known as the compatibility problem. This difficulty has motivated us to focus on the situation where all bivariate couplings are realized by a same copula [2], notwithstanding this might be an utterly artificial situation. In this specific setting, the transitivity property of the generated reciprocal relation can be expressed in the general framework of cycle-transitivity [1,5,6] if the chosen copula fulfills an infinite number of functional inequalities. Later on, we could prove that only copulas that belong to the Frank family of copulas and certain ordinal sums of Frank copulas satisfy all these inequalities [3,4].

Recall that the family of Frank copulas $(C_\lambda^F)_{\lambda \in [0,\infty]}$ consists of the copulas

$$ C_\lambda^F(x, y) = \log_\lambda\left[1 + \frac{(\lambda^x - 1)(\lambda^y - 1)}{\lambda - 1}\right], \quad \lambda \notin \{0, 1, \infty\}, $$

and the limits $C_0^F = M$, $C_1^F = \Pi$ and $C_{+\infty}^F = W$.

© Springer Nature Switzerland AG 2019
S. Destercke et al. (Eds.): SMPS 2018, AISC 832, pp. 54–58, 2019.
https://doi.org/10.1007/978-3-319-97547-4_8

Note that we use the parameterization common in the theory of t-norms, while for the limiting copulas we use the notation common in the framework of copulas. From here on, we will focus on random variables $X_1, \ldots, X_m$ that are pairwisely coupled by a same Frank copula C_λ^F, i.e. for which $F_{X_i, X_j}(x, y) = C_\lambda^F(F_{X_i}(x), F_{X_j}(y))$ for all $i, j \in \{1, \ldots, m\}$. The corresponding reciprocal relation is denoted by Q_λ. Note that Q_λ also depends on the marginal CDF F_{X_i} of the RV X_i, $i \in \{1, \ldots, m\}$, but in general, we are interested in properties that hold for any possible marginal CDFs.

2 Cycle-Free Cuts

In the present contribution, we are interested in obtaining minimum cutting levels c_λ^n such that the crisp relation obtained from the reciprocal relation Q_λ by setting to 0 its elements smaller than or equal to c_λ^n, and to 1 its other elements[1], is free from cycles of length n, irrespective of the marginal CDFs of the RVs. For instance, cutting Q_λ at level c_λ^3 yields a crisp relation of which the associated directed m-graph (a graph with m nodes) does not contain any 3-cycle (though it can contain cycles of length greater than 3). Clearly, cutting Q_λ at a level greater than or equal to c_λ^m guarantees that the corresponding graph is free of any cycles. The transitive closure of such graph clearly is the graph of a strict partial order.

The cutting level c_λ^3, i.e. the one at which 3-cycles disappear, can be straightforwardly computed from the inequality that characterizes the type of transitivity of the reciprocal relation Q_λ [2]. It turns out that c_λ^3 is the largest solution in $[0, 1]$ of the equation

$$C_{1/\lambda}^F(x, x) = 1 - x \, .$$

Defining $s = 1/\lambda$ and $t = s^x$, we have, given s, to solve for t the cubic equation

$$t^3 - 2t^2 + st - s(s - 1) = 0 \, .$$

This can be done with any symbolic computation package. However, for the copulas M, Π and W, the result is immediate:

(a) M ($\lambda = 0$): since $s = +\infty$, it holds that c_0^3 is a solution of $2x - 1 = 1 - x$, hence, $c_0^3 = 2/3$.
(b) Π ($\lambda = 1$): since $s = 1$, it holds that c_1^3 is a solution of $x^2 = 1 - x$, hence, $c_0^3 = (\sqrt{5} - 1)/2 = 0.628\ldots$, commonly known as the golden section.
(c) W ($\lambda = \infty$): since $s = 0$, it holds that c_0^3 is a solution of $x = 1 - x$, hence, $c_\infty^3 = 1/2$.

---

[1] This notion of cutting is known in fuzzy set theory as strict α-cuts with cutting level α.

3   Main Results

In our search for those cutting levels that, like c_λ^3, are computable in closed form, we found that this is possible for the following families: c_λ^4 for any $\lambda \in \mathbb{R}^+$, c_p^n for any $p \in \{0, 1, \infty\}$ and any $n > 3$, and c_λ^∞ for any $\lambda \in \mathbb{R}^+$. Surprisingly, the computation of c_λ^4, the cutting level at which 3-cycles and 4-cycles disappear, is even simpler than the computation of c_λ^3. The reason is that, using the same notations as before, the key equation to be solved is a quadratic equation in t (instead of a cubic equation), namely $t^2 + \sqrt{s}\, t - s(\sqrt{s} + 1) = 0$.

Proposition 1. *Let $X_1, \ldots, X_m$ be m RVs and consider the reciprocal relation Q_λ generated from the pairwise comparison of these RV through coupling with the same Frank copula C_λ^F. Then the cutting level c_λ^4 can be computed as*

$$c_\lambda^4 = \frac{1}{2} + \log_\lambda \left(\frac{\sqrt{\lambda} + \sqrt{5\lambda + 4\sqrt{\lambda}}}{2(\sqrt{\lambda} + 1)} \right).$$

For the limit cases $\lambda \to 0$, $\lambda \to 1$ and $\lambda \to +\infty$, we respectively obtain

$$c_0^4 = 3/4\,, \qquad c_1^4 = 2/3\,, \qquad c_\infty^4 = 1/2\,.$$

Note that it also holds that $c_\infty^n = 1/2$ for all $n \geq 3$.

Proposition 2.

(i) *Let $(X_1, \ldots, X_m)$ be a random vector with pairwise independent components (i.e., pairwisely coupled with Π) and let Q_1 be the reciprocal relation built from the pairwise winning probabilities. The cutting levels c_1^n can be computed as*

$$c_1^n = 1 - \frac{1}{4\cos^2\left(\frac{\pi}{n+2}\right)}.$$

In the limit of $n \to +\infty$, we obtain $c_1^\infty = 3/4$.

(ii) *Let $(X_1, \ldots, X_m)$ be a random vector with pairwise co-monotone components (i.e., pairwisely coupled with M, in other words, with underlying m-copula given by the co-monotone m-copula) and let Q_0 be the reciprocal relation built from the pairwise winning probabilities. The cutting levels c_0^n can be computed as*

$$c_0^n = \frac{n-1}{n}.$$

In the limit of $n \to +\infty$, we obtain $c_0^\infty = 1$.

Note that with the above formulas, we retrieve $c_1^3 = 1 - \frac{1}{4}\cos^2(\pi/5) = (\sqrt{5}-1)/2 = 0.628\ldots$ and $c_1^4 = 1 - \frac{1}{4}\cos^2(\pi/6) = 2/3$.

Proposition 3. *Let $X_1, \ldots, X_m$ be m RVs and consider the reciprocal relation Q_λ generated from the pairwise comparison of these RVs through coupling with the same Frank copula C_λ^F. Then the cutting level c_λ^∞ can be computed as*

$$c_\lambda^\infty = 1 - \log_\lambda \left(\frac{\lambda - 1}{2(\sqrt{\lambda} - 1)} \right) = 1 - \log_\lambda \left(\frac{\sqrt{\lambda} + 1}{2} \right)$$

with limits $c_0^\infty = 1$, $c_1^\infty = 3/4$ and $c_\infty^\infty = 1/2$.

4   Discussion

The cutting levels we have computed show regularities that allow us to draw the following conclusions.

1. In practice, we can use c_λ^∞ as cutting level to avoid cycles of any length. Note that this is too demanding as for any m RVs, it suffices to cut at the lower level c_λ^m. However, except for $\lambda \in \{0, 1, +\infty\}$, the cutting levels c_λ^n, $n > 4$, cannot be derived in closed form, but can only be computed by solving numerically a polynomial equation of high degree.
2. The closer the parameter λ of the Frank copula C_λ^F is to 0 (i.e., the more the RVs are treated as being pairwisely comonotone), the higher the cutting level c_λ^n is, for any fixed n, and the closer to 1 the winning probabilities in Q_λ must be to prevent the occurrence of n-cycles. Hence, the closer λ is to 0, the sparser the crisp relation that results from cutting Q_λ at level c_λ^m becomes, thus rendering more RVs incomparable.
3. At the other end of the scale, the more the parameter λ tends to ∞ (i.e., the more the RVs are treated as being pairwisely counter-monotone, a situation that is more and more artificial the larger $m > 2$), the closer the cutting level c_λ^n is to $1/2$ for any n, which means that the probability that the $(1/2)$-cut of Q_λ is cycle-free tends to 1, hence that the simple property of having more chance to win than to lose (winning probability $> 1/2$) suffices to induce a partial order on the RV.
4. We have assumed that all pairwise couplings are realised with the same Frank copula. However, due to the monotonicity of the cutting levels w.r.t. λ, when using different Frank copulas in the pairwise couplings, a cycle-free cut can be performed at the cutting level associated with the Frank copula with the smallest value of λ. If no information is available about the bivariate margins, one can safely cut at cutting level c_0^m.

References

1. De Baets, B., De Meyer, H.: Transitivity frameworks for reciprocal relations: cycle-transitivity versus FG-transitivity. Fuzzy Sets Syst. **152**, 249–270 (2005)
2. De Baets, B., De Meyer, H.: On the cycle-transitive comparison of artificially coupled random variables. Int. J. Approx. Reason. **47**, 306–322 (2008)

3. De Baets, B., De Meyer, H.: On a conjecture about the Frank copula family. Fuzzy Sets Syst. **228**, 15–28 (2013)
4. De Baets, B., De Meyer, H.: The Frank inequality. Fuzzy Sets Syst. **335**, 18–29 (2018)
5. De Baets, B., De Meyer, H., De Schuymer, B., Jenei, S.: Cyclic evaluation of transitivity of reciprocal relations. Soc. Choice Welfare **26**, 217–238 (2006)
6. De Schuymer, B., De Meyer, H., De Baets, B., Jenei, S.: On the cycle-transitivity of the dice model. Theor. Decis. **54**, 261–285 (2003)

Density Estimation with Imprecise Kernels: Application to Classification

Guillaume Dendievel[1,2(✉)], Sebastien Destercke[1,2], and Pierre Wachalski[1,2]

[1] Technologic University of Compiegne, CNRS, UMR 7253 - Heudiasyc,
Centre de Recherche de Royallieu, Compiègne, France
guillaume.dendievel@gmail.com, sebastien.destercke@utc.fr,
pierre.wachalski@gmail.com
[2] Openvalue, 58 Avenue Charles de Gaulle, 92200 Neuilly-sur-Seine, France

Abstract. In this paper, we explore the problem of estimating lower and upper densities from imprecisely defined families of parametric kernels. Such estimations allow to rely on a single bandwidth value, and we show that it provides good results on classification tasks when extending the naive Bayesian classifier.

Keywords: Density estimation · Kernel · Imprecision · Classification

1 Introduction

Estimating probability densities is a key task in many problems: signal filtering, classification, risk and uncertainty analysis, ... When the densities are known to belong to some parametric family, one can use efficient estimators of the parameters, yet when it is not the case, non-parametric methods such as kernel-based estimation must be used [3].

To perform this estimation, we need a kernel shape and a kernel bandwidth. It is commonly recognized that the resulting estimation will often not be sensitive to the kernel shape, but can be highly sensitive to the choice of the bandwidth [4]. A too low bandwidth will capture very local variations, while a too high bandwidth will provide a too smooth density. This is particularly true when the number of samples is low.

Except for specific cases, finding an optimal bandwidth for a finite sample of values is not doable. It may therefore be interesting to let the bandwidth vary in a pre-determined interval, obtaining upper and lower values of the estimated density. Such bounds can then be used in robustness analysis, ensuring that the inferences do not depend too much on the bandwidth value. For example, Destercke and Strauss [2] consider so-called cloudy kernels (pairs of possibility distributions) to perform signal filtering.

In this paper, we study how lower and upper density bounds can be obtained from imprecise bandwidth defined for a given family of kernels. The approach is described Sect. 2, and Sect. 3 deals with the practical problem of computing

S. Destercke et al. (Eds.): SMPS 2018, AISC 832, pp. 59–67, 2019.
https://doi.org/10.1007/978-3-319-97547-4_9

those bounds for specific kernels. We apply in Sect. 4 our findings to the naive Bayesian classifier, obtaining a non-parametric credal naive classifier that can deal with continuous data.

2 From Precise to Imprecise Kernel Density Estimation

A common problem when observing a sample $x_1, \ldots, x_N$ of a random variable $X \in \mathbb{R}$ is to estimate its density function $f : \mathbb{R}^+ \to \mathbb{R}$. When f can be assumed to follow some parametric model, estimating it comes down to estimate its parameters. There are cases, though, where f cannot be satisfactorily approximated by a simple parametric family.

In such cases, non-parametric kernel density estimation can be used to estimate density values without making a priori assumptions about its shape. Given a scaled kernel K with bandwidth h^1, the estimated density at point x is

$$\hat{f}_h(x) = \frac{1}{Nh} \sum_{i=1}^{N} K\left(\frac{x - x_{(i)}}{h}\right).$$

It is well known that the exact shape of K has in general a small influence on the end-result, while different choices of bandwidth h may lead to very different results.

This is why it could be interesting to develop tools that allow one to consider sets of bandwidth at once, thus providing a way to perform a global sensitivity analysis. The basic idea is the following: given an interval $H = [\underline{h}, \overline{h}]$ of possible values, how can we determine, for a point x, the upper and lower bounds of the corresponding density, i.e.,

$$\underline{\hat{f}}_H(x) = \inf_{h \in H} \hat{f}_h(x) \text{ and } \overline{\hat{f}}_H(x) = \sup_{h \in H} \hat{f}_h(x). \tag{1}$$

Finding the solutions to these equations is non-trivial in general, as the functions to optimize are usually non-convex in h. In practice, we should try to find kernels for which efficient algorithmic solutions exist. This is what we do next, for the cases of triangular and Epanechnikov kernels, recalled in Table 1.

Table 1. Triangular and Epanechnikov kernels

Name	K	Shape
Epanechnikov	$K(x) = \frac{3}{4}(1-x^2)I_{|x|\leq 1}$	-1 0 1 x
Triangular	$K(x) = (1-|x|)I_{|x|\leq 1}$	-1 0 1 x

---

[1] A kernel is here a symmetric, non-negative function with $\int_{\mathbb{R}} K(y)dy = 1$ and mean 0.

3   Particular Tractable Cases

In this section, we study how solutions for Eq. (1) can be found for some specific kernels, namely the Triangular and Epanechnikov ones, and for a specific value x. Since we focus on a particular value x, we will consider the re-indexing $x_{(1)}, \ldots, x_{(N)}$ of the sample in an increasing sequence with respect to their distance of x, that is such that $|x_{(i)} - x| \leq |x_{(i+1)} - x|$ for any $i = 1, \ldots, N - 1$.

We will also use the notations $D_{(i)} = |x_{(i)} - x|$ and $\mathbb{E}(D_{(i)}) = \left(\sum_{j=1}^{i} D_{(j)} \right) / i$ to simplify further proofs.

3.1   Triangular Kernel

To compute the bounds given by Eq. (1) over the interval $[\underline{h}, \overline{h}]$, we need to identify points that will reach this global optimum, and to do this we show that local optimums are easy to obtain within each interval $[D_{(i)}, D_{(i+1)}]$. This is shown below.

Proposition 1. *For a triangular kernel K and values of $h \in [D_{(i)}, D_{(i+1)}]$, we have:*

  – if $D_{(i+1)} < 2\mathbb{E}(D_{(i)})$ or $D_{(i)} > 2\mathbb{E}(D_{(i)})$,

$$\max_{h \in [D_{(i)}, D_{(i+1)}]} \hat{f}_h(x) = \max(\hat{f}_{D_{(i+1)}}(x), \hat{f}_{D_{(i)}}(x))$$

  – if $D_{(i)} \leq 2\mathbb{E}(D_{(i)}) \leq D_{(i+1)}$, $\hat{f}_h(x)$ has one maximal value in h and

$$\max_{h \in [D_{(i)}, D_{(i+1)}]} \hat{f}_h(x) = \hat{f}_{2\mathbb{E}(D_{(i)})}(x)$$

and the minimal value is given by

$$\min_{h \in [D_{(i)}, D_{(i+1)}]} \hat{f}_h(x) = \min(\hat{f}_{D_{(i)}}(x), \hat{f}_{D_{(i+1)}}(x))$$

Proof (sketch). If $h \in [D_{(i)}, D_{(i+1)}]$, we can write

$$\hat{f}_h(x) = \frac{1}{Nh} \sum_{j=1}^{i} K\left(\frac{x - x_{(j)}}{h} \right).$$

as $K\left(x - x_{(j)}/h \right)$ will be null for any $j \geq i + 1$, and non-negative for any $j < i + 1$. The derivative in h is $\partial \hat{f}_h(x)/\partial h = -i/Nh^2 + 2*\sum_{j=1}^{i} D_{(j)}/N*h^3$. The following table shows the sign variation of this function, which is sufficient to obtain the proposition

h	$< 2\mathbb{E}(D_{(i)})$	$= 2\mathbb{E}(D_{(i)})$	$> 2\mathbb{E}(D_{(i)})$
$\partial \hat{f}_h(x)/\partial h$	< 0	$= 0$	> 0

Next we show that $\hat{f}_h(x)$ is continuous in h, hence that going from $[\underline{h}, \overline{h}]$ to $[\underline{h} - \epsilon, \overline{h} + \epsilon]$ will not induce "jumps" in our results.

Proposition 2. $\hat{f}_h(x) = \frac{1}{Nh} \sum_{j=1}^{i} K\left(\frac{x - x_{(j)}}{h}\right)$ is piecewise continuous

Proof. Let us show the continuity in $D_{(i+1)}$. Consider first $h \in [D_{(i)}; D_{(i+1)}]$, we have

$$\lim_{\substack{h \to D_{(i+1)} \\ h < D_{(i+1)}}} \hat{f}_{h,i}(x) = \frac{i}{Nh} - \frac{(\sum_{j=1}^{i} D_{(j)})}{N * h^2} = \frac{i}{ND_{(i+1)}} - \frac{\sum_{j=1}^{i} D_{(j)}}{N * D_{(i+1)}^2}$$

with the last equality being obtained by taking $h = D_{(i+1)}$. Now consider $h \in [D_{(i+1)}; D_{(i+2)}]$

$$\lim_{\substack{h \to D_{(i+1)} \\ h > D_{(i+1)}}} \hat{f}_{h,i+1}(x) = \frac{i+1}{Nh} - \frac{\sum_{j=1}^{i+1} D_{(j)}}{N * h^2} = \frac{i+1}{ND_{(i+1)}} - \frac{\sum_{j=1}^{i+1} D_{(j)}}{N * D_{(i+1)}^2}$$

and we have that these two values are equal.

Algorithm 1. Find $\underline{\hat{f}}(x), \overline{\hat{f}}(x)$ in $H = [\underline{h}, \overline{h}]$

Input: $x_{(i)}$, $D_{(i)}$ sorted in ascending order, H, x, $i = 1$
Output: Bounds $\underline{\hat{f}}(x), \overline{\hat{f}}(x)$
while $i \neq N$ do
    if $[a, b] = H \cap [D_{(i)}, D_{(i+1)}] \neq \emptyset$ then
        $\underline{\hat{f}}(x) \leftarrow min(\underline{\hat{f}}(x), \hat{f}(a), \hat{f}(b))$;
        $\overline{\hat{f}}(x) \leftarrow max(\overline{\hat{f}}(x), \hat{f}(a), \hat{f}(b))$
    if $2 * \mathbb{E}(D_{(i)}) \in [a, b]$ then
        $\overline{\hat{f}}(x) \leftarrow max(\overline{\hat{f}}(x), \hat{f}(2 * \mathbb{E}(D_{(i)})))$
    $i \leftarrow i + 1$

3.2 Epanechnikov Kernel

Results similar to the previous case can be given for the Epanechnikov case. Due to page limit restriction, we only provide the main results.

Proposition 3. *For an Epanechnikov kernel K and values of $h \in [D_{(i)}, D_{(i+1)}]$, we have:*

– *if* $D_{(i+1)} < (3 * \mathbb{E}(D^2_{(i)}))^{\frac{1}{2}}$ *or* $D_{(i)} > (3 * \mathbb{E}(D^2_{(i)}))^{\frac{1}{2}}$,

$$\max_{h \in [D_{(i)}, D_{(i+1)}]} \hat{f}_h(x) = \max(\hat{f}_{D_{(i+1)}}(x), \hat{f}_{D_{(i)}}(x))$$

– *if* $D_{(i)} \leq (3 * \mathbb{E}(D^2_{(i)}))^{\frac{1}{2}} \leq D_{(i+1)}$, $\hat{f}_h(x)$ *has one maximal value in* h *and*

$$\max_{h \in [D_{(i)}, D_{(i+1)}]} \hat{f}_h(x) = \hat{f}_{(3 * \mathbb{E}(D^2_{(i)}))^{\frac{1}{2}}}(x)$$

and the minimal value is given by

$$\min_{h \in [D_{(i)}, D_{(i+1)}]} \hat{f}_h(x) = \min(\hat{f}_{D_{(i)}}(x), \hat{f}_{D_{(i+1)}}(x))$$

Proposition 4. $\hat{f}_h(x) = \frac{1}{Nh} \sum_{j=1}^{i} K\left(\frac{x - x_{(j)}}{h}\right)$ *is piecewise continuous*

3.3   Illustrative Experiments

To illustrate the results provided by imprecise kernels, we perform experiments where we modify the number of observations and the size of the interval H. To do that, we generated points from a bimodal mixture of Gaussian $X \sim 0.6\mathcal{N}(-1, 1) + 0.4\mathcal{N}(5, 1)$. To perform our experiments, we started from a reference bandwidth that corresponds to the optimal one for the normal case $h^* = (1.06 \cdot \hat{\sigma} \cdot N)^{-\frac{1}{5}}$.

Increasing sample. Figure 1 shows the results of the following experiments: we generate 1500 points and pick a fixed $H = [h^* - 0.2 * h^*, h^* + 0.2 * h^*]$. We then randomly shuffle the samples, and take each time the first n samples to achieve density estimation. From the picture, we can easily see that the more points we get, the less imprecise we are.

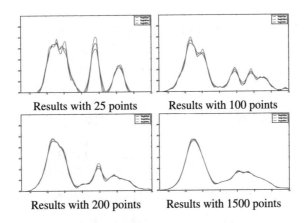

Results with 25 points   Results with 100 points

Results with 200 points   Results with 1500 points

Fig. 1. Imprecise estimation with varying sample sizes

Increasing bandwidth. we now fix the number of samples to 75, and make estimations using intervals going from $H = [h^* - 0.05 * h^*, h^* + 0.05 * h^*]$ to $H = [h^* - 0.90h^*, h^* + 0.90h^*]$. The results are shown in Fig. 2, and again we can easily see the increase of imprecision, as well as the increasing noise as the lower bound of H gets close to 0.

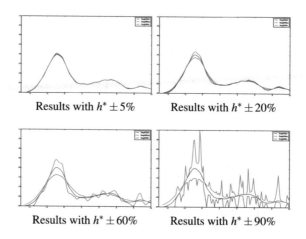

Results with $h^* \pm 5\%$ Results with $h^* \pm 20\%$

Results with $h^* \pm 60\%$ Results with $h^* \pm 90\%$

Fig. 2. Imprecise estimation with varying bandwidths

4 Application to Naive Credal Classification

As an illustration of our approach, we will apply it to the popular naive Bayes classifier, and will turn in into a credal naive classifier [5] on continuous variables, whereas most of the previous versions only accepts discrete variables [1].

4.1 Naive Credal Classification: Brief Reminder

The Naive Bayes model is a very popular classification model that considers inputs from a multivariate space $X = X^1 \times \ldots \times X^p$ and outputs in the form of a discrete class Y. The Naive Bayes model proposes to estimate the posterior probability $p(y|x)$ of class y given observation $x = (x^1, \ldots, x^p)$ by assuming that input variables are independent, given the class. That is, $p(y|x)$ can be rewritten

$$p(y|x) = \frac{p(x|y)p(y)}{\sum_{y \in Y} p(x|y)p(y)} = \frac{p(y) \prod_{i=1}^{p} p(x^i|y)}{\sum_{y \in Y} p(y) \prod_{i=1}^{p} p(x^i|y)}. \tag{2}$$

Given two classes y and y', checking that $p(y|x) \geq p(y'|x)$ comes down to check that

$$\frac{p(y|x)}{p(y'|x)} = \frac{p(y) \prod_{i=1}^{p} p(x^i|y)}{p(y') \prod_{i=1}^{p} p(x^i|y')} \tag{3}$$

is higher than one. In this case, we say that y is preferred to y', noted $y \succ y'$. When probabilities become imprecise, this comes down to test whether the infimum value of Eq. (3) is higher than 1. This becomes, when class prior probabilities are assumed precise

$$\inf_{p(x^i|y) \in [\underline{p}(x^i|y), \overline{p}(x^i|y)]} \frac{p(y|x)}{p(y'|x)} = \frac{p(y) \prod_{i=1}^{P} \underline{p}(x^i|y)}{p(y') \prod_{i=1}^{P} \overline{p}(x^i|y')}$$

This may result in a partial order, in which case our prediction consists in taking all the maximal elements of the resulting order.

4.2   Experimental Protocol

In our training, for any pair feature/class, we consider symmetric intervals around the estimation $h^* = (1.06 \cdot \hat{\sigma} \cdot N)^{-\frac{1}{5}}$. To avoid zero-probability of $\underline{p}(x^i|y)$ or of $\underline{p}(x^i|y)^2$, we take $\underline{p}(x^i|y) = \max(0, 0.1 * p(x^i|y))$, and we set $\underline{p}(x^i|y) = 10^{-3}$ if it is null.

The protocol adopted is the following: for each data set, we decide how imprecise our kernels will be by setting ϵ and taking $H = [h^* - \epsilon h^*, h^* + \epsilon h^*]$, and a split ratio of training/test. We then perform ten repetitions for each couple (ϵ, ratio). The selected data sets are summarised in Table 2.

Table 2. Selected data sets

#	a	b	c	d	e	f	g	h	i	j
Names	Breats	Iris	Wine	Automobile	Seed	Glass	Forest	Dermatology	Diabete	Segment
Instances	106	150	178	205	210	214	325	366	769	2310
Features	10	4	13	26	7	9	27	34	8	19
Labels	6	3	3	7	3	7	4	6	2	7

As we deal with imprecise results, we have chosen to use well-motivated utility-discounted accuracies u65 and u80 [6]

$$u_{65} = \frac{1}{T} \sum_{t=1}^{T} -0.6a_t^2 + 1.6a_t \quad \text{and} \quad u_{80} = \frac{1}{T} \sum_{t=1}^{T} -1.2a_t^2 + 2.2a_t$$

where $a_t = \mathbb{1}_{y_t \in Y_t}/|Y_t|$, Y_t being the predicted set. In practice, u_{80} rewards more imprecise predictions than u_{65}, hence should be more favourable to imprecise methods in comparisons. Table 3 summarizes the obtained results, showing that our method is clearly superior in terms of u_{80} for most configurations, and quite competitive in terms of u_{60}. Figure 3 shows the accuracy on those instances who were imprecisely predicted by our approach, on which we can notice that the precise model accuracy drops (e.g. for data set c, accuracy on those instances is less than 10%, while it is more than 60% in average), while our approach is almost systematically right.

---

[2] In some sense, to regularize our model.

Table 3. Experimental results

#	stats	SR = 50%		ε = 20%		#	SR = 50%		ε = 20%	
		ε = 10%	ε = 40%	SR=30%	SR=75%		ε = 10%	ε = 40%	SR=30%	SR=75%
a	precise	56.4	56.4	56.9	57.0	f	80.7	80.7	80.1	82.6
	u_{65}	61.4	49.0	55.2	55.9		52.7	39.6	45.9	46.3
	u_{80}	71.1	61.1	67.5	67.6		64.0	50.6	57.1	57.4
b	precise	97.1	97.1	96.3	95.8	g	87.4	87.4	87.2	87.3
	u_{65}	97.3	96.6	96.9	96.1		88.5	88.9	88.2	88.4
	u_{80}	97.5	97.1	97.3	96.3		89.1	90.9	89.4	89.4
c	precise	62.7	62.7	61.3	62.9	h	98.9	98.9	99.1	99.0
	u_{65}	86.2	82.0	84.9	85.9		96.9	78.3	92.2	92.8
	u_{80}	92.0	88.5	91.0	91.6		98.2	84.7	95.4	95.8
d	precise	80.0	80.0	79.6	79.8	i	79.2	79.2	79.7	79.7
	u_{65}	82.8	61.0	74.9	74.0		79.7	79.6	80.0	79.5
	u_{80}	86.2	71.6	81.5	80.9		81.5	85.7	83.5	83.1
e	precise	93.1	93.1	93.6	94.0	j	89.3	89.3	89.2	89.3
	u_{65}	92.4	91.6	92.2	92.2		61.7	50.1	56.7	56.3
	u_{80}	93.1	93.4	93.7	93.8		71.8	60.9	66.9	66.6

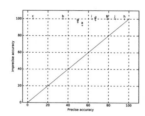

Fig. 3. Accuracy of imprecise predictions for $\epsilon = 0.2$ and split ratio $= 0.3$

5   Conclusion and Perspectives

In this paper, we have introduced the idea of using imprecise parametric kernels in order to estimate density bounds. We have shown that for some kernels, efficient algorithmic procedure can be developed, and that good results can be obtained in classification problems (at least for the naive credal classifier). Since density estimation plays an important role in many applications, we expect our approach to be of interest to many people.

Possible extensions to our paper include the study of more generic forms of kernels (e.g., polynomials), as well as the extension of the current study to multi-dimensional kernels and density estimation.

References

1. Corani, G., Zaffalon, M.: Learning reliable classifiers from small or incomplete data sets: the naive credal classifier 2. J. Mach. Learn. Res. **9**, 581–621 (2008)
2. Destercke, S., Strauss, O.: Filtering with clouds. Soft Comput. **16**(5), 821–831 (2012)
3. Silverman, B.W.: Density estimation for statistics and data analysis, vol. 26. CRC Press, Boca Raton (1986)

4. Wand, M.P., Jones, M.C.: Kernel smoothing. CRC Press, Boca Raton (1994)
5. Zaffalon, M.: The naive credal classifier. J. Probab. Plann. Infer. **105**, 105–122 (2002)
6. Zaffalon, M., Corani, G., Mauá, D.: Evaluating credal classifiers by utility-discounted predictive accuracy. Int. J. Approx. Reason. **53**(8), 1282–1301 (2012)

Z-numbers as Generalized Probability Boxes

Didier Dubois$^{(\boxtimes)}$ and Henri Prade

Institut de Recherche en Informatique de Toulouse (IRIT),
CNRS & Université de Toulouse, 118 Route de Narbonne,
31062 Toulouse Cedex 9, France
{dubois,prade}@irit.fr

Abstract. This paper proposes a new approach to the notion of Z-number, i.e., a pair (A, B) of fuzzy sets modeling a probability-qualified fuzzy statement, proposed by Zadeh. Originally, a Z-number is viewed as the fuzzy set of probability functions stemming from the flexible restriction of the probability of the fuzzy event A by the fuzzy interval B on the probability scale. However, a probability-qualified statement represented by a Z-number fails to come down to the original fuzzy statement when the attached probability is 1. This representation also leads to complex calculations. It is shown that simpler representations can be proposed, that avoid these pitfalls, starting from the remark that when both fuzzy sets A and B forming the Z-number are crisp, the generated set of probabilities is representable by a special kind of belief function that corresponds to a probability box (p-box). Then two proposals are made to generalize this approach when the two sets are fuzzy. One idea is to consider a Z-number as a weighted family of crisp Z-numbers, obtained by independent cuts of the two fuzzy sets, that can be averaged. In the other approach, a Z-number comes down to a pair of possibility distributions on the universe of A forming a generalized p-box. With our proposal, computation with Z-numbers come down to uncertainty propagation with random intervals.

Keywords: Fuzzy event · Belief function · p-box
Imprecise probability

1 Introduction

In order to account for uncertainty attached to fuzzy statements, Zadeh [1] introduced the notion of a Z-number. It was one of the last proposals made by him, that triggered a significant amount of literature, most noticeably by Aliev and colleagues, for instance [2]. A Z-number is a pair (A, B) where A is a fuzzy subset of U and B is a fuzzy subset of $[0, 1]$ modeling a fuzzy restriction on the probability of A. It tries to formalize the meaning of a statement of the form "$(X\ is\ A)$ has probability B", like "it is *probable* that my income will he *high* this year". Zadeh introduced this kind of probability-qualified statements much

© Springer Nature Switzerland AG 2019
S. Destercke et al. (Eds.): SMPS 2018, AISC 832, pp. 68–77, 2019.
https://doi.org/10.1007/978-3-319-97547-4_10

earlier, when introducing a general framework for the mathematical representation of linguistic statements (the PRUF language [3]). They are interpreted as a family of probability distributions obtained by flexibly constraining the probability of the fuzzy event A by the fuzzy set B. The latter is viewed as a possibility distribution restricting the possible value of an ill-known probability. So a Z-number can be viewed as a fuzzy set of probability measures, more specifically a possibility distribution over them. Zadeh [1] outlines a method to make computations with Z-numbers, a question taken up by other authors with a view to provide a practical computation tool. However, these methods seem to extensively use linear programming even for solving small problems.

This research trend seems to have developed on its own with no connection with other generalized probabilistic frameworks for uncertainty management, like belief functions [4] and imprecise probabilities [5]. Yet, in those frameworks, the idea of a second-order possibility distribution over probability functions has been envisaged as a finer representation of convex probability sets by Moral [6], or Walley [7], with a view to represent linguistic information [8]. A full-fledged behavioral approach to the fuzzy probability of a crisp event (understood as partial knowledge about an objective probability) has been even studied in this spirit by De Cooman [9].

Here, we reconsider Z-numbers and their interpretation by Zadeh in the light of belief functions and imprecise probabilities, questioning some choices made. First we characterize crisp Z-numbers where both A and B are crisp sets. This seems not to have been done yet. However it is quite important to do it so as to put Z-numbers in the general perspective of uncertainty modeling. We show that a crisp Z-number can be exactly represented by a belief function [4], more specifically a p-box [10]. On such a basis we first propose alternative interpretations of Z-numbers, which may sound more natural and make them easier to process in computations. Basically we show that a Z-number can be represented, or at least approximated, by a belief function. As a consequence, computing with Z-numbers come down to computing with random intervals.

The paper is organized as follows. Section 2 recalls the original definition of a Z-number and highlights some limitations of this representation. Section 3 shows that in the crisp case a Z-number is equivalently represented by a special kind of belief function. In the fourth section, we extend the belief function view to the case when both elements of the pair (A, B) are fuzzy sets, proposing two different approaches. Proofs of propositions are omitted due to length restrictions.

2   Z-numbers: Comments on Their Definition

Consider two fuzzy sets A and B where A is a fuzzy subset of the real line, typically a fuzzy interval, that stands for a fuzzy restriction on the value of some quantity X, and B is a fuzzy interval on $[0, 1]$ that stands for a fuzzy probability. After Zadeh [1], a Z-number (A, B) represents the fuzzy set of probability measures P such that the probability $P(A)$ of the fuzzy event A $(P(A) = \int_0^1 \mu_A(x)p(x)dx$ [11], where p is the density of probability measure P)

is fuzzily restricted by the fuzzy set B. Zadeh [1] calls Z^+-number, or bimodal distribution, the pair (A, P) where the probability distribution is made precise. It represents the statement "The probability that X is A is $P(A)$". So a Z-number (A, B) represents a second-order possibility distribution over probability measures defined by $\forall P : \pi_{(A,B)}(P) = \mu_B(P(A))$. On this basis one may compute the probability of some other fuzzy event C as the fuzzy probability $\tilde{P}(C)$ such that $\mu_{\tilde{P}(C)}(p) = \sup_{P:P(C)=p} \mu_B(P(A))$, which defines an inferred Z-number $(C, \tilde{P}(C))$.

This interpretation of Z-number meets a difficulty: it seems we cannot recover, as a special case, the statement X is A expressed with full certainty, which corresponds to the Z-number $(A, 1)$, i.e., $P(A) = 1$. Indeed if A is a genuine fuzzy interval, $P(A) = 1$ if and only if the support of P lies in the core of A, $\hat{A} = \{x : \mu_A(x) = 1\}$. Indeed, it is clear that without this restriction, we always have $P(A) < 1$. One should then accept the equivalence between $(A, 1)$ and $(\hat{A}, 1)$, the latter being crisp. In other words, the original interpretation of Z-numbers loses the membership function of A on the way when $P(A) = 1$.

Zadeh [1] puts an additional restriction on the set of probability functions compatible with a Z-number, namely that the mean-value of P be equal to the centroid of A. However this restriction is questionable because the centroid of A, of the form $cent(A) = \frac{\int_0^1 x\mu_A(x)}{\int_0^1 \mu_A(x)}$ comes down to considering the membership function as a probability distribution, the mean value of which is the centroid of A, thus doing away with the possibilistic understanding of A as representing incomplete probabilistic information.

Zadeh [1] also deals with the computation of a Z-number (A_Z, B_Z) such that $Z = f(X, Y)$ where the information on X is (A_X, B_X), and on Y is (A_Y, B_Y). Zadeh proposes to apply the extension principle to $A_Z = f(A_X, A_Y)$ to compute:

$$\mu_{f(A_X, A_Y)}(z) = \sup_{z=f(x,y)} \min(\mu_{A_X}(x), \mu_{A_Y}(y)),$$

and to define, again by the extension principle, a possibility distribution $\pi_{(A_Z, B_Z)}$ over the set of probability functions $P_Z = P_X \circ_f P_Y$ obtained by probabilistic f-convolution of P_X and P_Y:

$$\pi_{(A_Z, B_Z)}(P_Z) = \sup_{P_X, P_Y : P_Z = P_X \circ_f P_Y} \min(\mu_{B_X}(P_X(A_X)), \mu_{B_Y}(P_Y(A_Y)))$$

(again requesting that $E(P_X) = Cent(A_X)$, $E(P_Y) = Cent(A_Y)$). Then the recovering of a fuzzy probability B_Z is obtained by projection: $\mu_{B_Z}(b) = \sup_{P_Z : P_Z(A)=b} \pi_{(A_Z, B_Z)}(P_Z)$. This computation scheme is further studied by Yager [12] and Aliev et al. [2].

However the proposed approach looks a bit questionable as (i) it is very complicated to implement, (ii) its formal justification is questionable. On the latter point, the fuzzy interval A represents the family of compatible probability functions $\mathcal{P}(A) = \{P : P(C) \leq \Pi_A(C) = \sup_{u \in C} \mu_A(u), \forall C\}$ [13]. The above approach seems to consider that if probability measures P_1, P_2 are respectively compatible with fuzzy intervals A_1 and A_2 then $P = P_1 \circ_f P_2$ will be compatible with the fuzzy interval $A = f(A_1, A_2)$, which is quite unclear. For instance, it is well-known that the two conditions $P_1 \in \mathcal{P}(A_X), P_2 \in \mathcal{P}(A_Y)$ do not imply that $P_1 \circ P_2 \in \mathcal{P}(f(A_X, A_Y))$ for standard convolution $\circ$ [14,15]. It is even more dubious whether $E(P_1) = Cent(A_X)$ and $E(P_2) = Cent(A_Y)$ imply $E(P_1 \circ_f P_2) = Cent(f(A_X, A_Y))$. The above discussion motivates the search for a different interpretation of Z-numbers, where the idea that it represents a set or a fuzzy set of probability measures is kept, but the use of a probability of fuzzy events is given up.

3   Crisp Z-numbers

Define a crisp Z-number to be a Z-number (A, B) where A is an interval $[a^-, a^+]$ and B a probability interval $[b^-, b^+]$. It is clear that Zadeh's modeling of Z-numbers in this crisp case yields the convex probability family $\mathcal{P}(A, B) = \{P : b^- \leq P(A) \leq b^+\}$ and $\pi_{(A,B)}$ is the characteristic function of this probability family. This set of probabilities can actually be described by a belief function on $\mathbb{R}$.

A belief function on a set U is defined by a mass assignment $m : \mathcal{F} \to (0, 1]$ of positive numbers to a finite family $\mathcal{F}$ of subsets of U, whose elements E are called focal sets $(m(E) > 0)$, and such that $\sum_{E \in \mathcal{F}} m(E) = 1$, and $m(\emptyset) = 0$. m is a probability distribution over 2^U, namely over possible statements of the form X is E. The degree of belief $Bel(C)$ in an event C induced by m is defined by $Bel(C) = \sum_{E \in \mathcal{F}, E \subseteq C} m(E)$. It is the probability that X is C can be inferred from some X is E for $\bar{E} \in \mathcal{F}$. The degree of plausibility of an event C is $Pl(C) = 1 - Bel(\bar{C}) \geq Bel(C)$, where the overbar denotes set-complementation.

A belief function defined by m characterizes the convex set of probabilities (called credal set) $\mathcal{P}(m) = \{P : P(C) \geq Bel(C), \forall C \subseteq U\}$ in the sense that $Bel(C) = \inf\{P(C) : P \in \mathcal{P}(m)\}$ is the lower probability induced by $\mathcal{P}(m)$. Given a crisp Z-number $(A, [b^-, b^+])$, consider the mass function defined by

$$m_{(A,B)}(A) = b^-; \quad m_{(A,B)}(\bar{A}) = 1 - b^+; \quad m_{(A,B)}(U) = b^+ - b^-.$$

In this model, the weight $m_{(A,B)}(A)$ is the probability of knowing that X is A, the weight $m_{(A,B)}(\bar{A})$ is the probability of knowing that X is $\bar{A}$, the weight $m_{(A,B)}(U)$ is the probability of not knowing anything. The coincidence between this interpretation of the crisp Z-number with Zadeh's approach is summarized by the following easy-to-check proposition:

Proposition 1. $\mathcal{P}(m_{(A,B)}) = \mathcal{P}(A, B)$.

A belief function Bel_1 is said to be more committed (informative) than Bel_2 if and only if $Bel_1 \neq Bel_2$ and $Bel_1 \geq Bel_2$, which is clearly equivalent to $\mathcal{P}(m_1) \subset \mathcal{P}(m_2)$. The belief function induced by (A, B) via $m_{(A,B)}$ is clearly the least committed belief function such that $Bel(A) \geq b^-$ and $Pl(A) \leq b^+$. This is in conformity with Zadeh's intuition of Z-numbers. The Z-number is also akin to Shafer's view of belief functions as coming from an unreliable testimony [4]. The

latter can be represented as a special case of a Z-number with $b^+ = 1$, i.e., of the
form $(A, [b^-, 1])$, which corresponds to a simple support belief function. A crisp
Z-number generally corresponds to an unreliable testimony where the witness
may not only be irrelevant, but also has possibility of lying (See Pichon et al.
[16]).

The crisp Z-number on an interval $U = [u^-, u^+]$ can be actually represented
by a special family of belief functions called p-boxes [10]. A p-box is the set
of probability measures on a totally ordered set whose cumulative function is
limited by an upper and a lower cumulative function: there are distribution
functions $F^* \geq F_*$ such that $\mathcal{P}(m) = \{P : F^*(u) \geq P([u^-, x]) \geq F_*(u)\}$. Focal
sets are in the form $E_\alpha = [\inf\{u : F^*(u) \geq \alpha\}, \inf\{u : F_*(u) \geq \alpha\}]$. Given the
mass assignment $m_{(A,B)}$, suppose $u^- = a^- < a^+ < u^+$ and let:

$$F_*(x) = Bel([u^-, x]) = \begin{cases} 0 & \text{if } x < a^+ \\ b^- & \text{if } a^+ \leq x < u^+ \\ 1 & \text{if } x = u^+ \end{cases} ; F^*(x) = Pl([u^-, x]) = \begin{cases} b^+ & \text{if } x < a^+ \\ 1 & \text{if } a^+ \leq x \end{cases}.$$

We recover the three focal sets of $m_{(A,B)}$ as A for $0 \leq \alpha < b^-$, U for $b^- \leq \alpha < b^+$,
and $\bar{A}$ for $b^+ \leq \alpha < 1$. More generally, if A is a subset of a finite set U, it is
always possible to rank-order the n elements of U such that $A = \{u_1, \ldots, u_k\}$,
and $\bar{A} = \{u_{k+1}, \ldots, u_n\}$ so that the focal sets are of the form $A = [u_1, u_k]$,
$U = [u_1, u_n]$, $\bar{A} = [u_{k+1}, u_n]$.

Proposition 1 makes the computation with crisp p-boxes quite easy to perform. Let m_X and m_Y the mass functions induced by two independent crisp
Z-numbers (A_X, B_X), and (A_Y, B_Y), respectively. We can apply the random set
propagation method (Yager [17], Dubois and Prade [18]) to compute the mass
function for $f(X, Y)$ as

$$m_{f(X,Y)}(G) = \sum_{E_X \in \mathcal{F}_X, E_Y \in \mathcal{F}_Y : G = f(E_X, E_Y)} m_X(E_X) \cdot m_Y(E_Y).$$

This is not computationally extensive as each belief function only has 3 focal
sets. However, it is easy to figure out that in general the result cannot always
be represented by a Z-number as the result potentially has $3 \times 3 = 9$ focal sets.

Requesting the additional condition $E(P) = Cent(A)$ comes down to restricting to probability distributions in $\mathcal{P}(m_{(A,B)})$ such that the mean value $E(P)$ is
the midpoint of interval $[b^-, b^+]$. This linear constraint, notwithstanding its lack
of natural justification (it does not even imply the symmetry of the density of
P), leads to a smaller credal set that cannot generally be represented by a belief
function, which significantly increases the complexity of handling Z-numbers in
practice.

4   Interpreting Z-numbers: Two Proposals

In the following, we see to what extent general Z-numbers can be interpreted by
belief functions, and whether this point of view on Z-numbers is in agreement

or not with Zadeh's approach relying on probabilities of a fuzzy event. Suppose that both A and B are fuzzy intervals, i.e., their α-cuts $\{u : \mu_A(u) \geq \alpha\}$ and $\{u : \mu_B(u) \geq \beta\}$ are closed intervals of the form $[a^-(\alpha), a^+(\alpha)]$ and $[b^-(\beta), b^+(\beta)]$. We can suggest two approaches, different from the one of Zadeh, but that agree with the crisp case of the previous section.

The Average Belief Function Approach. The starting point is to view both A and B as sets of α-cuts. If the fuzzy set A has a finite number of membership grades $\alpha_1 = 1 > \alpha_2 > \cdots > \alpha_k > 0$, denote the α_i-cut of the focal set A by A_i. To each A_i is associated a weight $b^-(\beta)(\alpha_i - \alpha_{i+1})$, $\overline{A_i}$ is assigned weight $(1 - b^+(\beta))(\alpha_i - \alpha_{i+1})$, and U is assigned weight $b^+(\beta) - b^-(\beta)$. For each β, this approach coincides with the one by Yen [19] to belief functions with fuzzy focal sets. We then get a parameterized family of belief functions, with parameter β selecting a probability interval from B, which may sound hard to use in practice if there is no criterion to select a value β.

However we can average β out as well. Suppose that B is a discrete fuzzy set with membership levels $\beta_1 > \cdots > \beta_\ell > 0$. For each β_j the above approach yields a belief function Bel_j, and we can compute $Bel = \sum_{j=1}^{\ell}(\beta_j - \beta_{j+1})Bel_j$ as the representation of the Z-number. It could be proved that this approach comes down to interpreting (A, B) as a set of $k\ell$ crisp Z-numbers (A_i, B_j), where $B_j = [b^-(\beta_j), b^+(\beta_j)]$, each pair yielding a belief function Bel_{ij} with mass function

$$m_{ij}(A_i) = b^-(\beta_j), \quad m_{ij}(\overline{A_i}) = 1 - b^+(\beta_j), \quad m_{ij}(U) = b^+(\beta_j) - b^-(\beta_j)$$

and to computing the weighted average $Bel = \sum_{i=1}^{k} \sum_{j=1}^{\ell}(\alpha_i - \alpha_{i+1})(\beta_j - \beta_{j+1})Bel_{ij}$; (A, B) is then equated to a crisply qualified fuzzy set $(A, E(B))$ using the fuzzy belief structure approach of Yen [19], where the interval average $E(B)$ of B is computed as the Aumann integral $\int_0^1 B_\beta d\beta = [\int_0^1 \inf B_\beta d\beta, \int_0^1 \sup B_\beta d\beta]$ [20].

The p-box Approach. Another simple approach is to interpret (A, B) as a unique belief function representing a generalized p-box [21], i.e., a family of probability functions induced by upper and lower bounds on a nested family of sets. We again use A via its α-cuts, but simultaneously use the α-cuts of B to derive bounds on $P(A_\alpha)$. However since if $\alpha > \beta$ we get $P(A_\alpha) \leq P(A_\beta)$, it is fruitless to interpret (A, B) as the set of constraints $b^-(\alpha) \leq P(A_\alpha) \leq b^+(\alpha)$. Indeed, $b^-(\alpha)$ increases with α, so that the lower bounds $b^-(\alpha)$ are redundant when $\alpha < 1$ (as $b^-(\alpha) \leq b^-(1) \leq P(\hat{A}) \leq P(A_\alpha)$).

One way out is to consider $b^-(1 - \alpha)$ as the lower bound of $P(A_\alpha)$, that is we interpret B as a p-box on $[0, 1]$ using the pair of decumulative functions $\Pi([b, 1]) = \max_{x \geq b} \mu_B(x)$ and $N([b, 1]) = \min_{x < b} 1 - \mu_B(x)$ associated to B. Using a continuous membership function for B we have that $\Pi([b^+(\alpha), 1]) = \mu_B(b^+(\alpha)) = \alpha$ and $N([b^-(\alpha), 1]) = 1 - \mu_B(b^-(\alpha)) = 1 - \alpha$ (see Fig. 1, right).

The Z-number (A, B) is then viewed as the set of constraints $\mathcal{C} = \{b^-(1-\alpha) \leq P(A_\alpha) \leq b^+(\alpha) : \alpha \in (0, 1]\}$. It forms a generalized p-box on U, since it is a nested family of subsets whose probability is upper and lower bounded. It can be characterized by two possibility distributions (π^+ and π^-) built from A and B such that $1 - \pi^- \leq \pi^+$ and $1 - \pi^-, \pi^+$ are comonotone functions [22], as we shall detail below. The ordering on U for generating the cumulative distributions is the one induced by the membership function μ_A.

First, the set of constraints $b^-(1 - \alpha) \leq P(A_\alpha), \alpha > 0$ is representable by a least specific possibility distribution π^+ on U such that [13]

$$\pi^+(u) = \min_{u \notin A_\alpha} 1 - b^-(1 - \alpha)$$

In fact it is easy to see that $\pi^+(a^-(\alpha)) = \pi^+(a^+(\alpha)) = 1 - b^-(1 - \alpha)$.

Likewise the set of constraints $P(A_\alpha) \leq b^+(\alpha), \alpha \in (0, 1]$, once written as $P(\overline{A_\alpha}) \geq 1 - b^+(\alpha), \alpha \in (0, 1]$ is representable by a possibility distribution π^- on U such that

$$\pi^-(u) = \min_{u \in A_\alpha} b^+(\alpha).$$

Again, we have that $\pi^-(a^-(\alpha)) = \pi^-(a^+(\alpha)) = b^+(\alpha)$. We can describe π^+ and π^- more precisely:

Proposition 2. *If $\mu_A(u) = 0$, then $\pi^+(u) = 1 - b^-(1)$ and $\pi^-(u) = 1$. If the support of B is $[0, 1]$ and μ_B is continuous and concave (in the usual sense) then $\pi^+(u) \geq \mu_A(u) \geq 1 - \pi^-(u)$ and $\inf \pi^-(u) = b^+(1)$*

Note that the bracketing property $\pi^+(u) \geq \mu_A(u) \geq 1 - \pi^-(u)$ no longer holds if the support of B is not $[0, 1]$ (see Fig. 1). In particular, if there is a value α^* such that $1 - b^-(1 - \alpha^*) = \alpha^*$, then $\pi^+(a_{\alpha^*}^-) = \mu_A(a_{\alpha^*}^-) = \alpha^*$. Then we shall generally have $1 - b^-(1 - \alpha) > \alpha$ for $\alpha > \alpha^*$ so that the α-cut of π^+ may be contained in the α-cut of A

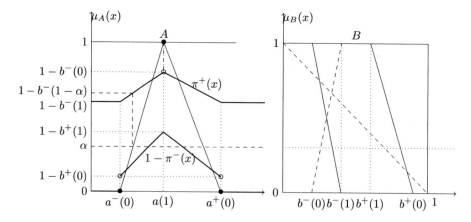

Fig. 1. p-box associated with Z-number (A, B) ($\bullet$: point included; $\circ$: point excluded)

We can express the two possibility distributions induced by (A, B) as follows.

$$\pi^+(u) = \begin{cases} 1 & \text{if } \mu_A(u) = 1, \\ 1 - b^-(1 - \mu_A(u)) & \text{if } 0 < \mu_A(u) < 1, \\ 1 - b^-(0) & \text{if } \mu_A(u) = 0. \end{cases} \qquad \pi^-(u) = \begin{cases} 1 & \text{if } \mu_A(u) = 0, \\ b^+(\mu_A(u)) & \text{if } 0 < \mu_A(u) < 1, \\ b^+(1) & \text{if } \mu_A(u) = 1. \end{cases}$$

Even if π^+ and $1 - \pi^-$ will not always bracket μ_A, we do have the inequality $\pi^+ \geq 1 - \pi^-$ since it comes down to $1 - b^-(1 - \alpha) \geq 1 - b^+(1 - \alpha)$. The following result can also be established:

Proposition 3. *The two functions π^+ and $\delta = 1 - \pi^-$ are comonotonic*

As explained in [22], the pair $(\pi^+, 1 - \pi^-(u))$ forms a co-monotonic cloud [23] corresponding to a set of probabilities $\mathcal{P} = \mathcal{P}(\pi^+) \cap \mathcal{P}(\pi^-)$, where $\mathcal{P}(\pi) = \{P : P(C) \leq \Pi(C), \forall C \text{ measurable}\}$. It yields a belief function whose focal sets are $E_\alpha = \{u : \pi^+(u) \geq \alpha\} \setminus \{u : 1 - \pi^-(u) \geq \alpha\}, \alpha > 0$. More specifically, in the case of a continuous Z-number, the focal sets obtained are of the form:

1. $A_1 = \hat{A}$ with mass $m(A_1) = b^-(0)$;
2. $\{u : \pi^+(u) \geq \alpha\}$ with (infinitesimal) mass $d\alpha$ for $1 - b^-(0) \geq \alpha > 1 - b^-(1)$;
3. U with mass $b^+(1) - b^-(1)$;
4. $\{u : 1 - \pi^-(u) < \alpha\}$ with (infinitesimal) mass $d\alpha$ for $1 - b^+(1) \geq \alpha > 1 - b^+(0)$;
5. $\overline{Supp(A)}$ with mass $1 - b^+(0)$.

What we obtain is a (partially) continuous belief function [24]. There are interesting special cases to be noticed.

– In the case A and B are crisp intervals, the result of the p-box approach degenerates in the p-box of Sect. 3 since $1 - b^-(0) = 1 - b^-(1)$ and $1 - b^+(1) = 1 - b^+(0)$;
– If the support of B is $[0, 1]$, some discrete parts of the mass assignment (cases 1 and 5) disappear, and the comonotonic cloud brackets μ_A.
– If $b^+(1) = 1$ (B expresses a form of certainty) then $1 - \pi^-(x) = 0$ and only the upper possibility distribution π^+ remains. If moreover $b^-(1) = 1$ then the support of π^+ is equal to the support of A (and cores are equal if $b^-(0) = 0$).
– If $\mu_B(x) = x$ (a genuine gradual representation of probabilistic certainty), it is easy to see that $\pi^+ = \mu_A$ and $1 - \pi^-(x) = 0$. In this case, (A, B) just reduces to the sure statement X is A. This is reminiscent of Zadeh's truth qualification (X is A is τ) by the fuzzy truth-value τ, he called "u-true" [3], where $\mu_\tau(x) = x$ and the result of truth-qualification is of the form $\mu_\tau(\mu_A)$.
– If $\mu_B(x) = 1 - x$ (a genuine gradual representation of negative probabilistic certainty), it is easy to see that $\pi^+ = 1$ (since $b^-(0) = b^-(1) = b^+(0) = 0$ and $b^+(1) = 1$) and $\pi^-(u) = 1 - \mu_A(u)$ since $b^+(\alpha) = 1 - \alpha$. It corresponds to the sure statement X is not $\hat{A}$. In turn, $\mu_B(x) = 1 - x$ is reminiscent of Zadeh's truth-qualifier "u-false" [3].
– If A is an interval and B is fuzzy, the constraints $\mathcal{C}$ reduce to $b^-(1) \leq P(A) \leq b^+(1)$, which is equivalent to the crisp Z-number $(A, \hat{B})$ using the core of B.

– If B is an interval and A is fuzzy, the constraints $\mathcal{C}$ reduce to $b^- \leq P(Supp(A)) \leq b^+$, which is equivalent to the crisp Z-number $(Supp(A), B)$ using the support of A. In particular if $B = [0, 1]$ (expressing ignorance), it is easy to see that $\pi^+ = \pi^- = 1$, which corresponds to complete ignorance about A.

The p-box approach seems to assume that it is not so natural to assign a fuzzy probability to a crisp event or a precise probability to a fuzzy event (in some sense the gradual nature of B reflects the gradual nature of A).

5   Conclusion

The notion of a Z-number is rather naturally found in human-originated information. It is thus important to propose faithful mathematical representations of their meaning. In this paper, Z-numbers have been examined in the light of imprecise probabilities, in order to provide more solid foundations to this concept. The main message is that it is possible to interpret a Z-number (A, B) as a special kind of belief function (or random set) on the universe of A (namely, a p-box). Indeed the original approach yields a convex (fuzzy) set of probabilities defined by linear constraints, that seems to be very hard to handle in practice. On the contrary, it is much easier to use random sets than convex sets of probabilities induced by any kind of linear constraints. Using the approaches described in this paper, we can compute the uncertainty pervading expressions of the form $f(X, Y)$ where X and Y are Z-numbers by means the random set propagation principle recalled in Sect. 3 and Monte-Carlo methods (see for instance [15]). Note that the result will not generally be equivalent to another Z-number, but a more general random set, contrary to what some works are presupposing, which does not prevent other Z-numbers on quantities of interest from being extracted from the resulting random set obtained via computation.

References

1. Zadeh, L.A.: A note on Z-numbers. Inf. Sci. **181**(14), 2923–2932 (2011)
2. Aliev, R.A., Alizadeh, A.V., Huseynov, O.H., Jabbarova, K.I.: Z-number-based linear programming. Int. J. Intell. Syst. **30**(5), 563–589 (2015)
3. Zadeh, L.A.: PRUF - a language for the representation of meaning in natural languages. Int. J. Man Mach. Stud. **10**, 395–460 (1978)
4. Shafer, G.: A Mathematical Theory of Evidence. Princeton University Press, Princeton (1976)
5. Walley, P.: Statistical Reasoning with Imprecise Probabilities. Chapman and Hall, London (1991)
6. Moral, S., de Campos, L.M.: Updating uncertain information. In: Bouchon-Meunier, B., Yager, R.R., Zadeh, L.A. (eds.) Uncertainty in Knowledge Bases, Proceedings of IPMU 1990, Paris. LNCS, vol. 521, pp. 58–67. Springer (1991)
7. Walley, P.: Statistical inferences based on a second-order possibility distribution. Int. J. Gen. Syst. **26**, 337–384 (1997)

8. Walley, P., De Cooman, G.: A behavioral model for linguistic uncertainty. Inf. Sci. **134**(1–4), 1–37 (2001)
9. De Cooman, G.: A behavioral model for vague probability assessments. Fuzzy Sets Syst. **154**(3), 305–358 (2005)
10. Ferson, S., Ginzburg, L., Kreinovich, V., Myers, D.M., Sentz, K.: Constructing probability boxes and Dempster-Shafer structures. Technical report, Sandia National Laboratories (2003)
11. Zadeh, L.A.: Probability measures of fuzzy events. J. Math. Anal. Appl. **23**(2), 421–427 (1968)
12. Yager, R.R.: On Z-valuations using Zadeh's Z-numbers. Int. J. Intell. Syst. **27**(3), 259–278 (2012)
13. Dubois, D., Prade, H.: When upper probabilities are possibility measures. Fuzzy Sets Syst. **49**, 65–74 (1992)
14. Ferson, S., Ginzburg, L.R.: Hybrid arithmetic. In: Proceedings of ISUMA/NAFIPS 1995, pp. 619–623 (1995)
15. Baudrit, C., Dubois, D., Guyonnet, D.: Joint propagation and exploitation of probabilistic and possibilistic information in risk assessment. IEEE Trans. Fuzzy Syst. **14**(5), 593–608 (2006)
16. Pichon, F., Denœux, T., Dubois, D.: Relevance and truthfulness in information correction and fusion. Int. J. Approx. Reason. **53**(2), 159–175 (2012)
17. Yager, R.R.: Arithmetic and other operations on Dempster-Shafer structures. Int. J. Man Mach. Stud. **25**(4), 357–366 (1986)
18. Dubois, D., Prade, H.: Random sets and fuzzy interval analysis. Fuzzy Sets Syst. **42**, 87–101 (1991)
19. Yen, J.: Generalizing the Dempster-Shafer theory to fuzzy sets. IEEE Trans. Syst. Man Cybern. **20**(3), 559–570 (1990)
20. Dubois, D., Prade, H.: The mean value of a fuzzy number. Fuzzy Sets Syst. **24**, 279–300 (1987)
21. Destercke, S., Dubois, D., Chojnacki, E.: Unifying practical uncertainty representations - I: generalized p-boxes. Int. J. Approx. Reason. **49**(3), 649–663 (2008)
22. Destercke, S., Dubois, D., Chojnacki, E.: Unifying practical uncertainty representations. II: clouds. Int. J. Approx. Reason. **49**(3), 664–677 (2008)
23. Neumaier, A.: Clouds, fuzzy sets and probability intervals. Reliab. Comput. **10**, 249–272 (2004)
24. Smets, P.: Belief functions on real numbers. Int. J. Approx. Reason. **40**(3), 181–223 (2005)

Computing Inferences for Large-Scale Continuous-Time Markov Chains by Combining Lumping with Imprecision

Alexander Erreygers$^{(\boxtimes)}$ and Jasper De Bock

SYSTeMS, ELIS, Ghent University, Ghent, Belgium
{alexander.erreygers,jasper.debock}@ugent.be

Abstract. If the state space of a homogeneous continuous-time Markov chain is too large, making inferences—here limited to determining marginal or limit expectations—becomes computationally infeasible. Fortunately, the state space of such a chain is usually too detailed for the inferences we are interested in, in the sense that a less detailed—smaller—state space suffices to unambiguously formalise the inference. However, in general this so-called lumped state space inhibits computing exact inferences because the corresponding dynamics are unknown and/or intractable to obtain. We address this issue by considering an imprecise continuous-time Markov chain. In this way, we are able to provide guaranteed lower and upper bounds for the inferences of interest, without suffering from the curse of dimensionality.

1 Introduction

State space explosion, or the exponential dependency of the size of a finite state space on a system's dimensions, is a frequently encountered inconvenience when constructing mathematical models of systems. In the setting of continuous-time Markov chains (CTMCs), this exponentially increasing number of states has as a consequence that using the model to perform inferences—for the sake of brevity here limited to marginal and limit expectations—about large-scale systems becomes computationally intractable. Fortunately, for many of the inferences we would like to make, a higher-level state description actually suffices, allowing for a reduced state space with considerably fewer states. However, unfortunately, the low-level description and its corresponding larger state space are necessary in order to accurately model the system's dynamics. Therefore, using the reduced state space to make inferences is generally impossible.

In this contribution, we address this problem using imprecise continuous-time Markov chains [5, 11, 15]. In particular, we outline an approach to determine guaranteed lower and upper bounds on marginal and limit expectations using the reduced state space. We introduced a preliminary version of this approach in [8, 14], but the current contribution is—to the best of our knowledge—its first fully general and theoretically justified exposition. Compared to other approaches [3,9] that also determine lower and upper bounds on expectations, ours has the

© Springer Nature Switzerland AG 2019
S. Destercke et al. (Eds.): SMPS 2018, AISC 832, pp. 78–86, 2019.
https://doi.org/10.1007/978-3-319-97547-4_11

advantage that it is not restricted to limit expectations. Furthermore, based on our preliminary experiments, our approach seems to produce tighter bounds.

2   Continuous-Time Markov Chains

We are interested in making inferences about a system, more specifically about the state of this system at some future time t, denoted by X_t. The complication is that we are unable to predict the temporal evolution of the state with certainty. Therefore, at all times $t \in \mathbb{R}_{\geq 0}$,[1] the state X_t of the system is a random variable that takes values—generically denoted by x, y or z—in the state space $\mathcal{X}$.

2.1   Homogeneous Continuous-Time Markov Chains

We assume that the stochastic process that models our beliefs about the system, denoted by $(X_t)_{t \in \mathbb{R}_{\geq 0}}$, is a *continuous-time Markov chain* (CTMC) that is *homogeneous*. For a thorough treatment of the terminology and notation concerning CTMCs, we refer to [1,11,13]. Due to length constraints, we here limit ourselves to the bare necessities.

The stochastic process $(X_t)_{t \in \mathbb{R}_{\geq 0}}$ is a CTMC if it satisfies the *Markov property*, which says that for all $t_1, \ldots, t_n, t, \Delta$ in $\mathbb{R}_{\geq 0}$ with $n \in \mathbb{N}$ and $t_1 < \cdots < t_n < t$, and all $x_1, \ldots, x_n, x, y$ in $\mathcal{X}$,

$$P(X_{t+\Delta} = y \mid X_{t_1} = x_1 \ldots, X_{t_n} = x_n, X_t = x) = P(X_{t+\Delta} = y \mid X_t = x). \quad (1)$$

The CTMC $(X_t)_{t \in \mathbb{R}_{\geq 0}}$ is *homogeneous* if for all t, Δ in $\mathbb{R}_{\geq 0}$ and all x, y in $\mathcal{X}$,

$$P(X_{t+\Delta} = y \mid X_t = x) = P(X_\Delta = y \mid X_0 = x). \quad (2)$$

It is well-known that—both in the classical measure-theoretic framework [1] and the full conditional framework [11]—a homogeneous continuous-time Markov chain is uniquely characterised by a triplet $(\mathcal{X}, \pi_0, Q)$, where $\mathcal{X}$ is a state space, π_0 an initial distribution and Q a transition rate matrix.

The state space $\mathcal{X}$ is taken to be a non-empty, finite and—without loss of generality—ordered set. This way, any real-valued function f on $\mathcal{X}$ can be identified with a column vector, the x-component of which is $f(x)$. The set containing all real-valued functions on $\mathcal{X}$ is denoted by $\mathcal{L}(\mathcal{X})$.

The initial distribution π_0 is defined by

$$\pi_0(x) := P(X_0 = x) \text{ for all } x \text{ in } \mathcal{X}, \quad (3)$$

and hence is a probability mass function on $\mathcal{X}$. We will (almost) exclusively be concerned with positive (initial) distributions, whom we collect in $\mathcal{D}(\mathcal{X})$ and will identify with row vectors.

---

[1] We use $\mathbb{R}_{\geq 0}$ and $\mathbb{R}_{>0}$ to denote the set of non-negative real numbers and positive real numbers, respectively. Furthermore, we use $\mathbb{N}$ to denote the natural numbers and write $\mathbb{N}_0$ when including zero.

The transition rate matrix Q is a real-valued $|\mathcal{X}| \times |\mathcal{X}|$ matrix—or equivalently, a linear map from $\mathcal{L}(\mathcal{X})$ to $\mathcal{L}(\mathcal{X})$—with non-negative off-diagonal entries and rows that sum up to zero. If for any t in $\mathbb{R}_{\geq 0}$ we define the *transition matrix over* t as

$$T_t := e^{tQ} = \lim_{n \to +\infty} \left(I + \frac{t}{n} Q \right)^n, \tag{4}$$

then for all t in $\mathbb{R}_{\geq 0}$ and all x, y in $\mathcal{X}$,

$$P(X_t = y \mid X_0 = x) = T_t(x, y). \tag{5}$$

Finally, we denote by E the expectation operator with respect to the homogeneous CTMC $(X_t)_{t \in \mathbb{R}_{\geq 0}}$ in the usual sense. It follows immediately from (3) and (5) that $E(f(X_t)) = \pi_0 T_t f$ for any f in $\mathcal{L}(\mathcal{X})$ and any t in $\mathbb{R}_{\geq 0}$.

2.2   Irreducibility

In order not to be tangled up in edge cases, in the remainder we are only concerned with irreducible transition rate matrices. Many equivalent necessary and sufficient conditions exist; see for instance [13, Theorem 3.2.1]. For the sake of brevity, we here say that a transition rate matrix Q is *irreducible* if, for all t in $\mathbb{R}_{>0}$ and x, y in $\mathcal{X}$, $T_t(x, y) > 0$.

Consider now a homogeneous CTMC that is characterised by $(\mathcal{X}, \pi_0, Q)$. It is then well-known that for any f in $\mathcal{L}(\mathcal{X})$, the limit $\lim_{t \to +\infty} E(f(X_t))$ exists. Even more, since we assume that Q is irreducible, this limit value is the same for all initial distributions π_0 [13, Theorem 3.6.2]! This common limit value, denoted by $E_\infty(f)$, is called the *limit expectation of* f. Furthermore, the irreducibility of Q also implies that there is a unique stationary distribution π_∞ in $\mathcal{D}(\mathcal{X})$ that satisfies the *equilibrium condition* $\pi_\infty Q = 0$. This unique distribution is called the *limit distribution*, as $E_\infty(f) = \pi_\infty f$.

In the remainder of this contribution, a *positive and irreducible CTMC* is any homogeneous CTMC characterised by a positive initial distribution π_0 and an irreducible transition rate matrix Q.

3   Lumping and the Induced (Imprecise) Process

In many practical applications—see for instance [3,8,9,14]—we have a positive and irreducible CTMC that models our system and we want to use this chain to make inferences of the form $E(f(X_t)) = \pi_0 T_t f$ or $E_\infty(f)$. As analytically evaluating the limit in (4) is often infeasible, we usually have to resort to one of the many available numerical methods—see for example [12]—that approximate T_t. However, unfortunately these numerical methods turn out to be computationally intractable when the state space becomes large. Similarly, determining the unique distribution π_∞ that satisfies the equilibrium condition also becomes intractable for large state spaces.

Fortunately, as previously mentioned in Sect. 1, the state space $\mathcal{X}$ is often unnecessarily detailed. Indeed, many interesting inferences can usually still be unambiguously defined using real-valued functions on a less detailed state space that corresponds to a higher-order description of the system, denoted by $\hat{\mathcal{X}}$. However, this provides no immediate solution as the motive behind using the detailed state space $\mathcal{X}$ in the first place is that this allows us to accurately model the (uncertain) dynamics of the system using a homogeneous CTMC; see [3, 8–10, 14] for practical examples. In contrast, the dynamics of the induced stochastic process on the reduced state space $\hat{\mathcal{X}}$ are often unknown and/or intractable to obtain, which inhibits us from making exact inferences using the induced stochastic process. We now set out to address this by allowing for imprecision.

3.1 Notation and Terminology Concerning Lumping

We assume that the lumped state space $\hat{\mathcal{X}}$ is obtained by *lumping*—sometimes called grouping or aggregating, see [2, 4]—states in $\mathcal{X}$, such that $1 < |\hat{\mathcal{X}}| \leq |\mathcal{X}|$. This lumping is formalised by the surjective *lumping map* $\Lambda \colon \mathcal{X} \to \hat{\mathcal{X}}$, which maps every state x in $\mathcal{X}$ to a state $\Lambda(x) = \hat{x}$ in $\hat{\mathcal{X}}$. In the remainder, we also use the inverse lumping map Γ, which maps every $\hat{x}$ in $\hat{\mathcal{X}}$ to a subset $\Gamma(\hat{x}) \coloneqq \{x \in \mathcal{X} \colon \Lambda(x) = \hat{x}\}$ of $\mathcal{X}$. Given such a lumping map Λ, a function f in $\mathcal{L}(\mathcal{X})$ is *lumpable with respect to* Λ if there is an $\hat{f}$ in $\mathcal{L}(\hat{\mathcal{X}})$ such that $f(x) = \hat{f}(\Lambda(x))$ for all x in $\mathcal{X}$. We use $\mathcal{L}_\Lambda(\mathcal{X}) \subseteq \mathcal{L}(\mathcal{X})$ to denote the set of all real-valued functions on $\mathcal{X}$ that are lumpable with respect to Λ.

As far as our results are concerned, it does not matter in which way the states are lumped. For a given f in $\mathcal{L}(\mathcal{X})$—recall that we are interested in the (limit) expectation of $f(X_t)$—a naive choice is to lump together all states that have the same image under f. However, this is not necessarily a good choice. One reason is that the resulting lumped state space can become very small, for example when f is an indicator, resulting in too much imprecision in the dynamics and/or the inference. Lumping-based methods therefore often let $\hat{\mathcal{X}}$ correspond to a natural higher-level description of the state of the system; see for example [3, 8, 9] for some positive results. An extra benefit of this approach is that the resulting model can be used to determine the (limit) expectation of multiple functions.

3.2 The Lumped Stochastic Process

Let $(X_t)_{t \in \mathbb{R}_{\geq 0}}$ be a positive and homogeneous continuous-time Markov chain. Then any lumping map $\Lambda \colon \mathcal{X} \to \hat{\mathcal{X}}$ unequivocally induces a *lumped stochastic process* $(\hat{X}_t)_{t \in \mathbb{R}_{\geq 0}}$. It has $\hat{\mathcal{X}}$ as state space and is defined by the relation

$$(\hat{X}_t = \hat{x}) \Leftrightarrow (X_t \in \Gamma(\hat{x})) \text{ for all } t \text{ in } \mathbb{R}_{\geq 0} \text{ and all } \hat{x} \text{ in } \hat{\mathcal{X}}. \tag{6}$$

In some cases, this lumped stochastic process is a homogeneous CTMC, and the inference of interest can then be computed using this reduced CTMC. See for example [2, Theorem 2.3(i)] for a necessary condition and [2, Theorem 2.4] or [4, Theorem 3] for a necessary and sufficient one. However, these conditions are very stringent. Indeed, in general, the lumped stochastic process is not homogeneous nor Markov. For this general case, we are not aware of any previous work that characterises the dynamics of the lumped stochastic process efficiently—i.e., directly from Λ, Q and π_0 and without ever determining T_t.

3.3   The Induced Imprecise Continuous-Time Markov Chain

Nevertheless, that is exactly what we now set out to do. Due to length constraints, we will here restrict ourselves to providing an intuitive explanation of our methodology, becoming formal only when stating our main results; see Theorems 1 and 2 further on. For a detailed exposition, we refer to the appendix of the extended preprint of this contribution [7].

The essential point is that, while we cannot exactly determine the dynamics of the lumped stochastic process $(\hat{X}_t)_{t \in \mathbb{R}_{\geq 0}}$, we can consider a *set of possible stochastic processes*, not necessarily homogeneous and/or Markovian but all with $\hat{\mathcal{X}}$ as state space, that definitely contains the lumped stochastic process $(\hat{X}_t)_{t \in \mathbb{R}_{\geq 0}}$. In the remainder, we will denote this set by $\mathbb{P}_{\pi_0,Q,\Lambda}$. As is indicated by our notation, $\mathbb{P}_{\pi_0,Q,\Lambda}$ is fully characterised by π_0, Q and Λ.

Crucially, it turns out that $\mathbb{P}_{\pi_0,Q,\Lambda}$ takes the form of a so-called *imprecise continuous-time Markov chain*. For a formal definition of general imprecise CTMCs, and an extensive study of their properties, we refer the reader to the work of Krak et al. [11] and De Bock [5]. For our present purposes, it suffices to know that tight lower and upper bounds on the expectations that correspond to the set of stochastic processes of an imprecise CTMC are relatively easy to obtain. In particular, they can be determined without having to explicitly optimise over this set of processes, thus mitigating the need to actually construct it.

There are many parallels between homogeneous CTMCs and imprecise CTMCs. For instance, the counterpart of a transition rate matrix is a *lower transition rate operator*. For our imprecise CTMC $\mathbb{P}_{\pi_0,Q,\Lambda}$, this lower transition rate operator is $\hat{\underline{Q}} \colon \mathcal{L}(\hat{\mathcal{X}}) \to \mathcal{L}(\hat{\mathcal{X}}) \colon g \mapsto \hat{\underline{Q}}g$ where, for every g in $\mathcal{L}(\hat{\mathcal{X}})$, $\hat{\underline{Q}}g$ is defined by

$$[\hat{\underline{Q}}g](\hat{x}) := \min \left\{ \sum_{\hat{y} \in \hat{\mathcal{X}}} g(\hat{y}) \sum_{y \in \Gamma(\hat{y})} Q(x,y) : x \in \Gamma(\hat{x}) \right\} \quad \text{for all } \hat{x} \text{ in } \hat{\mathcal{X}}. \quad (7)$$

Important to mention here is that in case the lumped state space corresponds to some higher-order state description, we often find that executing the optimisation in (7) is fairly straightforward, as is for instance observed in [8,14].

The counterpart of the transition matrix over t is now the *lower transition operator over t*, denoted by $\hat{\underline{T}}_t \colon \mathcal{L}(\hat{\mathcal{X}}) \to \mathcal{L}(\hat{\mathcal{X}})$ and defined for all g in $\mathcal{L}(\hat{\mathcal{X}})$ by

$$\hat{\underline{T}}_t g := \lim_{n \to +\infty} \left(I + \frac{t}{n}\hat{\underline{Q}} \right)^n g, \quad (8)$$

where the n-th power should be interpreted as consecutively applying the operator n times. Note how strikingly (8) resembles (4). Analogous to the precise case, one needs numerical methods—see for instance [6] or [11, Sect. 8.2]—to approximate $\underline{\hat{T}}_t g$ because analytically evaluating the limit in (8) is, at least in general, impossible.

4 Performing Inferences Using the Lumped Process

Everything is now set up to present our main results. Due to length constraints, we have relegated our proofs to the appendix of the extended arXiv version of this contribution [7].

4.1 Guaranteed Bounds on Marginal Expectations

We first turn to marginal expectations. Once we have $\mathbb{P}_{\pi_0, Q, \Lambda}$, the following result is a—not quite immediate—consequence of [11, Corollary 8.3].

Theorem 1. *Consider a positive and irreducible CTMC characterised by $(\mathcal{X}, \pi_0, Q)$ and a lumping map $\Lambda\colon \mathcal{X} \to \hat{\mathcal{X}}$. Let f in $\mathcal{L}(\mathcal{X})$ be lumpable with respect to Λ and let $\hat{f}$ be the corresponding element of $\mathcal{L}(\hat{\mathcal{X}})$. Then for any t in $\mathbb{R}_{\geq 0}$,*

$$\hat{\pi}_0 \underline{\hat{T}}_t \hat{f} \leq E(f(X_t)) = \pi_0 T_t f \leq -\hat{\pi}_0 \underline{\hat{T}}_t (-\hat{f}),$$

where $\hat{\pi}_0$ in $\mathcal{D}(\hat{\mathcal{X}})$ is defined by $\hat{\pi}_0(\hat{x}) := \sum_{x \in \Gamma(\hat{x})} \pi_0(x)$ for all $\hat{x}$ in $\hat{\mathcal{X}}$.

This result is highly useful in the setting that was outlined in Sect. 3. Indeed, for large systems we can use Theorem 1 to compute guaranteed lower and upper bounds on marginal expectations that cannot be computed exactly.

4.2 Guaranteed Bounds on Limit Expectations

Our second result provides guaranteed lower and upper bounds on limit expectations. This is extremely useful because the limit expectation is (almost surely) equal to the long-term temporal average due to the ergodic theorem [13, Theorem 3.8.1], and in practice—see for instance [8]—the inference one is interested in is often a long-term temporal average.

Theorem 2. *Consider an irreducible CTMC and a lumping map $\Lambda\colon \mathcal{X} \to \hat{\mathcal{X}}$. Let f in $\mathcal{L}(\mathcal{X})$ be lumpable with respect to Λ and let $\hat{f}$ be the corresponding element of $\mathcal{L}(\hat{\mathcal{X}})$. Then for all n in $\mathbb{N}_0$ and δ in $\mathbb{R}_{>0}$ such that $\delta \max\{|Q(x,x)|\colon x \in \mathcal{X}\} < 1$,*

$$\min(I + \delta \underline{\hat{Q}})^n \hat{f} \leq E_\infty(f) \leq -\min(I + \delta \underline{\hat{Q}})^n (-\hat{f}).$$

Furthermore, for fixed δ, the lower and upper bounds in this expression become monotonously tighter with increasing n, and each converges to a (possibly different) constant as n approaches $+\infty$.

84   A. Erreygers and J. De Bock

This result can be used to devise an approximation method similar to [8, Algorithm 1]: we fix some value for δ, set $g_0 = \hat{f}$ (or $g_0 = -\hat{f}$) and then repeatedly compute $g_i := (I + \delta\hat{Q})g_{i-1} = g_{i-1} + \delta\underline{Q}g_{i-1}$ until we empirically observe convergence of $\min g_i$ (or $-\min g_i$). In general, the lower and upper bounds obtained in this way are dependent on the choice of δ and this choice can therefore influence the tightness of the obtained bounds. Empirically, we have seen that smaller δ tend to yield tighter bounds, at the expense of requiring more iterations—that is, larger n—before empirical convergence.

4.3   Some Preliminary Numerical Results

Due to length constraints, we leave the numerical assessment of Theorem 1 for future work. For an extensive numerical assessment of—the method implied by—Theorem 2, we refer the reader to [8]. We believe that in this contribution, it is more fitting to compare our method to the only existing method—at least the only one that we are aware of—that also uses lumping to provide guaranteed lower and upper bounds on limit expectations. This method was first outlined by Franceschinis and Muntz [9], and then later improved by Buchholz [3]. In order to display the benefit of their methods, they use them to determine bounds on several performance measures for a closed queueing network that consists of a single server in series with multiple parallel servers. We use the method outlined in Sect. 4.2 to also compute bounds on these performance measures, as reported in Table 1. Note that our bounds are tighter than those of [9]. We would very much like to compare our method with the improved method of [3] as well. Unfortunately, the system parameters Buchholz uses do not—as far as we can tell—correspond to the number of states and the values for the performance measures he reports in [3, Fig. 3], thus preventing us from comparing our results.

Table 1. Comparison of the bounds obtained by using Theorem 2 with those obtained by the method presented in [9, Sect. 3.2] for the closed queueing network of [9].

	Exact	[9, Table 1]		Theorem 2	
		Lower	Upper	Lower	Upper
Mean queue length	1.2734	1.2507	1.3859	1.2664	1.2802
Throughput	0.9828	0.9676	0.9835	0.9826	0.9831

5   Conclusion

Broadly speaking, the conclusion of this contribution is that imprecise CTMCs are not only a robust uncertainty model—as they were originally intended to be—but also a useful computational tool for determining bounds on inferences for large-scale CTMCs. More concretely, the first important observation of this contribution is that lumping states in a homogeneous CTMC inevitably introduces imprecision, in the sense that we cannot exactly determine the parameters that describe the dynamics of the lumped stochastic process without also explicitly determining the original process. The second is that we can easily characterise a set of processes that definitely contains the lumped process, in the form of an imprecise CTMC. Using this imprecise CTMC, we can then determine guaranteed lower and upper bounds on marginal and limit expectations with respect to the original chain. From a practical point of view, these results are helpful in cases where state space explosion occurs: they allow us to determine guaranteed lower and upper bounds on inferences that we otherwise could not determine at all.

Regarding future work, we envision the following. For starters, a more thorough numerical assessment of the methods outlined in Sect. 4 is necessary. Furthermore, it would be of theoretical as well as practical interest to determine bounds on the *conditional* expectation of a lumpable function, or to consider functions that depend on the state at *multiple* time points. Finally, we are developing a method to determine lower and upper bounds on limit expectations that only requires the solution of a simple linear program.

Acknowledgements. Jasper De Bock's research was partially funded by H2020-MSCA-ITN-2016 UTOPIAE, grant agreement 722734. Furthermore, the authors are grateful to the reviewers for their constructive feedback and useful suggestions.

References

1. Anderson, W.J.: Continuous-Time Markov Chains. Springer-Verlag (1991)
2. Ball, F., Yeo, G.F.: Lumpability and marginalisability for continuous-time Markov chains. J. Appl. Probab. **30**(3), 518–528 (1993)
3. Buchholz, P.: An improved method for bounding stationary measures of finite Markov processes. Perform. Eval. **62**(1), 349–365 (2005)
4. Burke, C.J., Rosenblatt, M.: A Markovian function of a Markov chain. Ann. Math. Stat. **29**(4), 1112–1122 (1958)
5. De Bock, J.: The limit behaviour of imprecise continuous-time Markov chains. J. Nonlinear Sci. **27**(1), 159–196 (2017)
6. Erreygers, A., De Bock, J.: Imprecise continuous-time Markov chains: efficient computational methods with guaranteed error bounds. In: Proceedings of ISIPTA 2017, pp. 145–156. PMLR (2017). extended pre-print: arXiv:1702.07150
7. Erreygers, A., De Bock, J.: Computing inferences for large-scale continuous-time Markov chains by combining lumping with imprecision (2018). arXiv:1804.01020
8. Erreygers, A., Rottondi, C., Verticale, G., De Bock, J.: Imprecise Markov models for scalable and robust performance evaluation of flexi-grid spectrum allocation policies (2018, submitted). arXiv:1801.05700
9. Franceschinis, G., Muntz, R.R.: Bounds for quasi-lumpable Markov chains. Perform. Eval. **20**(1), 223–243 (1994)
10. Ganguly, A., Petrov, T., Koeppl, H.: Markov chain aggregation and its applications to combinatorial reaction networks. J. Math. Biol. **69**(3), 767–797 (2014)
11. Krak, T., De Bock, J., Siebes, A.: Imprecise continuous-time Markov chains. Int. J. Approx. Reason. **88**, 452–528 (2017)

Robust Fuzzy Relational Clustering of Non-linear Data

Maria Brigida Ferraro$^{(\boxtimes)}$ and Paolo Giordani

Department of Statistical Sciences, Sapienza University of Rome, Rome, Italy
{mariabrigida.ferraro,paolo.giordani}@uniroma1.it

Abstract. In many practical situations data may be characterized by non-linear structures. Classical (hard or fuzzy) algorithms, usually based on the Euclidean distance, implicitly lead to spherical shape clusters and, therefore, do not identify clusters properly. In this paper we deal with non-linear structures in clustering by means of the geodesic distance, able to capture and preserve the intrinsic geometry of the data. We introduce a new fuzzy relational clustering algorithm based on the geodesic distance. Furthermore, to improve its adequacy, a robust version is proposed in order to take into account the presence of outliers.

1 Introduction

Cluster analysis is a well-known class of (exploratory) techniques allowing us to partition n objects in a limited number of clusters such that objects belonging to the same cluster are similar as much as possible according to a given distance measure. In standard clustering methods, clusters are usually obtained by considering the Euclidean distance. It implies that clusters have spherical shape and the similarities among objects are computed linearly. Although there exist several variants of these algorithms for coping with arbitrary shape clusters, it frequently occurs that the linearity assumption may be inadequate. When data contain non-linear structures, the Euclidean distance is no longer appropriate. In order to compute the dissimilarities between objects reasonably well, the so-called geodesic distance should be considered. The aim of this paper is to propose some fuzzy relational clustering methods based on the geodesic distance. The paper is organized as follows. In the next section we introduce non-linear data and the geodesic distance. Then, in Sect. 3, the relational fuzzy clustering methods for non-linear data are proposed.

2 Non-linear Data and Geodesic Distance

In case of non-linear data, the Euclidean distance is not helpful to compute the dissimilarities between objects. An example of such data is reported in Fig. 1. By looking at the figure, two groups of data points are intuitively observed. Each group defines the so-called manifold. In mathematics, a manifold is a topological

© Springer Nature Switzerland AG 2019
S. Destercke et al. (Eds.): SMPS 2018, AISC 832, pp. 87–90, 2019.
https://doi.org/10.1007/978-3-319-97547-4_12

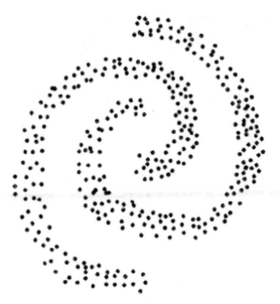

Fig. 1. Example of non-linear data

space which is globally non-linear, but locally linear. It follows that the dissimilarity between two neighboring points belonging to the manifold can be evaluated through, for instance, the Euclidean distance, whilst the one between two points far from each other cannot be measured by their straight-line Euclidean distance. In this case, the so-called geodesic distance should be adopted. In contrast with the Euclidean distance, the geodesic one is able to capture and, therefore, preserve the intrinsic geometry of the manifold. We can interpret a manifold as a graph with edges connecting neighboring data points (see, e.g., [1]). The graph can be constructed according to two approaches differing on how the neighbors of each object are determined. In the K-neighboring approach, each object is connected by an edge to the K nearest neighbors, whilst, in the ϵ-neighboring approach, the number of neighbors is not fixed and each object is connected to all the other objects lying in an ϵ-radius environment. In both cases, the Euclidean distance can be used to compute the length of the edges taking into account the local linearity. Once the neighbors are determined, the Floyd-Warshall algorithm [2,3] can be used to compute the geodesic distances for all the objects pertaining to a manifold. The intuition of the algorithm is that, in case of two faraway objects, the geodesic distance is equal to the shortest path in the graph connecting these two objects. The Floyd-Warshall algorithm consists in the following steps.

1. Let N_i be the set of neighbors for object i, $i = 1, \ldots, n$. For each object $j \in N_i$, set the edge length between i and j equal to, for instance, their Euclidean distance, say $d_E(i,j)$.

2. By denoting the geodesic distance for objects i and j by $d_G(i,j)$, set

$$d_G(i,j) = \begin{cases} d_E(i,j) & \text{if } i \text{ and } j \text{ are linked by an edge,} \\ +\infty & \text{otherwise.} \end{cases} \quad (1)$$

3. For each $k \in \{1,\ldots,n\}$, if $d_G(i,j) > d_G(i,k) + d_G(k,j)$, then recursively set

$$d_G(i,j) = d_G(i,k) + d_G(k,j). \quad (2)$$

Following [4], the computation of the geodesic distances between pair of objects allows us to determine the dissimilarity matrix $\mathbf{D}$, whose generic element is

$$d(i,j) = \begin{cases} +\infty & \text{if there is no path between } i \text{ and } j, \\ d_E(i,j) = d_G(i,j) & \text{if } i \text{ and } j \text{ are neighbors,} \\ d_G(i,j) & \text{otherwise,} \end{cases} \quad (3)$$

$i,j = 1,\ldots,n.$

3 Fuzzy Relational Clustering for Non-linear Data

Although several procedures for clustering non-linear data are available in the literature (examples can be found in, for instance, [4–7]), robust clustering methods for such data have not been proposed. This is done in this paper by considering fuzzy clustering algorithms for relational data (see [8] for an overview). Relational data consists in the pairwise relations (similarities or dissimilarities) between objects. In the present case, such relations are stored in the dissimilarity matrix $\mathbf{D}$. It is therefore fundamental to consider fuzzy clustering algorithms suitable for relational data matrices, which do not contain Euclidean distances, such as $\mathbf{D}$. The Non-Euclidean Fuzzy Relational data clustering algorithm (NE-FRC), proposed in [9], can be used. The NE-FRC for Non-Linear data (NE-FRC-NL) can be formulated as:

$$\min_{\mathbf{U}} \sum_{g=1}^{c} \frac{\sum_{i=1}^{n}\sum_{j=1}^{n} u_{ig}^m u_{jg}^m d(i,j)}{2\sum_{t=1}^{n} u_{tg}^m},$$

$$\text{s.t.} \sum_{g=1}^{c} u_{ig} = 1, \qquad u_{ig} \in [0,1], \quad (4)$$

where c denotes the number of clusters, $m > 1$ is the fuzziness parameter (usually $m = 2$) and u_{ig}, the generic element of the membership degree matrix $\mathbf{U}$, expresses the degree of membership of object i to cluster g.

By NE-FRC-NL we detect c clusters taking into account the non-linear structure of the n objects. However, it may suffer from the presence of outliers. A noise resistant version of NE-FRC-NL can be built by using the concept of noise cluster [10]. It is an extra cluster such that the membership degree of objects belonging to the noise cluster increases with their distance to the c regular clusters. In this way, the influence of outliers on the regular clusters is reduced. In order to define the noise cluster, we fix the so-called noise dissimilarity δ and assume that, in the noise cluster, all the dissimilarities are equal to δ. By means of δ, objects such that their dissimilarities with even one object in every cluster are larger than δ will be classified as outliers [9]. Taking into account [9], we get the Robust NE-FRC for Non-Linear data (Robust-NE-FRC-NL):

$$
\min_{\mathbf{U}} \sum_{g=1}^{c} \frac{\sum_{i=1}^{n}\sum_{j=1}^{n} u_{ig}^{m} u_{jg}^{m} d(i,j)}{2\sum_{t=1}^{n} u_{tg}^{m}} + \frac{\sum_{i=1}^{n}\sum_{j=1}^{n} u_{i*}^{m} u_{j*}^{m}\delta}{2\sum_{t=1}^{n} u_{t*}^{m}},
\tag{5}
$$
$$
\text{s.t.} \sum_{g=1}^{c} u_{ig} \leq 1, \qquad u_{ig} \in [0,1],
$$

being u_{i*} the membership degree of object i to the noise cluster. The Robust-NE-FRC-NL cost function is equal to the NE-FRC-NL one plus an additional term concerning the noise cluster. By comparing (4) and (5) one more difference emerges. For each object, the sum of the membership degrees to the regular clusters is not equal to one. The complement of this sum,

$$
u_{i*} = 1 - \sum_{g=1}^{c} u_{ig},
\tag{6}
$$

gives the membership degree of object i to the noise cluster. Outliers will have high values of u_{i*} and, therefore, have arbitrarily small membership values to regular clusters.

References

1. Tenenbaum, J.B., de Silva, V., Langford, J.C.: A global geometric framework for nonlinear dimensionality reduction. Science **290**, 2319–2323 (2000)
2. Floyd, R.W.: Algorithm 97: shortest path. Commun. ACM **5**, 345 (1962)
3. Warshall, S.: A theorem on Boolean matrices. J. ACM. **9**, 11–12 (1962)
4. Király, A., Vathy-Fogarassy, Á., Abonyi, J.: Geodesic distance based fuzzy c-medoid clustering - searching forcentral points in graphs and high dimensional data. Fuzzy Sets Syst. **286**, 157–172 (2016)
5. Asgharbeygi, N., Maleki, A.: Geodesic k-means clustering. In: 19th International Conference on Pattern Recognition, ICPR 2008, Tampa, Florida, pp. 1–4 (2008)
6. Babaeian, A., Babaee, M., Bayestehtashk, A., Bandarabadi, M.: Nonlinear subspace clustering using curvature constrained distances. Pattern Recognit. Lett. **68**, 118–125 (2015)
7. Karygianni, S., Frossard, P.: Tangent-based manifold approximation with locally linear models. Signal Process. **104**, 232–247 (2014)
8. Runkler, T.A.: Relational fuzzy clustering. In: de Oliveira, J.V., Pedrycz, W. (eds.) Advances in Fuzzy Clustering and Its Applications, pp. 31–51. Wiley, Chichester (2007)
9. Davé, R.N., Sen, S.: Robust fuzzy clustering of relational data. IEEE Trans. Fuzzy Syst. **10**, 713–727 (2002)
10. Davé, R.N.: Characterization and detection of noise in clustering. Pattern Recog. Lett. **12**, 657–664 (1991)

Measures of Dispersion for Interval Data

Przemyslaw Grzegorzewski[1,2(✉)]

[1] Systems Research Institute, Polish Academy of Sciences, Warsaw, Poland
pgrzeg@ibspan.waw.pl
[2] Faculty of Mathematics and Information Science,
Warsaw University of Technology, Warsaw, Poland

Abstract. Almost all experiments reveal variability of their results. In this contribution we consider the measures of dispersion for sample of random intervals. In particular, we suggest a generalization of two well-known classical measures of dispersion, i.e. the range and the interquartile range, for interval-valued samples.

Keywords: Dispersion · Interquartile range · Interval data
Random interval · Range · Spread · Variability

1 Introduction

Besides measures of location, measures of dispersion play a key role both in descriptive and inferential statistics. Many tools have been proposed to characterize a dispersion of data, like the range, interquartile range, sample variance, standard deviation and so on. However, most of the contributions devoted to measures of dispersion are focused on univariate real data. In the case of multidimensional data one may apply univariate dispersion measures to each attribute separately, but this way we loose information on possible relations between variables. A general study on dispersion measure for p-dimensional real observations have been conducted by Kolacz and Grzegorzewski [8].

Recently, the interval-valued data have drawn an increasing interest. Quite often a real random variable is imprecisely observed or is so uncertain that the results are recorded as the real intervals containing the precise outcomes of the experiment. In all such cases intervals are considered as disjunctive sets representing incomplete information (*epistemic view* according to [3]). However, sometimes the experimental data appear as essentially interval-valued data describing a precise information (e.g. ranges of fluctuations of some physical measurements, time interval spanned by an activity). Such intervals are called conjunctive and correspond to the *ontic view* (see [3]). Interval data appear also as a kind in the so-called *Symbolic Data Analysis* summarizing the information stored in large data sets (see [1]).

In the epistemic approach we assume that the interval-valued observations are perceptions of the unknown, not observed, true outcomes of a real-valued random variable. In the ontic approach, contrary to the epistemic approach,

© Springer Nature Switzerland AG 2019
S. Destercke et al. (Eds.): SMPS 2018, AISC 832, pp. 91–98, 2019.
https://doi.org/10.1007/978-3-319-97547-4_13

we deal no longer with usual real-valued random variables but with random intervals. Our attention in this paper is restricted to the ontic approach. Thus, further on, we assume that a given sample is just a sequence of random intervals.

Central tendency measures for interval-valued data have been extensively examined in the literature (see [4,10–13]). On the other hand, it seems that the sample variance and standard deviation are the only measures of dispersion for random intervals considered in the literature (for some exceptions see [5]). The lack of measures based on quantiles, like the range or interquartile range, can be somehow explained by the fact that interval data are not linearly ordered. However, using a suitable interpretation of the aforementioned measures we can generalize them also for random intervals. Hence, our goal is to propose the concept of the range and the interquartile range that could be applied for characterizing the dispersion in the sample of random intervals.

The paper is organized as follows. In Sect. 2 we introduce basic notation related to the interval data. Next, in Sect. 3, we recall some information on random intervals and their characteristics, especially on the median for a sample of random intervals. Sect. 4 contains main results of our contribution. Starting from general remarks (Sect. 4.1), we introduce the generalizations of the range (Sect. 4.2) and interquartile range for the samples of random intervals (Sect. 4.3).

2   Interval-Valued Data

Let $\mathcal{K}_c(\mathbb{R})$ denote a family of all non-empty closed and bounded intervals in the real line $\mathbb{R}$. Each interval $A \in \mathcal{K}_c(\mathbb{R})$ can be expressed by means of its endpoints, $(\inf A, \sup A) \in \mathbb{R}^2$ with $\inf A \leq \sup A$. Other way for describing interval, which is in some situations more operative, is based on the point $(\text{mid } A, \text{spr } A) \in \mathbb{R} \times \mathbb{R}^+$, where $\text{mid } A = (\sup A + \inf A)/2$ is the mid-point (center) of the interval A, and $\text{spr } A = (\sup A - \inf A)/2$ denotes the spread (radius). Thus, A can be represented as $A = [\inf A, \sup A] = [\text{mid } A \pm \text{spr } A]$.

When dealing with intervals, a natural arithmetic is defined on $\mathcal{K}_c(\mathbb{R})$ based on the Minkowski addition and the product by scalars. These operations are settled as $A + B = \{a + b : a \in A, b \in B\}$ and $\lambda A = \{\lambda a : a \in A\}$, for all $A, B \in \mathcal{K}_c(\mathbb{R})$ and $\lambda \in \mathbb{R}$, respectively. Using the mid/spr notation the above operations can be jointly expressed as $A + \lambda B = [(\text{mid } A + \lambda \text{mid } B) \pm (\text{spr } A + |\lambda| \text{spr } B)]$.

It should be noted that the space $(\mathcal{K}_c(\mathbb{R}), +, \cdot)$ is not linear but semilinear, due to the lack of the inverse element with respect to the Minkowski addition: in general, $A + (-1)A \neq \{0\}$, unless $A = \{a\}$ is a singleton. To overcome this problem, sometimes it is possible to consider the so-called Hukuhara difference $A -_H B$ between the intervals A and B, defined by such an interval $C \in \mathcal{K}_c(\mathbb{R})$ that $B + C = A$. Unfortunately, the Hukuhara difference does not exist for any two intervals $A, B \in \mathcal{K}_c(\mathbb{R})$ but only for such $A, B \in \mathcal{K}_c(\mathbb{R})$ that $\text{spr } A \geqslant \text{spr } B$.

3   Random Intervals

As we have mentioned in Sect. 1, in the ontic approach we deal no longer with usual real-valued random variables but with random intervals defined as follows.

Definition 1. *Given a probability space $(\Omega, \mathcal{A}, P)$, a mapping $X : \Omega \longrightarrow \mathcal{K}_c(\mathbb{R})$ is said to be a **random interval** (interval-valued random set) if it is Borel-measurable with the Borel σ-field generated by the topology associated with by the Hausdorff metric on $\mathcal{K}_c(\mathbb{R})$.*

Equivalently, a mapping $X : \Omega \longrightarrow \mathcal{K}_c(\mathbb{R})$ is a random interval if mid $X :$ $\Omega \to \mathbb{R}$ and spr $X : \Omega \to \mathbb{R}^+$ are (real-valued) random variables defined as the mid-point and the spread of the interval $X(\omega)$, respectively, for each $\omega \in \Omega$.

If X is a random interval and mid X, spr $X \in L^1(\Omega, \mathcal{A}, P)$, then the *Aumann mean of* X is given by the set $\mathbb{E}[X] = \{\int_\Omega f dP : f \in X \text{ a.s.}[P]\}$, which leads to the following interval $\mathbb{E}[X] = [\mathbb{E}(\text{mid } X) \pm \mathbb{E}(\text{spr } X)]$.

The dispersion of a random interval can be measured by means of a distance between X and $\mathbb{E}[X]$. Consider a generalized family of L^2-type metric between intervals, introduced by Gil et al. [6] for any $A, B \in \mathcal{K}_c(\mathbb{R})$ as

$$d_{T,\theta}(A, B) = \sqrt{(\text{mid } A - \text{mid } B)^2 + \theta(\text{spr } A - \text{spr } B)^2}, \tag{1}$$

were $\theta > 0$ determines the relative weight of the distance between the spreads against the distance between the mids. It should be noted that a value of θ closer to 0 gives more importance to the midpoint, while a high value of θ gives more importance to the spread of the interval. Such a metric appears to be useful for defining the Fréchet-variance of a random interval X as $\sigma_X^2 = \mathbb{E}\left(d_{T,\theta}^2(X, \mathbb{E}[X])\right)$. As it is detailed in [2] this variance can be also expressed in terms of the classical variances of the mid and spr variables, namely, $\sigma_X^2 = \sigma_{\text{mid } X}^2 + \theta \sigma_{\text{spr } X}^2$.

Another useful metric on $\mathcal{K}_c(\mathbb{R})$ is the Hausdorff metric defined as follows

$$d_H(A, B) = |\text{mid } A - \text{mid } B| + |\text{spr } A - \text{spr } B|. \tag{2}$$

The Hausdorff metric can be applied for defining the median of a random interval as the interval value minimizing the expected distance $\mathbb{E}\left(d_H(X, A)\right)$ over $A \in \mathcal{K}_c(\mathbb{R})$. In this way we obtain the following definition (see [11]):

Definition 2. *Given a probability space $(\Omega, \mathcal{A}, P)$ and a random interval X associated with it, the **median** of the distribution of X is the nonempty compact interval $\widetilde{\text{me}}_H[X] = [\text{me}(\text{mid } X) \pm \text{me}(\text{spr } X)]$, where $\text{me}(\cdot)$ stands for the population median of a real-valued random variable.*

Note, that as for the real-valued case, the median of a random interval can be either defined by a unique interval or not, depending on the medians of mid X and spr X being both defined in a unique way or not.

Now, given a sample $\mathbb{X} = (X_1, \ldots, X_n)$ of random intervals the sample median $\widetilde{\text{Me}}_H[\mathbb{X}] = \widehat{\text{me}_H[X]}$ is defined as follows

$$\widetilde{\text{Me}}_H[\mathbb{X}] = \left[\text{Me}(\text{mid } X_1, \ldots, \text{mid } X_n) \pm \text{Me}(\text{spr } X_1, \ldots, \text{spr } X_n)\right], \tag{3}$$

where $\text{Me}(\text{mid } X_1, \ldots, \text{mid } X_n)$ and $\text{Me}(\text{spr } X_1, \ldots, \text{spr } X_n)$ stand for the sample median of a real-valued sample of the centers and of spreads, respectively. Sinova et al. [12] introduced also the population d_θ-median on $\mathcal{K}_c(\mathbb{R})$.

Definition 3. *Given a probability space* $(\Omega, \mathcal{A}, P)$ *and a random interval* X *associated with it, the* d_θ-**median** *of* X *is the nonempty compact interval* $\mathrm{me}_\theta[X]$ *such that*

$$\mathrm{me}_\theta[X] = \arg \min_{A \in \mathcal{K}_c(\mathbb{R})} \mathbb{E}\left(d_\theta(X, A)\right), \tag{4}$$

whenever the involved expectation exists.

Consequently, the sample d_θ-median $\widetilde{\mathrm{Me}}_\theta[\mathbb{X}] = \widehat{\mathrm{me}_\theta[X]}$ can be defined as

$$\widetilde{\mathrm{Me}}_\theta[\mathbb{X}] = \arg \min_{A \in \mathcal{K}_c(\mathbb{R})} \frac{1}{n} \sum_{i=1}^{n} d_\theta(X_i, A) \tag{5}$$

$$= \arg \min_{(y,z) \in \mathbb{R} \times [0,\infty)} \frac{1}{n} \sum_{i=1}^{n} \sqrt{(\mathrm{mid}\, X_i - y)^2 + (\mathrm{spr}\, X_i - z)^2}.$$

Sample medians discussed above are strongly consistent whenever the corresponding population medians exist and are actually unique. However, in contrast to L^1-type median, the sample d_θ-median always exists and, moreover, it is unique for any sample realization $(x_1, \ldots, x_n)$ for which the two-dimensional sample points $\left\{(\mathrm{mid}\, x_i, \mathrm{spr}\, x_i)\right\}_{i=1}^{n}$ are not all collinear [10, 13].

4   Dispersion Measures for Interval Data

4.1   General Remarks

Central tendency measures do not characterize a sample sufficiently. The most important type of summary statistics, besides measures of location, are measures of dispersion (variability or scale) such as the range, interquartile range, sample variance, standard deviation, etc. Kołacz and Grzegorzewski [8] have recently proposed an axiomatic definition of measure of dispersion for a sample in $\mathbb{R}^n$. According to them, a function $\Delta : \bigcup_{n=1}^{\infty} (\mathbb{R}^p)^n \to [0, \infty)$ is called a *measure of dispersion* if Δ is a non-identically zero function satisfying the following axioms for any $(\boldsymbol{x}_1, \ldots, \boldsymbol{x}_n) \in (\mathbb{R}^p)^n$:

(A1) $\Delta(\boldsymbol{x}_1, \ldots, \boldsymbol{x}_n) = 0$ if $\boldsymbol{x}_1 = \ldots = \boldsymbol{x}_n$.
(A2) Δ is symmetric, i.e. $\Delta(\boldsymbol{x}_{\pi(1)}, \ldots, \boldsymbol{x}_{\pi(n)}) = \Delta(\boldsymbol{x}_1, \ldots, \boldsymbol{x}_n)$ for any permutation $\pi : \{1, \ldots, n\} \to \{1, \ldots, n\}$.
(A3) Δ is translation invariant, i.e. $\Delta(\boldsymbol{x}_1 + \boldsymbol{t}, \ldots, \boldsymbol{x}_n + \boldsymbol{t}) = \Delta(\boldsymbol{x}_1, \ldots, \boldsymbol{x}_n)$ for all $\boldsymbol{t} \in \mathbb{R}^p$.
(A4) Δ is rotation invariant, i.e. $\Delta(\boldsymbol{R}\boldsymbol{x}_1, \ldots, \boldsymbol{R}\boldsymbol{x}_n) = \Delta(\boldsymbol{x}_1, \ldots, \boldsymbol{x}_n)$ for any rotation p-matrix $\boldsymbol{R}$.

Sometimes one more axiom is also considered:

(A5) there exists a function $\rho : \mathbb{R} \to [0, \infty)$ such that $\Delta(a\boldsymbol{x}_1, \ldots, a\boldsymbol{x}_n) = \rho(a)\Delta(\boldsymbol{x}_1, \ldots, \boldsymbol{x}_n), \forall a \in \mathbb{R}^+$.

All the above properties seem reasonable and most absolute measures of dispersion, like the variance, standard deviation, range and interquartile range, satisfy all of them. On the other hand, such relative measures of spread like the coefficient of variation or the Gini index satisfy all of them but A3.

It seems that the investigations on the dispersion measures for the interval-valued samples, until now, are mostly reduced (for some exceptions see [5]) to the generalization of the sample variance. Given a random sample $\mathbb{X} = (X_1, \ldots, X_n)$ of random intervals the d_θ-sample variance is defined as follows (see [2])

$$\widetilde{S}_\theta^2[\mathbb{X}] = \frac{1}{n} \sum_{i=1}^{n} d_\theta^2(X_i, \overline{X}),\tag{6}$$

where $\overline{X} = \frac{1}{n}\sum_{i=1}^{n} X_i = \left[\operatorname{mid} \overline{X} \pm \operatorname{spr} \overline{X}\right]$. $\widetilde{S}_\theta^2[\mathbb{X}]$ is a consistent estimator of the population variance σ_X^2. Obviously, the standard deviation is given by $\widetilde{S}_\theta[\mathbb{X}] = \sqrt{\widetilde{S}_\theta^2[\mathbb{X}]}$.

Contrary to the sample variance (6), which is a direct generalization of the classical sample variance, the range and interquartile range cannot be generalized so straightforward. Indeed, both measures are based on the natural linear ordering of real observations, but intervals are not linearly ordered. Therefore, to suggest a reasonable analogue of the range and interquartile range for interval data we have to utilize their properties to avoid ranking intervals.

4.2   Range for Interval Data

The range for a real-valued sample $(Y_1, \ldots, Y_n)$ is defined as $R = Y_{n:n} - Y_{1:n}$, where $Y_{i:n}$ denotes the i-th order statistic, i.e. the i-th biggest observation in the sample. Thus the range is as the distance between the biggest and the smallest observation, which is actually the biggest distance between any two observations in the sample, i.e.

$$R = \max\left\{|Y_i - Y_j| : i, j = 1, \ldots, n\right\}.\tag{7}$$

This remark lies at the bottom of our definition. Let d denote an arbitrary distance (d_H or d_θ) defined in Sect. 2.

Definition 4. *Given a sample* $\mathbb{X} = (X_1, \ldots, X_n)$ *of random intervals the **d-range** is defined as follows*

$$\widetilde{R}_d[\mathbb{X}] = \max\left\{d(X_i, X_j) : i, j = 1, \ldots, n\right\}.\tag{8}$$

One can prove the following theorem.

Proposition 1. *The d-range is a dispersion measure.*

A distance is, roughly speaking, a two-argument function measuring how much any two points of the given space are separated. Martín and Mayor [9] extended the conventional definition of distance between two points so that it might be applied to collections of more than two elements. Their multi-argument "distance" function, called a *multidistance*, is defined as follows.

Definition 5. *A function* $D : \bigcup_{n \geqslant 1}(\mathbb{R}^p)^n \to [0, \infty)$ *is called a* **multidistance** *if it satisfies the following conditions for all* $\mathbf{x}_1, \ldots, \mathbf{x}_n, \mathbf{y} \in \mathbb{R}^p$:

(md1) $D(\mathbf{x}_1, \ldots, \mathbf{x}_n) = 0$ *if and only if* $\mathbf{x}_1 = \ldots = \mathbf{x}_n$.
(md2) $D(\mathbf{x}_{\pi(1)}, \ldots, \mathbf{x}_{\pi(n)}) = D(\mathbf{x}_1, \ldots, \mathbf{x}_n)$ *for any permutation* π *of the numbers* $1, \ldots, n$.
(md3) $D(\mathbf{x}_1, \ldots, \mathbf{x}_n) \leqslant D(\mathbf{x}_1, \mathbf{y}) + \ldots + D(\mathbf{x}_n, \mathbf{y})$.

The following proposition could be easily proved.

Proposition 2. *The d-range is a multidistance.*

Proposition 2 indicates another approach for defining the range in the framework of random intervals.

Definition 6. *Given a sample* $\mathbb{X} = (X_1, \ldots, X_n)$ *of random intervals the* **range based on a multidistance** D *is defined as follows*

$$\tilde{R}_D[\mathbb{X}] = D(X_1, \ldots, X_n). \tag{9}$$

Thus, by a suitable choice of a multidistance, like the diameter of the smallest ball containing all observations [9], we may obtain alternative definitions of the range. For more considerations on the relation between multidistances and measures of dispersion we refer the reader to [8].

4.3   Interquartile Range for Interval Data

The range is rather rarely applied in practice because of its high sensitivity to outliers. Alternatively, one may use the interquartile range IQR $= Q_3 - Q_1$ defined as the difference between the upper Q_3 and lower quartile Q_1, i.e. the 75th and 25th percentiles, respectively. This simple definition is not actually so straightforward since one may determine percentiles in many ways [7]. To be more strict, given a real-valued sample $(Y_1, \ldots, Y_n)$ we may specify a general formula for the interquartile range as

$$\text{IQR} = (1 - \gamma)(Y_{k:n} - Y_{l:n}) + \gamma(Y_{k+1:n} - Y_{l+1:n}),$$

where $k = \lfloor 0.75n + m \rfloor$ and $l = \lfloor 0.25n + m \rfloor$, while m and γ are some constants which depend on a particular method for determining percentiles (for more details we refer the reader to [7,8]). Thus, IQR like the range, depends on order statistics and hence it cannot be directly generalized for the interval data. However, we may notice that the interquartile range is equal to the width of the interval containing about 50% of the middle observations (we write "about 50%" instead of 50% because it may happen that depending on the sample size and a particular method for determining percentiles IQR actually contains 50% $\pm$ 1 or 2 of the middle observations). This way we omit about 50% of observations which are relatively far from the "more typical" observations. So we also get rid of possible outliers. For the real data outliers are either too big or too small against the majority, we omit about 25% of the smallest and about 25% of the biggest observations in the sample.

Now, turning back to interval data, we should notice that the meaning of the outlier should be modified. Indeed, since each interval is characterized by its location and width, one can consider as the outlier both an interval which lies far from the "typical" interval as well as the interval which is too narrow or too wide with respect to the "typical" one. Following these clues we may define the interquartile range for a sample of intervals as the biggest distance between intervals which do not differ too much from the "typical" one. It seems that the median would be a quite reasonable candidate for such "typical" interval.

Let us consider the following measure construction. Given a sample $\mathbb{X} = (X_1, \ldots, X_n)$ of random intervals let us compute the sample median $\widetilde{\mathrm{Me}}[\mathbb{X}]$, where $\widetilde{\mathrm{Me}}[\mathbb{X}] \in \{\widetilde{\mathrm{Me}}_H[\mathbb{X}], \widetilde{\mathrm{Me}}_\theta[\mathbb{X}]\}$ denotes one of the medians defined in Sect. 2. Secondly, we determine a distance between each observation X_i and $\widetilde{\mathrm{Me}}[\mathbb{X}]$, i.e. we obtain the following sequence $(\xi_1, \ldots, \xi_n)$, where $\xi_i = d(X_i, \widetilde{\mathrm{Me}}[\mathbb{X}])$. Next, we order this sequence $(\xi_1, \ldots, \xi_n)$, so we obtain $(\xi_{1:n}, \ldots, \xi_{n:n})$.

Let $\beta = \xi_{k:n}$, where $k = \lfloor \frac{n}{2} \rfloor$, denote the k-th biggest distance from the median and let us define the following subset $\mathbb{X}_{1/2}$ of the original sample $\mathbb{X}$

$$\mathbb{X}_{1/2} = \left\{ X_i \in \mathbb{X} : d(X_i, \widetilde{\mathrm{Me}}[\mathbb{X}]) \leqslant \beta \right\}. \tag{10}$$

One may notice that $\mathbb{X}_{1/2}$ contains "about" 50% of the "central" observations, i.e. such intervals which do not differ too much with respect to the sample median. Now we are able to define the interquartile range as the biggest distance in $\mathbb{X}_{1/2}$. More formally, we may consider the following definition.

Definition 7. *Given a sample* $\mathbb{X} = (X_1, \ldots, X_n)$ *of random intervals the **d-interquartile range** is defined as follows*

$$\widetilde{\mathrm{IQR}}_d[\mathbb{X}] = \max \left\{ d(X_i, X_j) : X_i, X_j \in \mathbb{X}_{1/2} \right\}. \tag{11}$$

One can proof the following theorem.

Proposition 3. *The d_θ-interquartile range is a dispersion measure.*

Similarly, as in the case of the range we may also define the interquartile range with respect to a multidistance.

Definition 8. *Given a sample* $\mathbb{X} = (X_1, \ldots, X_n)$ *of random intervals the **interquartile range based on a multidistance** D is defined as follows*

$$\widetilde{\mathrm{IQR}}_D[\mathbb{X}] = D(\mathbb{X}_{1/2}). \tag{12}$$

5   Conclusion and Further Perspectives

In this contribution we have considered measures of dispersion for random intervals. In particular, we have suggested a generalization of the range and the interquartile range for interval-valued samples. Because of the limited space, the contribution contains just a sketch of the idea connected with the introduced notions and many questions remain open. Firstly, one may ask which median $\widetilde{\mathrm{Me}}_H[\mathbb{X}]$ or $\widetilde{\mathrm{Me}}_\theta[\mathbb{X}]$ works better as a reference interval applied in the d-interquartile range. Moreover, even if we restrict our attention to d_θ-interquartile ranges one may be interested in possible relations between properties of the corresponding interquartile range and the choice of the parameter θ. Another important problem concerns the robustness of the suggested measures. It is obvious that the breakdown point of $\widetilde{\mathrm{R}}_d[\mathbb{X}]$ is low but the breakdown point for the interquartile range $\widetilde{\mathrm{IQR}}_d[\mathbb{X}]$ based on different distances should be examined. Finally, measures of dispersion based on various multidistances and their relation to the notion of statistical depth are worth further study.

References

1. Billard, L., Diday, E.: From the statistics of data to the statistics of knowledge: symbolic data analysis. J. Am. Stat. Assoc. **98**, 470–487 (2003)
2. Blanco-Fernández, A., Corral, N., González-Rodríguez, G.: Estimation of a flexible simple linear model for interval data based on set arithmetic. Comput. Stat. Data Anal. **55**, 2568–2578 (2011)
3. Couso, I., Dubois, D.: Statistical reasoning with set-valued information: Ontic vs. epistemic views. Int. J. Approx. Reason. **55**, 1502–1518 (2014)
4. De Carvalho, F.A.T., De Souza, R.M.C.R., Chavent, M., Lechevallier, Y.: Adaptive Hausdorff distances and dynamic clustering of symbolic interval data. Pattern Recogn. Lett. **27**, 167–179 (2006)
5. de la Rosa, S., de Sáa, M.A., Lubiano, B., Sinova, P.F.: Robust scale estimators for fuzzy data. Adv. Data Anal. Classif. **11**, 731–758 (2017)
6. Gil, M.A., Lubiano, M.A., Montenegro, M., López, M.T.: Least squares fitting of an affine function and strenght of association for interval-valued data. Metrika **56**, 97–111 (2002)
7. Hyndman, R.J., Fan, Y.: Sample quantiles in statistical packages. Am. Stat. **50**, 361–365 (1996)
8. Kolacz, A., Grzegorzewski, P.: Measures of dispersion for multidimensional data. Eur. J. Oper. Res. **251**, 930–937 (2016)
9. Martín, J., Mayor, G.: How separated Palma, Inca and Manacor are? In: Proceedings of the AGOP 2009, pp. 195–200 (2009)
10. Sinova, B.: M-estimators of location for interval-valued random elements. Chemometr. Intell. Lab. Syst. **156**, 115–127 (2016)
11. Sinova, B., Casals, M.A., Colubi, A., Gil, M.A. : The median of a random interval. In: Borgelt, C., et al. (eds.) Combining Soft Computing & Statistical Methods, pp. 575–583. Springer, Heidelberg (2010)
12. Sinova, B., González-Rodríguez, G., Van Aelst, S.: An alternative approach to the median of a random interval using an L^2 metric. In: Kruse, R., et al. (eds.) Synergies of Soft Computing and Statistics for Intelligent Data Analysis, pp. 273–281. Springer, Heidelberg (2013)
13. Sinova, B., Van Aelst, S.: On the consistency of a spatial-type interval-valued median for random intervals. Stat. Probab. Lett. **100**, 130–136 (2015)

A Maximum Likelihood Approach to Inference Under Coarse Data Based on Minimax Regret

Romain Guillaume and Didier Dubois[(✉)]

IRIT, CNRS and Université de Toulouse,
118 Route de Narbonne, 31062 Toulouse Cedex 9, France
{Romain.Guillaume,dubois}@irit.fr

Abstract. Various methods have been proposed to express and solve maximum likelihood problems with incomplete data. In some of these approaches, the idea is that incompleteness makes the likelihood function imprecise. Two proposals can be found to cope with this situation: maximize the maximal likelihood induced by precise datasets compatible with the incomplete observations, or maximize the minimal such likelihood. These approaches prove to be extremist, the maximax approach having a tendency to disambiguate the data, while the maximin approach favors uniform distributions. In this paper we propose an alternative approach consisting in minimax relative regret criterion with respect to maximal likelihood solutions obtained for all precise datasets compatible with the coarse data. It uses relative likelihood and seems to achieve a trade-off between the maximax and the maximin methods.

1 Introduction

Maximum likelihood is a standard approach to finding a probabilistic model based on data. It maximizes the probability of obtaining the observations (supposed to belong to a set of mutually exclusive outcomes) [3]. When observations are coarse and may overlap, it is not clear how to define the likelihood function: several options exist, according to whether we take into account the measurement process governing the incompleteness or not. In this paper, we focus on optimizing the likelihood function that we should have observed, had observations been precise. Due to incomplete observations, this likelihood function is imprecise, since there are several possible precise datasets compatible with the coarse observations. Two approaches have been proposed: one considers an optimistic point of view aiming to disambiguate the data, by maximizing the maximum likelihood value across candidate datasets [5]. Another more cautious one tries to maximize the minimum likelihood value across candidate datasets thus adopting a robust optimization approach [6]. Both approaches have their weaknesses and can be criticized as being extreme ones, yielding either too deterministic or too dispersed distribution functions.

© Springer Nature Switzerland AG 2019
S. Destercke et al. (Eds.): SMPS 2018, AISC 832, pp. 99–106, 2019.
https://doi.org/10.1007/978-3-319-97547-4_14

In this paper we propose an alternative criterion that tries to define a trade-off between the two previous approaches, and can be seen as minimizing a maximal regret criterion. We provide the results of some preliminary investigations of this approach, first by considering examples in the discrete setting, such as ill-observed coin flipping. Optimizing this criterion seems to pose challenging computational problems.

2 Definition of the Problem

We consider the setting of a random experiment where a variable X is incompletely observed via a measurement device providing values of a variable Y, namely [1]:

- $X : \Omega \to \mathcal{X}$ represents the outcome of a certain random experiment. In the finite case we assume $\mathcal{X} = \{a_1, \ldots, a_m\}$.
- $Y : \Omega \to \mathcal{Y} \subseteq \wp(\mathcal{X})$ that models the reports of the measurement device, where $\mathcal{Y} = \{A_1, \ldots, A_r\}$, $\wp(\mathcal{X})$ is the set of subsets of $\mathcal{X}$, and $A_i \subseteq \mathcal{X}$.

The information about the joint probability distribution $P(X, Y)$ on $\mathcal{X} \times \mathcal{Y}$ of the random vector (X, Y) can be obtained by modeling the random variable X ($P(X)$) and its measurement process ($P(Y|X)$). However, in this paper, we shall just ignore the measurement process and try to figure out what is the best choice for $P(X)$ based on the available data. In general, this probability function depends on a model parameter θ, and we write it $P(X|\theta)$.

Let $\mathbf{y} = (G_1, \ldots G_N)$ be a sample of coarse observations, where $G_j \in \mathcal{Y}$ denote the results of observing X through the measurement device $Y = G_j$ means that $x_j \in G_j$, where x_j is the jth (unobserved) outcome of in the sample $\mathbf{x} = (x_1, \ldots x_N)$ of the random process governing X. Let $\mathcal{X}^{\mathbf{y}}$ be the set of samples of X compatible with $\mathbf{y}$. We may consider three different likelihood functions depending on whether we refer to [1]:

1. the observed sample in $\mathcal{Y}$: $P(\mathbf{y}|\theta) = \prod_{i=1}^{N} p(y_i|\theta)$.
2. the hidden sample in $\mathcal{X}$: $P(\mathbf{x}|\theta) = \prod_{i=1}^{N} p(x_i|\theta)$.
3. the complete sample in $\mathcal{Z} = \mathcal{X} \times \mathcal{Y}$: $L^{\mathbf{z}}(\theta) = \prod_{i=1}^{N} p(z_i|\theta)$, where $z_i = (x_i, G_i)$, $\mathbf{z} = ((x_1, G_1), \ldots (x_N, G_N))$, $x_j \in G_j, \forall j = 1, \ldots, N$.

In the following we focus on the likelihood function based on the hidden sample $\mathbf{x}$. Clearly this likelihood function is ill-known and belongs to the set $\{P(\mathbf{x}|\theta) : \mathbf{x} \in \mathcal{X}^{\mathbf{y}}\}$. There are two strategies of likelihood maximization, based on a sequence of imprecise observations $\mathbf{y} = (y_1, \ldots, y_N) \in \mathcal{Y}^N$:

1. The maximax strategy: it aims at finding $(\mathbf{x}^*, \theta^*)$ that maximizes $\overline{\Lambda}(\theta) = \max_{\mathbf{x} \in \mathcal{X}^{\mathbf{y}}} P(\mathbf{x}|\theta)$.
2. The maximin strategy: it aims at finding $\theta_* \in \Theta$ that maximizes $\underline{\Lambda}(\theta) = \min_{\mathbf{x} \in \mathcal{X}^{\mathbf{y}}} P(\mathbf{x}|\theta)$.

The above setting for inference with incomplete data differs from the older, more standard approach by Heitjan and Rubin [4], which relies on an extensive use of partitions and a different view of the measurement process: an observation G_j is viewed the unique element, such that $x_j \in G_j$, of a partition of $\mathcal{X}$ that is selected at random. Here we rather adopt the framework of Dempster [2] where the defect in the measurement process is modelled by means of a multimapping from the sample space to the outcome space $\mathcal{X}$.

These optimization problems take the following form in the discrete case where $\mathcal{X}$ is finite [7], assuming exchangeability. Let n_k be the number of appearances of value a_k in the virtual sample $\mathbf{x}$, and q_j be the number of observations of A_j in the observed sample $\mathbf{y}$, and let n_{ik} be the number of times (a_i, A_k) is obtained in the complete sample $\mathbf{z}$; we have that $\mathbf{x} \in \mathcal{X}^{\mathbf{y}}$ if and only if:

$$\begin{cases} (a) & \sum_{k=1}^{r} n_k = \sum_{j=1}^{r} q_j = N \\ (b) & n_k = \sum_{j=1}^{r} n_{kj}, \forall k = 1, \ldots, m \\ (c) & q_j = \sum_{k=1}^{m} n_{kj}, \forall j = 1, \ldots, r. \\ (d) & n_{kj} = 0 \text{ if } a_k \notin A_j, \forall k, j. \end{cases} \tag{1}$$

Let $\mathcal{N}^{\mathbf{y}}$ be the set of tuples $\mathbf{n} = (n_1, \ldots, n_m)$ that are compatible with $\mathbf{y}$, namely they satisfy (1). The maximax, resp. maximin, strategy then takes the following form, using log-likelihoods, and defining $p_k^\theta = P(X = a_k | \theta)$:

$$\max_{\mathbf{p}} \max_{\mathbf{n}} \sum_{k=1}^{m} n_k \cdot \log p_k^\theta$$

or

$$\max_{\mathbf{p}} \min_{\mathbf{n}} \sum_{k=1}^{m} n_k \cdot \log p_k^\theta$$

s.t. constraints (1) hold and

$$(e) \qquad \sum_{k=1}^{m} p_k^\theta = 1$$
$$(f) \qquad n_k, n_{jk} \in \mathbb{N}, p_k^\theta > 0, \qquad \forall k = 1, \ldots, m,$$

In the finite case, when all discrete distributions are possible ($\theta = (p_1, \ldots p_{m-1})$), the set of probabilistic models is the credal set associated to the belief function defined by the mass assignment $m(A_j) = q_j/N$, for $j = 1, \ldots, r$. Then the optimal solution to the maximin likelihood problem is the distribution with maximal entropy, namely the solution to: $\max_{\mathbf{n}} - \sum_{k=1}^{m} \frac{n_k}{N} \cdot \log \frac{n_k}{N}$ under conditions (a, b, c), and $n_k \in \mathbb{R}^+$, i.e. $\mathbf{n}$ in the convex hull of $\mathcal{N}^{\mathbf{y}}$ [1,7].

In contrast, the optimal solution to the maximax likelihood problem (2) is the solution with minimal entropy, namely the solution to: $\min_{\mathbf{n}} - \sum_{k=1}^{m} \frac{n_k}{N} \cdot \log \frac{n_k}{N}$ under conditions (a, b, c) [1,7].

As a consequence the two approaches to handling incomplete observations look somewhat extreme. On the one hand, the max-max approach tends to strongly disambiguate the data, yielding Dirac distributions consistent with the coarse data when they are feasible. If the measured quantity is deterministic in nature, this is natural. However it becomes questionable if the measured quantity is tainted with variability (like the ill-known overlapping observed outcomes of tossing a die). On the other hand, the max-min approach yields distributions with high variances interpreting incomplete information as the result of extreme randomness: the less information the larger the variance, which is not fully satisfactory either.

3   A Criterion Based on a Likelihood Ratio Trade-Off

In this section we propose a new criterion which tries to maximize the confidence level in the fact that the parameters are acceptable for all possible realizations. It is well-known that the likelihood function represents the plausibility of parameter θ in view of some precise results $\mathbf{x}$ of observing a variable X, in relative value only. Namely the likelihood of θ is proportional to $P(\mathbf{x}|\theta)$ but the proportionality coefficient is arbitrary. We cannot evaluate the extent to which data $\mathbf{x}$ supports θ more that data $\mathbf{x}'$ supports θ' by comparing $P(\mathbf{x}|\theta)$ and $P'(\mathbf{x}'|\theta')$ We must compare the likelihood ratio $\frac{P(\mathbf{x}|\theta)}{P(\mathbf{x}|\theta')}$ with $\frac{P(\mathbf{x}'|\theta)}{P(\mathbf{x}'|\theta')}$. In other words, only likelihood ratios can be used to choose the right parameter value.

The new criterion we propose consists in comparing the likelihood ratios $\frac{P(\mathbf{x}|\theta)}{P(\mathbf{x}|\hat{\theta}_\mathbf{x})}$ across observations $\mathbf{x} \in \mathcal{X}^\mathbf{y}$, where $\hat{\theta}_\mathbf{x}$ is the maximum likelihood estimate of the distribution parameter θ for observation $\mathbf{x}$, and adopt a variant of the minmax regret approach (we could call minmax *relative* regret) of the form

$$\max_{\theta \in \Theta} \min_{\mathbf{x} \in \mathcal{X}^\mathbf{y}} \frac{P(\mathbf{x}|\theta)}{P(\mathbf{x}|\hat{\theta}_\mathbf{x})}. \tag{2}$$

The idea is that since the ideal parameter value θ for observation $\mathbf{x}$ is $\hat{\theta}_\mathbf{x}$, and X is ill-observed, we try to find the value of θ that reaches a best compromise between the various ideal values $\hat{\theta}_\mathbf{x}$ for all $\mathbf{x}$ in agreement with the incomplete data $\mathbf{y}$. The value $\frac{P(\mathbf{x}|\theta)}{P(\mathbf{x}|\hat{\theta}_\mathbf{x})}$ lies in $[0, 1]$, and can be viewed as the degree of membership of θ to the fuzzy set $\tilde{\theta}_\mathbf{x}$ of good parameter estimates based on observing $\mathbf{x}$. Then the problem (2) is a standard fuzzy multicriteria optimisation problem (finding θ with the maximal membership values in the intersection of fuzzy sets $\tilde{\theta}_\mathbf{x}$). Note that the problem (2) is very similar to the one induced by the maximin strategy. The latter can be seen as a fuzzy optimisation problem, albeit with non normalized fuzzy sets.

Using the log-likelihood $L(\mathbf{x}|\theta) = \log P(\mathbf{x}|\theta)$, the above problem can be formulated as one of minimizing the maximal regret:

$$\min_{\theta \in \Theta} \max_{\mathbf{x} \in \mathcal{X}^\mathbf{y}} L(\mathbf{x}|\hat{\theta}_\mathbf{x}) - L(\mathbf{x}|\theta) \tag{3}$$

Or yet, in the finite case ($\mathcal{X}^\mathbf{y} = \{\mathbf{x}^1, \dots, \mathbf{x}^h\}$), it can be expressed as a goal programming problem, minimizing the L^∞ distance $\max_{i=1}^h |L(\mathbf{x}^i|\hat{\theta}_{\mathbf{x}^i}) - L(\mathbf{x}^i|\theta)|$ between the vector $(\theta, \dots, \theta)$ with length h and the ideal point $(\hat{\theta}_{\mathbf{x}^1}, \dots, \hat{\theta}_{\mathbf{x}^h})$.

Remark. An alternative choice of probabilistic model in the presence of coarse data consists in choosing the pignistic probability of Smets [9]. The observation vector $(q_1, \ldots, q_r)$ on $\mathcal{Y}$ can be viewed as a Dempster-Shafer mass assignment m on $\wp(\mathcal{X})$ [8], letting $m(A_j) = q_j/N$ for $j = 1, \ldots r$ inducing lower probabilities in the sense of [2] in the form of a belief function $Bel(A) = \sum_{E \subseteq A} m(A)$. This Dempster-Shafer mass assignment defines a convex set $\{P_X : \bar{P}_X(A) \geq Bel(A), \forall A \subseteq \mathcal{X}\}$ of probabilities on $\mathcal{X}$, hence of joint probabilities on $\mathcal{X} \times \mathcal{Y}$ with known marginals $\hat{q}_j = q_j/N$ for $j = 1, \ldots, r$ on $\mathcal{Y}$. Knowing the distribution on $\mathcal{Y}$ (which can be estimated from $\mathbf{y}$), the pignistic probability induced by m can be obtained as:

$$p_k^P = \sum_{j:A_j \ni a_k} \frac{q_j}{N \cdot |A_j|}.$$

where $|A_j|$ is the cardinality of A_j, which yields a marginal distribution on $\mathcal{X}$.

4   Resolution Method

In this section we propose a mathematical programming approach to solving problem (3) in the discrete case. Let $c(\mathcal{N}^y)$ be the convex hull of $\mathcal{N}^y$. Elements of $c(\mathcal{N}^y)$ are denoted by $\mathbf{w} = (w_1, \ldots w_m)$. To build this model, we show (see Proposition 1) that the maximization part of Problem (3) can be reduced to maximization over the set of vertices of $c(\mathcal{N}^y)$, denoted by $V(\mathcal{N}^y)$, that actually lies in $\mathcal{N}^y$. The efficiency of the approach presupposes $V(\mathcal{N}^y)$ is small enough. Note that $L(\mathbf{x}|\theta) = \sum_{k=1}^{m} n_k \cdot \log p_k^\theta$, which can be extended to $c(\mathcal{N}^y)$, in the form $L(\mathbf{w}|\theta) = \sum_{k=1}^{m} w_k \cdot \log p_k^\theta$, where $\mathbf{w}$ is a vector of non-negative reals.

Proposition 1. $\max_{\mathbf{x} \in \mathcal{X}^y} L(\mathbf{x}|\hat{\theta}_\mathbf{x}) - L(\mathbf{x}|\theta) = \max_{\mathbf{w} \in c(\mathcal{N}^y)} L(\mathbf{w}|\hat{\theta}_\mathbf{w}) - L(\mathbf{w}|\theta) = \max_{\mathbf{n} \in V(\mathcal{N}^y)} L(\mathbf{n}|\hat{\theta}_\mathbf{n}) - L(\mathbf{n}|\theta)$.

Proof: $L(\mathbf{x}|\hat{\theta}_\mathbf{x}) - L(\mathbf{x}|\theta) = \sum_{k=1}^{m} n_k \cdot \log n_k - \sum_{k=1}^{m} n_k \cdot \log p_k^\theta$ with $(n_1, \ldots n_m) \in \mathcal{N}^y$, the term $\log p_k^\theta$'s being constant. Consider maximizing $L(\mathbf{w}|\hat{\theta}_\mathbf{x}) - L(\mathbf{w}|\theta)$ for $\mathbf{w} \in c(\mathcal{N}^y)$ instead. The function $\sum_{k=1}^{m} w_k \cdot \log w_k$ is convex, $\sum_{k=1}^{m} w_k \cdot \log p_k^\theta$ is convex, and clearly $L(\mathbf{w}|\hat{\theta}_\mathbf{x}) - L(\mathbf{w}|\theta)$ is convex too. We also know that $c(\mathcal{N}^y)$ is a convex polyhedron with integer vertices, so the maximal value of $L(\mathbf{w}|\hat{\theta}_\mathbf{x}) - L(\mathbf{w}|\theta)$ is reached at one of these vertices, i.e., for some $\mathbf{w} = \mathbf{n} \in V(\mathcal{N}^y) \subseteq \mathcal{N}^y$ corresponding to some sample $\mathbf{x} \in \mathcal{X}^y$. $\square$

From Proposition 1, we can solve the maximin relative regret using the following mathematical programming model, introducing an additional variable $\alpha \geq 0$:

$$\min_{\mathbf{p}} \alpha \tag{4}$$

s.t.

(a) $\sum_{k=1}^{m} n_k \cdot \log n_k - \sum_{k=1}^{m} n_k \cdot \log p_k^\theta \leq \alpha, \forall \mathbf{x} \in V(\mathcal{X}^y)$

(b) $\qquad \sum_{k=1}^{m} p_k^\theta = 1$

(c) $\qquad p_k^\theta > 0, \qquad\qquad \forall k = 1, \ldots, m,$

This model has an exponential number of constraints (a). As perspective, we intend to develop an iterative algorithm to compute the optimal probability distribution.

5 Examples

In this section, we discuss the differences between the maximin and maximax approaches and the new one proposed in this paper by means of examples.

Let us suppose that you want to estimate the probability to observe heads or tails in a coin flipping experiment, where $\mathcal{X} = \{h, t\}$. But the only precise observations made yielded heads and we could not see the other outcomes: so we have $\mathcal{Y} = \{\{h\}, \{h, t\}\}$. Suppose that we observed 30 times $\{h\}$ and 30 times nothing ($\{h, t\}$). To estimate the minmax regret probability distribution on $\mathcal{X}$ (noted p^R) we solved the mathematical formulation (4) given in the previous section using the solver SQP of software Octave.[1] To discuss the result, let us compare it with the probability distribution obtained using the maximax approach (denoted by p^M) and the maximin approach (denoted by p^m) and Smets' pignistic probability distribution [9] (denoted by p^P) induced by the belief function whose family of focal sets is $\mathcal{Y}$. The results are given in Table 1.

Table 1. Probability distribution with 30 times $\{h\}$ and 30 times $\{h, t\}$

$\mathcal{X}$	$\{h\}$	$\{t\}$
$p^M(X = a_i) =$	1	0
$p^m(X = a_i) =$	0.5	0.5
$p^R(X = a_i) =$	0.8	0.2
$p^P(X = a_i) =$	0.75	0.25

Firstly we can see that the solution p^R lies between the maximax and the maximin solutions. The maximax one assumes that the observation $\{h, t\}$ are $\{h\}$, in contrast with the maximin solution, which assumes that the outcomes behind $\{h, t\}$ are $\{t\}$. The criterion proposed in this paper achieves a trade-off between the two extreme samples S1: (30 $\{h\}$, 30 $\{t\}$) and S2: (60 $\{h\}$, 0 $\{t\}$) compatible with the observations. It selects parameter values which are the least incompatible with both samples, here $p_h = 0.8$ and $p_t = 0.2$. Noted that it is different from the pignistic distribution here, $p_h^P = 0.75$ and $p_t^P = 0.25$.

Let us now study, on this simple example, the impact of the imprecise data on the optimal parameter found by the 4 methods. This is provided on Table 2. The extreme behavior of maximin and maximax methods is patent. The maximax approach disambiguates the data assuming the coin always yield heads. If it is known that the measured process is random (a regular coin flipping experiment), this approach sounds totally counter-intuitive. However if the (ill-)observed process is known to be deterministic (e.g., it consists in repetitively reading h on the visible side of a coin that is not flipped), concluding h with probability 1, if it has been observed at least once, is natural. In contrast, the maximin approach (like the other ones) interprets the lack of precise observations as pure randomness,

---

[1] https://www.gnu.org/software/octave/.

Table 2. Impact of imprecision on the optimal parameter

Number of $\{h\}$	60	50	40	30	20	10	0
Number of $\{h,t\}$	0	10	20	30	40	50	60
$p^M(X = h) =$	1	1	1	1	1	1	1 or 0
$p^m(X = h) =$	1	$\frac{5}{6}$	$\frac{4}{6}$	0.5	0.5	0.5	0.5
$p^R(X = h) \approx$	1	0.937	0.87	0.8	0.72	0.63	0.5
$p^P(X = h) \approx$	1	0.92	0.83	0.75	0.66	0.58	0.5

which may sound questionable both when the underlying phenomenon is known to be deterministic, and when it is not. When no outcome can be observed, the maximax approach expresses pure ignorance, while the uniform distribution found by other approaches is an instance of Laplace principle of insufficient reason.

Observe that the maximin regret approach yields a smoother variation of the optimal value, in terms of the amount of dataset imprecision than the maximin approach (the latter produces a uniform distribution as soon as it can). The same smooth behavior is observed for the pignistic probability. However, the latter is the center of gravity of the set of probability assignments $\mathbf{p} = (p_1, \ldots, p_m)$ such that $N\mathbf{p} \in c(\mathcal{N}^\times)$, hence based on Euclidean distance, while the minmax regret uses an L^∞ distance.

Let us analyze the value of likelihood ratio in (2) for optimal parameters p^R obtained by model (4), as provided on Table 3. This ratio measures our confidence in the obtained model, i.e., the extent to which parameter $\theta = (p_h^R, p_t^R)$ is acceptable for all possible samples in $\mathcal{N}^\times$.

Table 3. Confidence on the parameter θ

Number of $\{h\}$	60	50	40	30	20	10	0
Number of $\{h,t\}$	0	10	20	30	40	50	60
θ	(1,0)	(0.937,0.063)	(0.87,0.13)	(0.8,0.2)	(0.72,0.28)	(0.63,0.37)	(0.5,0.5)
$\min_{\mathbf{x} \in \mathcal{XY}} \frac{P(\mathbf{x}\mid\theta)}{P(\mathbf{x}\mid\hat{\theta}_{\mathbf{x}})}$	1	0.68	0.44	0.26	0.14	0.06	0.015

We can see that when the imprecision of the dataset increases, the extent to which the optimal parameter is acceptable for all possible samples decreases. In other words, when all observations are imprecise we are bound to choose $(0.5, 0.5)$ but we know that our confidence on this choice is low.

6   Conclusion

This paper is a preliminary step towards a more robust approach to statistical estimation in the presence of coarse data. This approach is more compatible with the usual view of likelihood functions as defined up to multiplicative constants, while the maximax and maximin approaches compare absolute likelihood values, albeit on equal datasets. Rather than selecting a single probabilistic model we could also use our criterion to build a small range of suboptimal parameter values that guarantee a given confidence level. Another issue is to compare our approach with more traditional ones to coarse data using partitions [4]. In these approaches, it is assumed that the imprecision generation process stems from the random choice of a coarsening of $\mathcal{X}$ (a partition) such that if $x = a$ occurs, the corresponding element of the random partition is observed. In contrast, our setting, based on [1], relies on the multimapping representation of incomplete information due to Dempster [2]. The latter framework sounds more natural, while the partition-based framework seems to require the specification of more parameters (there are more partitions of a finite set of size m than non-empty subsets thereof, when m is large enough). In other words, if no specific knowledge is available on the measurement process, there are more parameters to be set in the partition-based framework, than the number of conditional probabilities defining $P(Y|X)$.

References

1. Couso, I., Dubois, D.: A general framework for maximizing likelihood under incomplete data. Int. J. Approx. Reason. **93**, 238–260 (2018)
2. Dempster, A.P.: Upper and lower probabilities induced by a multivalued mapping. Ann. Math. Stat. **38**, 325–339 (1967)
3. Edwards, A.W.F.: Likelihood. Cambridge University Press, Cambridge (1972)
4. Heitjan, D.F., Rubin, D.B.: Ignorability and coarse data. Ann. Stat. **19**, 2244–2253 (1991)
5. Hüllermeier, E.: Learning from imprecise and fuzzy observations: data disambiguation through generalized loss minimization. Int. J. Approx. Reason. **55**, 1519–1534 (2014)
6. Guillaume, R., Dubois, D.: Robust parameter estimation of density functions under fuzzy interval observations. In: 9th ISIPTA Symposium, Pescara, Italy, pp. 147–156 (2015)
7. Guillaume, R., Couso, I., Dubois, D.: Maximum likelihood with coarse data based on robust optimisation. In: 10th ISIPTA Symposium, Lugano, Switzerland, pp. 169–180 (2017)
8. Shafer, G.: A Mathematical Theory of Evidence. Princeton University Press, Princeton (1976)
9. Smets, P.: Constructing the pignistic probability function in a context of uncertainty. In: Henrion, M., et al., (eds.) Uncertainty in Artificial Intelligence, North-Holland, Amsterdam, vol. 5, pp. 29–39 (1990)

Monitoring of Time Series Using Fuzzy Weighted Prediction Models

Olgierd Hryniewicz$^{(\boxtimes)}$ and Katarzyna Kaczmarek-Majer

Systems Research Institute, Newelska 6, 01-447 Warsaw, Poland
{hryniewi,K.Kaczmarek}@ibspan.waw.pl

abstract
Abstract. Monitoring of processes described by autocorrelated time series data has been considered. For this purpose, we propose to use the Shewhart control chart for residuals, designed using the $WAM*$ approach introduced by the authors. The main problem, that has to be solved in the design of the proposed control chart, is the choice of the weight w_0 assigned to the model estimated directly from data. In this paper, we propose a method for choosing the optimal value of w_0 using fuzzy values of parameters describing a monitored process.

Keywords: Control chart · Autocorrelated observations
Fuzzy weighted process model

1 Introduction

Monitoring of processes is the main aim of Statistical Process Control (SPC). In many applications, the most important feature of a monitored process is its stability. A process is considered stable, or under control, when its uncontrolled variation is purely random (e.g., due to random measurement errors). In 1924, W. Shewhart, at that time working for Western Electric Company, introduced a simple tool for monitoring stable processes - a control chart. According to Shewhart's proposal, in an initial stage of a monitored process, which is assumed to be in-control, process characteristics are measured, and their mean value and standard deviation are recorded. These recorded values are used for the design of a control chart, known as the Shewhart control chart, which consists of three control lines: central line, and two (or one, when only deviation of a process level in one direction is interesting) control lines. The process is considered stable when its observations are located inside control lines (limits). When an observation falls outside the control lines, an alarm signal is generated, and the process is considered as being possibly out of control (unstable).

Since the time of Shewhart's invention, many other control charts have been proposed and applied in many areas, such as industry, insurance, banking, health services, and many others. In his original proposal, Shewhart assumed that recorded values of process characteristics (variables) are mutually independent and described by the normal distribution. This assumption is still widely used

© Springer Nature Switzerland AG 2019
S. Destercke et al. (Eds.): SMPS 2018, AISC 832, pp. 107–114, 2019.
https://doi.org/10.1007/978-3-319-97547-4_15

in many univariate and multivariate control charts. However, in many practical cases, especially when individual process observations are monitored, these assumptions are not fulfilled. Thus, since the years 1970s, many inspection procedures that do not rely on these assumptions have been proposed (for more information see, [6]). Some of these procedures, such as control charts, have been applied in practice. For example, Sohn et al. [7] used a control chart for autocorrelated data for monitoring structural health of aerospace, civil, and mechanical engineering systems using methods of vibration-based damage diagnosis. References to other important papers can be found, e.g., in [5].

In this paper, we are interested in monitoring individual observations that are autocorrelated. It is a well known fact that in presence of autocorrelations design methods for control charts, developed under the assumption of independence, fail. As the result of this failure, traditionally designed control charts are inefficient, i.e., either generate too many false alarms (when the area between control lines is too narrow), or are not able to generate such alarms when a monitored process goes out of control. Therefore, several methods have been proposed with the common aim to take into account existing dependencies in process data. In their design, it is usually assumed that several hundreds of observations have to be available. Unfortunately, in many practical cases, such large amount of data is either not available or its collection is too costly (or takes too long time).

It has been observed by many authors (see [5], for more information) that control charts for autocorrelated data, especially those designed using small samples of observations, have elevated false alarm rates. Hryniewicz and Kaczmarek-Majer [5] have noted that this rather unfavorable property is somewhat related to the problem of bad predictability in short time series (e.g., having less than 100 observations). Inspired by the very good properties of their prediction algorithm for short time series [3], they proposed in [5] a new control chart for the so called residuals, named the XWAM control chart. The chart proposed by them is based on the concept of model averaging, well known in the literature devoted to the analysis of time series.

The performance of the XWAM control chart, and its further modifications, strongly depends upon certain weights assigned to mathematical models used for prediction purposes. These weights influence different rates of false alarms. Unfortunately, in some cases, the change of the assigned weight may increase the rate of alarms of one type (e.g., false), but at the same time decreases the rate of alarms of other types (e.g., true). Moreover, decision maker's preferences related to this change of behavior of a control chart may be imprecisely perceived. In order to cope with this problem, and this is the main aim of this paper, we propose a method for choosing appropriate weights of models using simple methods of computational intelligence, and the analysis of fuzzy preferences.

The paper is organized as follows. In the second section, we present methods that are used in monitoring processes, especially when consecutive observations are autocorrelated. In the third section, we present the $WAM*$ approach, recently proposed by us for the design of control charts using small sample sizes. The properties of control charts designed using this methodology have been investigated in extensive simulation experiments. They are used in the fourth section of the paper, where we propose an original method for finding the weights used in the $WAM*$ approach. The paper is concluded in its last section.

2 Monitoring of Processes with Control Charts

The design of the Shewhart control chart is very simple when the number of observations collected when a monitored process is stable is sufficiently large, and these observations are statistically independent. For these observations, treated as a sample from the monitored process, one has to calculate the average value $\bar{x}$ and the standard deviation σ_x, and set the control limits, upper (UCL) and lower (LCL), to $UCL = \bar{x} + 3\sigma_x$ and $LCL = \bar{x} - 3\sigma_x$, respectively.

When process deterioration is related only to increase (decrease) of a process level, one can use one-sided control charts with respective upper (lower) control limits. Usually, it is assumed that the monitored characteristic is normally distributed, and in this case the probability of observing the observation outside control limits when the monitored process is stable (i.e., observing a false alarm) is very low, and equals 0.0027. It means, that the expected number of observations between consecutive false alarms $(ARL0)$, is approximately equal to 370. In many practical situations, the maximal run length is limited (curtailed) from above. This curtailment decreases the value of $ARL0$. For example, when the maximal length of a monitored process is equal to 1000 observations, the value of $ARL0$ is approximately equal to 345. A chart is considered better when the value of $ARL0$ is larger. However, the main purpose of a chart is to trigger alarms when the process average increases (decreases) by a certain value s, usually expressed as a multiplicity of the process standard deviation. The average waiting time to alarm $ARL(s)$, i.e. the time between process deterioration and an alarm, is another important characteristic of a chart. A chart is considered efficient, if this value is possibly low. For example, when the process average shifts upwards by 3 standard deviations, the average time to alarm $ARL(3)$ is expected to be very low (e.g., close to 2). One has to note, however, that requirements of large $ARL0$ and small $ARL(s)$ are contradictory. Therefore, in the design of a control chart one has to use an appropriate balance between them.

When consecutive observations of a monitored process are statistically dependent, the situation becomes much more complicated. For example, when sample data are autocorrelated, the properties of a control chart designed using a simple algorithm described above may be completely different from those observed for independent data. In this paper, we consider one of the oldest, and the most popular among practitioners, model - an autoregression process of $AR(p)$, and defined by the following equation

$$X_t - \mu - a_1 X_{t-1} - \cdots - a_p X_{t-p} = Y_t, \tag{1}$$

where $X_i, i = t, t - 1, \ldots$ are random observations of a process, $Y_i, i = t, t - 1, \ldots$ is a series of independent random variables with constant variance and null expected value, $a_1, \ldots, a_p$ are process parameters describing its correlation structure, and μ is a constant describing a process level.

To cope with the problem of autocorrelation, statisticians have proposed two general approaches. In the first one, we chart the original data, but control limits are adjusted using the knowledge about the type of dependence. In the second general approach, originally introduced by Alwan and Roberts [1], a control chart is used for monitoring residuals, i.e., differences between actual observed value and its prediction based on previous observations. Their methodology is applicable for any class of processes, so it is also applicable for the $AR(p)$ process considered in this paper. According to the methodology proposed by Alwan and Roberts [1], process parameters are estimated from sample data of n elements, and used for the calculation of residuals. Then, these residuals are used for the construction of our control chart according to the algorithm described above.

There are many methods used for estimation of parameters of the $AR(p)$ process. One of the most popular is based on sample autocorrelations defined as

$$r_i = \frac{n \sum_{t=1}^{n-i}(x_t - \bar{x})(x_{t+i} - \bar{x})}{(n - i) \sum_{t=1}^{n}(x_t - \bar{x})^2}, i = 1, \ldots, p, \tag{2}$$

where n is the number of observations in the sample (usually, it is assumed that $n \geq 4p$), and $\bar{x}$ is the sample average. Then, the parameters $a_1, \ldots, a_p$ of the $AR(p)$ model are calculated by solving the Yule-Walker equations (see, [2])

$$\begin{aligned} r_1 &= a_1 + a_2 r_1 + \ldots + a_p r_{p-1} \\ r_2 &= a_1 r_1 + a_2 + \ldots + a_p r_{p-2} \\ &\ldots \\ r_p &= a_1 r_{p-1} + a_2 r_{p-2} + \ldots + a_p \end{aligned} \tag{3}$$

The constant term μ in (1) may be estimated using the following equation

$$\hat{\mu} = \bar{x}(1 - a_1 - \cdots - a_p). \tag{4}$$

3   Control Charts for Autocorrelated Data - The $WAM*$ Approach

When the process model is correctly estimated, calculated residuals are independent, and described by the distribution of Y_t in (1). In practice, however, this property may be assured when the sample size is very large (e.g., greater than 1000). Unfortunately, in many processes which need monitoring, such large samples are not attainable. For example, in [4] a control chart for monitoring a certain health-recovery process has to be designed using, due to medical circumstances, only first 20 observations. It has been observed by many authors (see [5], for more information) that imprecise (due to insufficient sample size) estimation of the process model leads to autocorrelated residuals, and poor properties of control charts based on these residuals. Inspired by the very good properties of their prediction algorithm for short time series [3], Hryniewicz and Kaczmarek-Majer proposed in [5] a new control chart for residuals, named the XWAM control chart, based on the concept of model averaging.

Let us denote by M_0 the model of a monitored process estimated from a (usually) small sample, and describe its parameters by a vector $(a_{1,0}, \ldots . a_{p_0,0})$. We assign to this estimated model a certain weight $w_0 \in [0, 1]$. We also consider k alternative models $M_j, j = 1, \ldots, k$, each described by a vector of parameters $(a^0_{1,j}, \ldots, a^0_{p_j,j})$. In general, any model with known parameters can be used as an alternative one, but in this paper, we restrict ourselves to the $AR(p)$ models. Let $w'_1, \ldots, w'_k$ denote the weights assigned to models $M_1, \ldots, M_k$ by this algorithm when only alternative models are considered. Because the total weight of the chosen alternative models is $1 - w_0$, to the estimated model we assign the weight w_0, and to each chosen alternative model we will assign a weight $w_j = (1 - w_0)w'_j, j = 1, \ldots, k$.

When we model our process using $k + 1$ models (one estimated from data, and k alternatives) each process observation generates $k + 1$ residuals. In the case of the $AR(p)$ process considered in this paper, they are calculated using the following formula

$$z_{i,j} = x_i - (\mu_j + a_{1,j}x_{i-1} + \ldots + a_{p_j,j}x_{i-p_j}), j = 0, \ldots, k; i = p_j + 1, \ldots. \quad (5)$$

In (5), we have assumed that for a model with $p_j, j = 0, \ldots, k$ parameters we need exactly p_j previous consecutive observations in order to calculate the first residual. Therefore, we need $i_{min} = \max(p_0, \ldots, p_k) + 1$ observations for the calculation of all residuals in the sample. For the calculation of the parameters of the XWAM control chart we use $n - i_{min} + 1$ weighted residuals calculated from the formula

$$z^*_i = \sum_{j=0}^{k} w_j z_{i,j}, i = i_{min}, \ldots, n. \quad (6)$$

The central line of the chart is calculated as the mean of z^*_i, and the control limits are equal to the mean plus/minus three standard deviations of z^*_i, respectively. The operation of the XWAM control chart is a classical one. First decision is made after i_{min} observations. The weighted residual for the considered observation is calculated according to (6), and compared to the control limits. An alarm is generated when the weighted residual falls beyond the control limits.

The method for the calculation of weights, assigned to alternative models, was firstly proposed by Hryniewicz and Kaczmarek-Majer in [5]. Their algorithm is based on the methods of computational intelligence, namely the DTW (Dynamic Time Warping) algorithm for comparison of time series. Unfortunately, this algorithm is computationally demanding, so in their further publications, the authors have proposed other variants of their original algorithm that are based on more simple methods for the comparison of time series. In this paper, we apply one of them, coined $WAM*$, and proposed in a recent paper [4]. In this approach, Hryniewicz and Kaczmarek propose to compare autocorrelation (ACF) and partial autocorrelation $(PACF)$ functions estimated from sample observations with respective ACF and $PACF$ functions calculated for possible alternative models. As the measure of distance between the estimated ACF and $PACF$ functions and their theoretical counterparts calculated for alternative models Hryniewicz and Kaczmarek-Majer [4] use a simple sum of absolute differences (called the Manhattan distance in the community of data mining). As alternative models used in the design of a control chart according to the $WAM*$ methodology, they consider the models with k lowest distances from the estimated one. Their weights, after some standardization, are inversely proportional to these distances. However, one still needs to find an appropriate value of the weight w_0. In the next section of this paper, we propose a method of doing this using an analysis of fuzzy preferences related to statistical properties of the proposed control chart.

4   Choice of the $WAM*$ Weights Using Fuzzy Preferences

In Table 1, we present the values of ARL of the Shewhart control chart for residuals when the monitored process is $AR(-0.3)$ (with the null constant term), and the size of the sample used for the design of this chart is equal to 30. As possible alternative models we have considered, in this case, all autoregression models of first and second order. In the rightmost column of this table, we present the values of ARL for the same process, when the sample size is very large (50000 in our example), i.e., when we know the process model nearly perfectly. We can consider these values as the theoretical values of $ARLs$. We can easily notice, that the necessity to design a chart using a small sample drastically deteriorates statistical properties of the chart. We can also notice, that the usage of a chart designed using the $WAM*$ approach improves its characteristics for shifted process (smaller values of ARL). However, at the same time, for non-shifted processes these characteristics are even worse (larger values of $ARL0$). Therefore, one can think about an optimal choice of the weight w_0 in order to design a chart with the minimal loss of efficiency, in comparison to chart designed using a very large sample.

Table 1. Average ARL for different weights assigned to the estimated model and different shifts of the process level, $\rho = -0,3$, n $= 30$.

Shift/w_0	1.0	0.9	0.8	0.7	0.6	0.5	0.4	0.3	0.2	0.1	0.0	ARL-T
-3	1.64	1.63	1.63	1.62	1.62	1.62	1.62	1.62	1.62	1.62	1.62	**1.61**
-2	3.96	3.89	3.84	3.79	3.77	3.75	3.74	3.73	3.74	3.75	3.78	**3.44**
-1	33.24	32.25	31.40	30.74	30.20	29.76	29.45	29.20	29.08	29.02	29.04	**22.91**
0	278.0	273.9	270.0	266.3	262.7	259.4	256.3	253.4	250.7	248.4	246.3	**345.6**
1	32.73	31.74	30.90	30.23	29.68	29.26	28.93	28,69	28.54	28.48	28.51	**22.91**
2	3.91	3.85	3.79	3.75	3.72	3.70	3.69	3.69	3.69	3.71	3.73	**3.44**
3	1.64	1.63	1.63	1.62	1.62	1.61	1.61	1.61	1.61	1.61	1.62	**1.61**

Let us denote tables like Table 1, with the row describing a non-shifted process deleted, by $g(i,j), i = 1, \ldots, n_s, j = 1, \ldots, k_w$, where n_s is the number of considered shifts, and k_w is the number of considered weights w_0. Then, let us create a table $h(i, j), i = 1, \ldots, n_s, j = 1, \ldots, k_w$, such that $h(i, j) = g(i, j) - g^*(i), i = 1, \ldots, n_s, j = 1, \ldots, k_w$, where $g^*(i), i = 1, \ldots, n_s$ denotes the theoretical values of $ARLs$ (such as those given in the last column of Table 1). The entries of this table represent profits (or losses) from using the $WAM*$ approach in chart's design. In the next step of the optimization process, we will treat separately shifted processes described by values of $h(i, j)$. To each process, characterized by a shift s, we assign the conditional probability $p_s(i)$ of observing such a process, and the utility (gain or loss) $v_s(i)$ attributed to the change of its ARL by one unit. Then, for each considered value of w_0, we can calculate the expected utility $u(j) = \sum h(i, j) p_s(i) v_s(i), j = 1, \ldots, k_w$, where this sum is calculated over all shifted processes. In the next step, we assign unit utilities v_0 attributed to the change of $ARL0$ by one unit, and calculate utilities $u_0(j) = v_0 * (g_0(j) - g_0)$, where $g_0(j)$ is the respective value of $ARL0$, and g_0 is its theoretical value. Note, that unit utilities v_0 must have a different sign than unit utilities $v_s(i)$ used in the evaluation of shifted processes. The optimal value of w_0 is that one, for which the sum of $u_T(j) = u(j) + u_0(j)$ is maximal.

In practice, the values of probabilities $p_s(i)$, and unit utilities $v_s(i)$ and v_0 are never known precisely. Let us assume that their values are given by respective fuzzy numbers $\tilde{p}_s(i), \tilde{v}_s(i), \tilde{v}_0$, represented by the α-cuts: $(p_{s,L}^\alpha(i), p_{s,U}^\alpha(i))$, $(v_{s,L}^\alpha(i), v_{s,U}^\alpha(i))$, and $(v_{0,L}^\alpha, v_{0,U}^\alpha)$. Then, using Zadeh's extension principle, we can calculate the fuzzy values of $\tilde{u}_T(j)$. In the last step of our optimization procedure, we can use a chosen method for ranking of fuzzy numbers, and find the optimal value of the weight w_0.

Let us illustrate this procedure using data presented in Table 1. We have six shifted processes, and let us assume that the probabilities of the occurrence of shifts $3, 2, 1, -1, -2, -3$ are described by the following triangular fuzzy numbers: $(0.03, 0.05, 0.08)$, $(0.1, 0.15, 0.2)$, $(0.2, 0.3, 0.4)$, $(0.2, 0.3, 0.4)$, $(0.1, 0.15, 0.2)$, $(0.03, 0.05, 0.08)$, respectively. The unit utilities for shifted processes are assumed to be, respectively, the following triangular fuzzy numbers: $(-1, -1, -0.8)$, $(-0.8, -0.6, -0.6)$, $(-0.4, -0.3, -0.2)$, $(-0.4, -0.3, -0.2)$, $(-0.8, -0.6, -0.6)$, $(-1, -1, -0.8)$. Finally, for the non-shifted process, we assume that the unit utility is given by the fuzzy triangular number $(0.01, 0.02, 0.03)$. The computed fuzzy numbers $u_T(j)$ are in this case, unfortunately, not triangular. However, their membership functions are only slightly convex, and may be well approximated by triangular numbers. For example, for $w_0 = 0.5$ the membership function may be well approximated by the triangular membership function $(-4.639, -2.962, -1.509)$. The value of the core of this fuzzy number is the largest one. Moreover, the Carlsson-Fuller expected value, and the center of gravity, for this fuzzy number are also the largest. Therefore, we may say that the optimal value of the weight w_0 is equal to 0.5.

5   Conclusions

In the paper, we have considered the problem of monitoring of processes represented by autocorrelated time series data. For monitoring purposes, we propose to use the Shewhart control chart for residuals, designed using the $WAM*$ approach introduced by the authors in [4]. The main problem that has to be solved in the design procedure is the choice of the weight w_0 assigned to the model estimated directly from data. In the paper, we propose a method for choosing the optimal value of w_0 using fuzzy values of parameters describing a monitored process. In the paper, we have considered processes described by simple autoregression models. In future research, more complicated models of monitored processes should be considered. For example, we can consider models describing non-stationary time series, such as that considered in [4] for monitoring of health-recovery processes. Moreover, we can directly utilize the fuzziness of the chosen weight w_0, and thus the fuzziness of plotted observations, arriving at control charts for residuals for fuzzy data.

References

1. Alwan, L.C., Roberts, H.V.: Time-series modeling for statistical process control. J. Bus. Econ. Stat. **6**, 87–95 (1988)
2. Brockwell, P.J., Davis, R.A.: Introduction to Time Series and Forecasting, 2nd edn. Springer, New York (2002)
3. Hryniewicz, O., Kaczmarek, K.: Bayesian analysis of time series using granular computing approach. Appl. Soft Comput. **47**, 644–652 (2016)
4. Hryniewicz, O., Kaczmarek-Majer, K.: Monitoring of health-recovery processes with control charts. In: Lorenz, P., Trinekens, J. (eds.) Proceedings of ACCSE 2017, IARIA, pp. 6–11 (2017)
5. Hryniewicz, O., Kaczmarek-Majer, K.: Monitoring of short series of dependent observations using a XWAM control chart. In: Knoth, S., Schmid, W. (eds.) Frontiers in Statistical Quality Control 12. Springer, Cham (2018, in print)
6. Montgomery, D.C.: Introduction To Statistical Quality Control, 6th edn. Wiley, New York (2011)
7. Sohn, H., Czarnecki, J.A., Farrar, C.R.: Structural health monitoring using statistical process control. J. Struct. Eng. **126**(11), 1356–1363 (2000)

Control Charts Designed Using Model Averaging Approach for Phase Change Detection in Bipolar Disorder

Katarzyna Kaczmarek-Majer$^{(\boxtimes)}$, Olgierd Hryniewicz, Karol R. Opara,
Weronika Radziszewska, Anna Olwert, Jan W. Owsiński,
and Sławomir Zadrożny

Systems Research Institute, Polish Academy of Sciences,
Newelska 6, 01-447 Warsaw, Poland
{K.Kaczmarek,Olgierd.Hryniewicz,Karol.Opara,
Weronika.Radziszewska,AOlwert,Jan.Owsinski,
Slawomir.Zadrozny}@ibspan.waw.pl

Abstract. Bipolar disorder is a mental illness affecting over 1% of the world's population. In the course of disease there are episodic fluctuations between different mood phases, ranging from depression to manic episodes and mixed states. Early detection and treatment of prodromal symptoms of affective episode recurrence is crucial since it reduces the conversion rates to full-blown illness and decreases the symptoms severity. This can be achieved by monitoring the mood stability with the use of data collected from patients' smartphones. We provide an illustrative example of the application of control charts to early and reliably generate notifications about the change of the bipolarity phase. Our charts are designed with the weighted model averaging approaches WAM* and WAMs for the detection of disturbances in the stability of the monitored processes. The models are selected in a novel way using the autocorrelation functions. The proposed approach delivers results that have clear, psychiatric interpretation. Control charts based on weighted model averaging are a promising tool for monitoring patients suffering from bipolar disorder, especially in case of limited amount of diagnostic data.

Keywords: Control charts · Forecasting · Short time series
Model averaging · Bipolar disorder · Monitoring stability

1 Introduction

This research reports on preliminary results from an ongoing clinical trial aiming at the early diagnosis of the bipolarity phase change. Bipolar disorder is "a mental disorder characterized by manic episodes of elevated mood and overactivity, interspersed with periods of depression" [1]. It's a chronic and recurrent disease with the highest rate of suicide of all the psychiatric disorders and affecting more than 1% of the world's population [2]. The disorder has a serious impact

© Springer Nature Switzerland AG 2019
S. Destercke et al. (Eds.): SMPS 2018, AISC 832, pp. 115–123, 2019.
https://doi.org/10.1007/978-3-319-97547-4_16

on psychosocial functioning, cognition and quality of life [3]. Recurrence rates are of 70% at five years [4]. Continuous monitoring of illness activity may be clinically relevant since it would allow for early detection of prodromal symptoms of affective episode recurrence.

Conus et al. [5] state that there is an urgent need for the development of early intervention strategies aimed at prompt detection and more specific treatment of an occurring phase change. Apart from monitoring the episodes of mania or depression, monitoring the mood stability becomes increasingly important for the understanding of the bipolar disorder [6]. Early pharmacological intervention is crucial because it reduces conversion rates to full-blown illness and decreases the symptoms severity.

There is a growing number of applications that support mood charting and monitoring of the bipolar disorder, a good example of which is MONARCA [7]. However, they require active patient participation, which is difficult to obtain, it especially concerns manic phases. It is also known that bipolar disorder patients often have a lack of insight about their symptoms and the need of monitor and treatment. Moreover, the challenges remain regarding the proper analysis and interpretation of the self-reported data due to the frequent changes related to the nature of bipolar disorder.

Within this research, we discuss the design of a Shewhart control chart to analyse the stability of a process with the self-reported mood measurements of a patient suffering from bipolar disorder. Despite the common use of control charts in industry and finance they are only entering the area of medicine. The probable reason for this is the incompatibility of basic assumptions used for their construction. In particular, the consecutive observations of the characteristics related to the mental disorder are seldom independent. Moreover, they are often described by non-stationary random processes, and the runs of observations useful for prediction are short (for instance three weeks of daily observations). Small samples are inevitable at the beginning of the data collection process. They also result from the invalidation of the recent observations if the phase change occurred. Recently, Vazquez-Montes et al. [8] provide study that investigates the ability of control chart methodology to support diagnostics of the bipolar disorder (data are collected weekly). Within this research, we apply control charts for daily monitoring of mood stability. We modify the XWAM control chart that was first introduced in [9] and developed further in [10,11]. Weighted model averaging and autoregressive models are combined to reflect the uncertainty about the considered small and non-stationary samples. The main advantage of the proposed approach is its accuracy and simplicity.

2   Design of a Shewhart Control Chart

The design of a Shewhart control chart is simple in the case of a sufficiently large number of individual and mutually independent observations. After collecting a sample of data from a period during which the monitored process is stable (its uncontrolled variation is purely random [12]), it suffices to calculate average value $\bar{x}$ and standard deviation σ_x, and set the control limits, upper (UCL) and lower (LCL), to

$$\begin{aligned} \text{UCL} &= \bar{x} + 3\sigma_x \\ \text{LCL} &= \bar{x} - 3\sigma_x. \end{aligned} \tag{1}$$

When a new observation falls outside the control lines, an alarm signal is generated, and the process is considered as being possibly out of control (unstable). In such a case the user should diagnose the reason of this behaviour and take appropriate actions to revert it to the in-control state.

Consider an example of daily measurements by a bipolar patient. Figure 1(a) presents a control chart constructed using 21 observations for the self-rated mood score. Data from the remaining 19 days (to the right from the dotted vertical line) are treated as new samples, whose stability is to be analysed.

The upper and lower control limits depend significantly on the amount of the training data. Moreover, the characteristics of the process depicted in Fig. 1(a) change in its final phase (days 30 to 40). More importantly, the process is non-stationary because the value of the Kwiatkowski-Philips-Schmidt-Shin (KPSS) statistic [13] amounts to 2.27. Thus, insignificant (from a medical point of view) shifts of the monitored process may cause false alarms. However, already the series of the first order differences, that is presented in Fig. 1(b), turns out to be approximately stationary with the KPSS statistic 0.06 and the corresponding p-value of 0.1. Although, the alarm will be generated in (b) for the final phase with abnormal observations, the established control limits are rather wide leading to a risk of missing a phase change. The improper choice of the control limits is especially frequent when the charted data are autocorrelated.

There are two general approaches towards handling autocorrelation. In the first one, the original data are charted, but control limits are adjusted using the knowledge about the type of dependence. In the second approach, originally introduced by Alwan and Roberts [14], a chart for monitoring residuals is constructed. To obtain the residuals, predictions are commonly calculated according to the autoregressive model AR of the most appropriate order p

$$X_t - \mu - a_1 X_{t-1} - \cdots - a_p X_{t-p} = Z_t, \tag{2}$$

where $X_i, i = t, t-1, \ldots$ are random observations of a process, $Z_i, i = t, t-1, \ldots$ is a series of independent random variables with constant variance and zero expectation, $a_1, \ldots, a_p$ are process parameters describing its correlation structure, and μ is a constant describing a process level.

One of the most common criteria of choosing the order of the AR model is the Akaike's criterion (AIC). However, Hryniewicz and Kaczmarek-Majer [10] showed that such a control chart performs poorly for short series. Ineffectiveness of the residual control chart is a consequence of a more general problem of inaccurate model selection for small samples. Emiliano et al. [15] state that the indiscriminate use of the information criteria for small samples ($n = 100$) in selecting the best predictive model leads to inaccurate results.

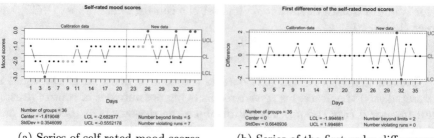

(a) Series of self-rated mood scores   (b) Series of the first order differences

Fig. 1. Control charts for the self-rated mood scores

3   Weighted Model Averaging for Monitoring Residuals

We follow Alwan and Roberts' [14] approach of monitoring residuals. However, we improved the model selection procedure. Instead of estimating the residuals with one autoregressive model, we combine multiple autoregressive models. Weighted Model Averaging (WAM*) algorithm [10] is adapted to select the k models from a set of alternative models $\{M_1, M_2, ..., M_J\}$. As a qualitative model selection tool, we compare the sample autocorrelation (ACF) and partial autocorrelation (PACF) against their known, theoretical values for the j predictive models. The sample lag-i autocorrelation r_i is given by

$$r_i = \frac{n \sum_{t=1}^{n-i}(x_t - \bar{x})(x_{t+i} - \bar{x})}{(n-i) \sum_{t=1}^{n}(x_t - \bar{x})^2}, i = 1, \ldots, p, \tag{3}$$

where n is the number of observations in the sample (usually, it is required that $n \geq 4p$), and $\bar{x}$ is the sample average. Furthermore, we improve the WAM* approach by calculating not only the sum of the autocorrelation distances, but also their standard deviations. We call this approach WAMs.

The WAMs algorithm, similarly to WAM*, is designed for the stationary time series. Thus, a preprocessing phase that ensures the transition of a time series to stationarity is necessary. Fortunately, in many practical contexts, already the series of first or second order differences are stationary.

After preprocessing, the WAMs algorithm consists of the following steps:

1. **Selection of alternative models.** Let us denote by M_0 the estimated model of a monitored process (e.g., with the AIC), and describe its parameters by a vector $(a_{1,0}, \ldots . a_{p_0,0})$. We assign to this estimated model a certain weight $w_0 \in [0, 1]$. We also consider k alternative models $M_j, j = 1, \ldots, k$, each described by a vector of parameters $(a_{1,j}, \ldots, a_{p_j,j})$. In general, any model with known parameters can be used as an alternative one, but in this paper, we restrict ourselves to the $AR(p)$ models. In practice, we take all combinations of coefficients $a_{p_j,j}$ from a grid with some small step α (e.g., $\alpha = 0.1$).
2. **Calculation of distances** for the sample ACF and PACF against known theoretical autocorrelation functions for the considered predictive models.

For the comparison of autocorrelation functions, Hryniewicz and Kaczmarek-Majer [10] use the sum of absolute differences (known also as L1 metric or the Manhattan distance). The sum of both distances, for the ACF(p) and the PACF(p) functions, is used for choosing the best alternative models and their weights. We modify this step to better handle the selection of the top performing models and instead of calculating only the sum of both distances, we also consider their standard deviations

$$s_{\phi,j} = \sqrt{\frac{\sum_{k=1}^{p}(d_{\phi_p(k),j} - \bar{d}_{\phi_p,j})^2}{p-1}}, s_{\psi,j} = \sqrt{\frac{\sum_{k=1}^{p}(d_{\psi_p(k),j} - \bar{d}_{\psi_p,j})^2}{p-1}} \quad (4)$$

where $d_{\phi_p(k),j}$ $(d_{\psi_p(k),j})$ is the difference between the jth model's kth autocorrelation (partial autocorrelation) coefficient and the kth coefficient estimated from the sample;

3. **Weighted model averaging** is applied for the selected k alternative models. Model weights w'_j, $j = 1, \ldots, k$ are calculated using the sum of the mean value and the standard deviation. Let $w'_1, \ldots, w'_k$ denote the weights assigned to models $M_1, \ldots, M_k$ by this algorithm when only alternative models are considered, and $w'_j = \frac{1}{\bar{d}_{\psi_p,j}+s_{\psi,j}+\bar{d}_{\phi_p,j}+s_{\phi,j}}$. Because the total weight of the chosen alternative models is $1 - w_0$, to the estimated model we assign the weight w_0, and to each chosen alternative model we will assign a weight $w_j = (1 - w_0)w'_j, j = 1, \ldots, k$. Individual forecasts f_i of the selected k models are calculated according to definitions in M. The final forecast $f(t) = \sum_{i=0}^{k} w_i f_i(t)$ is calculated as the weighted average of the individual forecasts $f_i(t)$ using the corresponding weights $w_0, w_1, ..., w_k$. More detailed discussion about the construction and performance of the WAM* algorithm is given e.g., in [10]. Also, the problem of the weights selection for the alternative models has been separately handled by Hryniewicz and Kaczmarek-Majer in [16].

4   Results

Although, the complete validation of the proposed methodology will be possible only when the clinical trail is finished, we can present some preliminary results. We assume that the model is estimated using a small sample ($n = 21$), and then, it's verified for 14 step-ahead forecasts ($h = 1, 2, ..., 14$). The original time series is divided to create multiple, overlapping fragments. The first 21 observations of every sample are used to estimate the predictive model, and the remaining 14 observations are applied to verify its predictive performance.

We compare the prediction errors obtained for the proposed model averaging approach in with the errors of the forecasts calculated from the M_0 process selected with the AIC criterion and estimated using the Yule-Walker equations (referenced as 'est AR' forecast). Table 1 shows the accuracy measured with the mean squared error (MSE) depending on the forecast horizon. The best results (and those that are not statistically significantly worse from the best) are typed

Table 1. Forecast accuracy (MSE and Relative change of MSE) according to the proposed WAMs, WAM* and est. AR.

Forecast horizon	MSE WAM*	MSE WAMs	MSE est AR	R-MSE WAM*	R-MSE WAMs
1	**0.693**	0.707	0.717	3.31%	1.41%
2	**0.807**	0.808	0.846	4.70%	4.55%
3	0.878	**0.875**	0.923	4.87%	5.18%
4	0.909	**0.907**	0.953	4.55%	4.85%
5	0.915	**0.912**	0.954	4.14%	4.39%
6	0.907	**0.905**	0.944	3.94%	4.16%
7	0.892	**0.890**	0.926	3.74%	3.93%
8	0.872	**0.870**	0.904	3.54%	3.71%
9	0.856	**0.855**	0.886	3.30%	3.46%
10	0.837	**0.836**	0.865	3.21%	3.36%
11	0.810	**0.809**	0.837	3.29%	3.43%
12	0.776	**0.775**	0.803	3.39%	3.52%
13	0.742	**0.741**	0.769	3.55%	3.68%
14	0.708	**0.707**	0.735	3.75%	3.87%
Avg	**0.829**	0.828	**0.862**	**3.81%**	**3.82%**

in bold. In all cases, the proposed WAMs method delivers forecasts that are more accurate than the forecasts of the estimated AR process.

As observed in Table 1, the average MSE amounts to 0.829, 0.828 and 0.862 for the WAM*, WAMs, and the referenced AR process respectively. Thus, the average relative improvement on using model averaging is 3.81%–3.82% and is very similar for both approaches. If we study the detailed results for shorter horizons, for example $h = 7$, we also observe that the proposed model averaging WAM* and WAMs approaches deliver improvement according to the MSE. Then, the relative change amounts to 3.74% and 3.93% for WAM* and WAMs, respectively.

Now, let us illustrate how the novel WAMs method of prediction is applied for the construction of a control chart for monitoring residuals. Suppose, that our control chart has to be put into operation after observing the first 21 measurements of the self-rated mood scores. The model M_0 is the following: $\mu = 0.014$ and $a_{0,1} = -0.409$. When we take into account only this model, the mean value of the calculated residuals equals 0.033, and sample standard deviation is 0.49. Hence, the Shewhart two-sided control chart for the mean value, calculated using the estimated model only, has the following control limits: LCL $= -1.45$, and UCL $= 1.51$. We used the estimated model for the calculation of residuals for the next 14 measurements, obtaining a control chart presented in Fig. 2(a). Two alarms are generated. However, both control lines are rather distant from the charted process values. Thus, this chart may be ineffective in the detection of the process instability.

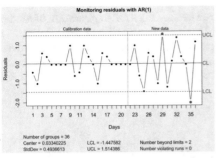

(a) Residuals of the est. AR model (M_0)

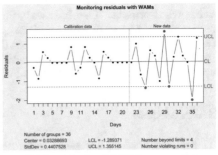

(b) Residuals of WAMs, $w_0 = 0.5$

Fig. 2. Control charts for the series of residuals estimated for the process with self-rated mood scores

Finally, let us use the WAMs approach for the calculation of residuals using the first 21 measurements. The three best alternative models are the following: M_1 being AR(1) where $a_{1,1} = -0.3$ with weight $w_1 = 0.342$; $M_2 : $ AR(1) : $a_{1,2} = -0.4$ with weight $w_2 = 0.337$ and $M_3 : $ AR(1) : $a_{1,3} = -0.2$ with weight $w_3 = 0.32$. For $w_0 = 0.5$ the mean value of calculated residuals is equal to 0.0328, and sample standard deviation is equal to 0.44. Hence, the Shewhart two-sided control chart for the mean value, calculated using the WAMs approach, has the following control limits: LCL $= -1.289$, and UCL $= 1.355$. The resulting control chart is presented in Fig. 2(b). The standard deviation is 10% smaller, and as a consequence, the control limits are closer to the observed values. Four alarms are generated. This control chart represents a compromise between a chart designed using only an estimated model, and a chart designed using only the alternative models.

5   Conclusions

In this paper, we monitor the stability of short and non-stationary processes using the Shewhart control chart. We propose a new way of computing the control limits through weighted model averaging called WAMs. Its main strength consists in taking into account the uncertainty about the data at the modeling stage due to the averaging approach. Illustrative examples of the control charts designed in a novel way have been applied for an early detection of stability disturbance in the monitoring of patients with bipolar disorder.

Our preliminary results suggest that, the proposed approach is highly competitive with the state-of-the-art approach of the statistical process control in terms of both, the accuracy and simplicity. Notably, this approach is suitable for detecting changes even in case of no medical diagnoses (similarly to the idea of unsupervised learning).

In future work we plan to validate the proposed approach with the complete data from the clinical trial. Moreover, we plan further developments of the forecasting methodology, so that it includes other predictive models.

Acknowledgments. The study was submitted to the Office for Registration of Medicinal Products, Medical Devices and Biocidal Products in accordance with Polish law. This project is financed from EU funds (Regional Operational Program for Mazovia) - a project entitled "Smartphone-based diagnostics of phase changes in the course of bipolar disorder" (RPMA.01.02.00-14-5706/16-00).

References

1. American Psychiatric Association: Bipolar and related disorders in diagnostic. In: Diagnostic and Statistical Manual of Mental Disorders, 5th ed., Washington, D.C. (2000)
2. Grande, I., Berk, M., Birmaher, B., Vieta, E.: Bipolar disorder. Lancet **387**(10027), 1561–1572 (2016). http://www.sciencedirect.com/science/article/pii/S014067361500241X
3. Catala-Lopez, F., Genova-Maleras, R., Vieta, E., Tabares-Seisdedos, R.: The increasing burden of mental and neurological disorders. Eur. Neuropsychopharmacol.: J. Eur. Coll. Neuropsychopharmacol. **23**, 1337–1339 (2013)
4. Gitlin, M., Swendsen, J., Heller, T., Hammen, C.: Relapse and impairment in bipolar disorder. Am. J. Psychiatry **152**, 1635–1640 (1995)
5. Conus, P., Macneil, C., McGorry, P.: Public health significance of bipolar disorder: implications for early intervention and prevention. Bipolar Disord. **16**, 548–556 (2014)
6. Bonsall, M.B., Wallace-Hadrill, S.M.A., Geddes, J.R., Goodwin, G.M., Holmes, E.A.: Nonlinear time-series approaches in characterizing mood stability and mood instability in bipolar disorder. Proc. R. Soc. Lond. Ser. B: Biol. Sci. **279**, 916–924 (2012)
7. Faurholt-Jepsen, M., Frost, M., Ritz, C., Christensen, E.M., Jacoby, A., Mikkelsen, R., Knorr, U., Bardram, J., Vinberg, M., Kessing, L.V.: Daily electronic self-monitoring in bipolar disorder using smartphones-the MONARCA I trial: a randomized, placebo-controlled, single-blind, parallel group trial. Psychol. Med. **45**(13), 2691–2704 (2015)
8. Vazquez-Montes, M., Stevens, R., Perera, R., Saunders, K., Geddes, J.R.: Control charts for monitoring mood stability as a predictor of severe episodes in patients with bipolar disorder. Int. J. Bipolar Disorder **6**, 7 (2018)
9. Hryniewicz, O., Kaczmarek-Majer, K.: Monitoring of short series of dependent observations using a XWAM control chart. In: Frontiers in Statistical Quality Control, vol. 12. Springer, Cham (2018)
10. Hryniewicz, O., Kaczmarek-Majer, K.: Monitoring of non-stationary health-recovery processes with control charts. Int. J. Adv. Life Sci. (2018)
11. Kaczmarek-Majer, K., Hryniewicz, O.: Data-mining approach to finding weights in the model averaging for forecasting of short time series. In: Proceedings of EUSFLAT 2017 (2017)
12. Montgomery, D.C.: Introduction To Statistical Quality Control, 6th edn. Wiley, New York (2011)
13. Kwiatkowski, D., Philips, P., Schmidt, P., Shin, Y.: Testing the null hypothesis of stationarity against the alternative of a unit root. J. Econom. **54**, 159–178 (1992)
14. Alwan, L.C., Roberts, H.V.: Time-series modeling for statistical process control. J. Bus. Econ. Stat. **6**, 87–95 (1988)

15. Emiliano, P.C., Vivanco, M.J., de Menezes, F.S.: Information criteria: how do they behave in different models. Comput. Stat. Data Anal. **69**, 141–153 (2014)
16. Hryniewicz, O., Kaczmarek-Majer, K.: Monitoring of time series using fuzzy weighted prediction models. In: Destercke, S., Denoeux, T., Gil, M., Grzegorzewski, P., Hryniewicz, O. (eds.) Uncertainty Modelling in Data Science (2018)

An Imprecise Probabilistic Estimator for the Transition Rate Matrix of a Continuous-Time Markov Chain

Thomas Krak[✉], Alexander Erreygers, and Jasper De Bock

ELIS, SYSTeMS, Ghent University, Ghent, Belgium
{thomas.krak,alexander.erreygers,jasper.debock}@ugent.be

Abstract. We consider the problem of estimating the transition rate matrix of a continuous-time Markov chain from a finite-duration realisation of this process. We approach this problem in an imprecise probabilistic framework, using a set of prior distributions on the unknown transition rate matrix. The resulting estimator is a set of transition rate matrices that, for reasons of conjugacy, is easy to find. To determine the hyperparameters for our set of priors, we reconsider the problem in discrete time, where we can use the well-known Imprecise Dirichlet Model. In particular, we show how the limit of the resulting discrete-time estimators is a continuous-time estimator. It corresponds to a specific choice of hyperparameters and has an exceptionally simple closed-form expression.

1 Introduction

Continuous-time Markov chains (CTMCs) are mathematical models that describe the evolution of dynamical systems under (stochastic) uncertainty [9]. They are pervasive throughout science and engineering, finding applications in areas as disparate as medicine, mathematical finance, epidemiology, queueing theory, and others. We here consider time-homogeneous CTMCs that can only be in a finite number of states.

The dynamics of these models are uniquely characterised by a single *transition rate matrix* Q. This Q describes the (locally) linearised dynamics of the model, and is the generator of the semi-group of transition matrices $T_t = \exp(Qt)$ that determines the conditional probabilities $P(X_t = y \,|\, X_0 = x) = T_t(x, y)$. In this expression, X_t denotes the uncertain state of the system at time t, and so T_t contains the probabilities for the system to move from any state x at time zero to any state y at time t.

In this work, we consider the problem of estimating the matrix Q from a single realisation of the system up to some finite point in time. This problem is easily solved in both the classical frequentist and Bayesian frameworks, due to the likelihood of the corresponding CTMC belonging to an exponential family; see e.g. the introductions of [3,7]. The novelty of the present paper is that we instead consider the estimation of Q in an *imprecise probabilistic* [1,14] context.

© Springer Nature Switzerland AG 2019
S. Destercke et al. (Eds.): SMPS 2018, AISC 832, pp. 124–132, 2019.
https://doi.org/10.1007/978-3-319-97547-4_17

Specifically, we approach this problem by considering an entire *set* of Bayesian priors on the likelihood of Q, leading to a *set-valued* estimator for Q. In order to obtain well-founded hyperparameter settings for this set of priors, we recast the problem by interpreting a continuous-time Markov chain as a limit of *discrete*-time Markov chains. This allows us to consider the imprecise-probabilistic estimators of these discrete-time Markov chains, which are described by the popular Imprecise Dirichlet Model (IDM) [10]. The upshot of this approach is that the IDM has well-known prior hyperparameter settings which can be motivated from first principles [4,15].

This leads us to the two main results of this work. First of all, we show that the limit of these IDM estimators is a set of transition rate matrices that can be described in closed-form using a very simple formula. Secondly, we identify the hyperparameters of our imprecise CTMC prior such that the resulting estimator is equivalent to the estimator obtained from this discrete-time limit. For reasons of brevity, the proofs of our results are omitted. They are available in the appendix of an online extended version[1].

The immediate usefulness of our results is two-fold. From a domain-analysis point of view, where we are interested in the parameter values of the process dynamics, our imprecise estimator provides prior-insensitive information about these values based on the data. If we are instead interested in robust inference about the future behaviour of the system, our imprecise estimator can be used as the main parameter of an *imprecise continuous-time Markov chain* [5,6,8,13].

2   A Brief Refresher on Stochastic Processes

Intuitively, a stochastic process describes the uncertainty in a stochastic system's behaviour as it moves through some state space $\mathcal{X}$ as time t progresses over some time dimension $\mathbb{T}$. A fundamental choice is whether we are considering processes in discrete time, in which case typically $\mathbb{T} = \mathbb{N}_0$, or in continuous time, in which case $\mathbb{T} = \mathbb{R}_{\geq 0}$. Here we write $\mathbb{N}$ for the natural numbers, and let $\mathbb{N}_0 := \mathbb{N} \cup \{0\}$. The real numbers are denoted by $\mathbb{R}$, the positive reals by $\mathbb{R}_{>0}$, and the non-negative reals by $\mathbb{R}_{\geq 0}$. We briefly recall the basic definitions of stochastic processes below; for an introductory work we refer to e.g. [9].

Formally, a realisation of a stochastic process is a *sample path*, which is a map $\omega : \mathbb{T} \to \mathcal{X}$. Here $\omega(t) \in \mathcal{X}$ represents the state of the process at time $t \in \mathbb{T}$. We collect all sample paths in the set Ω and, when $\mathbb{T} = \mathbb{R}_{\geq 0}$, these paths are assumed to be càdlàg under the discrete topology on $\mathcal{X}$. With this domain in place, we then consider some abstract underlying probability space $(\Omega, \mathcal{F}, P)$, where $\mathcal{F}$ is some appropriate $(\sigma\text{-})$algebra on Ω, and where P is a (countably-)additive probability measure.

The stochastic process can now finally be defined as a family of random variables $\{X_t\}_{t \in \mathbb{T}}$ associated with this probability space. In particular, for fixed $t \in \mathbb{T}$, the quantity X_t is a random variable $\Omega \to \mathcal{X} : \omega \mapsto \omega(t)$. Conversely, for a fixed realisation $\omega \in \Omega$, $X_t(\omega)$ is a deterministic map $\mathbb{T} \to \mathcal{X} : t \mapsto \omega(t)$.

---

[1] URL: https://arxiv.org/abs/1804.01330.

Well-known and popular kinds of stochastic processes are *Markov chains*:

Definition 1 (Markov Chain). *Fix* $\mathbb{T} \in \{\mathbb{N}_0, \mathbb{R}_{\geq 0}\}$, *and let* $\{X_t\}_{t \in \mathbb{T}}$ *be a stochastic process. We call this process a* Markov *chain if, for all* $s_0, \dots, s_n, s, t \in \mathbb{T}$ *for which* $s_0 < \cdots < s_n < s < t$, *it holds that* $P(X_t = x_t \mid X_{s_0} = x_{s_0}, \dots, X_{s_n} = x_{s_n}, X_s = x_s) = P(X_t = x_t \mid X_s = x_s)$ *for all* $x_{s_0}, \dots, x_{s_n}, x_s, x_t \in \mathcal{X}$. *If then* $\mathbb{T} = \mathbb{N}_0$, *we call* $\{X_t\}_{t \in \mathbb{T}}$ *a* discrete-time Markov chain *(DTMC). If instead* $\mathbb{T} = \mathbb{R}_{\geq 0}$, *we call it a* continuous-time Markov chain *(CTMC).*

Furthermore, attention is often restricted to *homogenous* Markov chains:

Definition 2 (Homogeneous Markov Chain). *Let* $\{X_t\}_{t \in \mathbb{T}}$ *be a Markov chain. We call this Markov chain* (time-)homogeneous *if, for all* $s, t \in \mathbb{T}$, $s \leq t$, *and all* $x, y \in \mathcal{X}$, *it holds that* $P(X_t = y \mid X_s = x) = P(X_{(t-s)} = y \mid X_0 = x)$.

This homogeneity property makes such processes particularly easy to describe.

In what follows, we will say that a $|\mathcal{X}| \times |\mathcal{X}|$ matrix T is a *transition matrix*, if it is a real-valued and row stochastic matrix, i.e. if $T(x, y) \geq 0$ and $\sum_{z \in \mathcal{X}} T(x, z) = 1$ for all $x, y \in \mathcal{X}$. We write $\mathfrak{T}$ for the space of all transition matrices. The elements T of $\mathfrak{T}$ can be used to describe the single-step conditional probabilities of a (homogeneous) DTMC:

Proposition 3 ([9]). *Let* $\{X_t\}_{t \in \mathbb{N}_0}$ *be a homogeneous DTMC. Then this process is completely and uniquely characterised by a probability mass function p on $\mathcal{X}$ and some $T \in \mathfrak{T}$. In particular, $P(X_0) = p$ and, for all $t \in \mathbb{N}_0$ and all $x, y \in \mathcal{X}$, $P(X_t = y \mid X_0 = x) = T^t(x, y)$, where T^t is the tth matrix power of T.*

On the other hand, to describe CTMCs we need the concept of a *(transition) rate matrix*: a $|\mathcal{X}| \times |\mathcal{X}|$ real-valued matrix Q with non-negative off-diagonal elements and zero row-sums, i.e. $Q(x, y) \geq 0$ and $\sum_{z \in \mathcal{X}} Q(x, z) = 0$ for all $x, y \in \mathcal{X}$ such that $x \neq y$. We write $\mathfrak{Q}$ for their entire space. A rate matrix describes the "speed" with which a CTMC moves between its states:

Proposition 4 ([9]). *Let* $\{X_t\}_{t \in \mathbb{R}_{\geq 0}}$ *be a homogeneous CTMC. Then this process is completely and uniquely characterised by a probability mass function p on $\mathcal{X}$ and some $Q \in \mathfrak{Q}$. In particular, $P(X_0) = p$ and, for all $t \in \mathbb{R}_{\geq 0}$ and all $x, y \in \mathcal{X}$, $P(X_t = y \mid X_0 = x) = \exp(Qt)(x, y)$, where $\exp(Qt)$ is the matrix exponential of Qt. Furthermore, for small enough $\Delta \in \mathbb{R}_{\geq 0}$ and all $x, y \in \mathcal{X}$, it holds that $P(X_\Delta = y \mid X_0 = x) \approx (I + \Delta Q)(x, y)$, where I is the identity matrix.*

3 Estimation of a CTMC's Rate Matrix

In what follows, we will derive methods to estimate the rate matrix Q of a homogeneous CTMC from a realisation $\omega \in \Omega$ that was observed up to some finite point in time $t_{\max} \in \mathbb{R}_{\geq 0}$. We denote with $\widetilde{\omega}$ the restriction of ω to this interval $[0, t_{\max}] \subset \mathbb{R}_{\geq 0}$, and we consider this (finite-duration) observation to be fixed throughout the remainder of this paper.

For any $x, y \in \mathcal{X}$ such that $x \neq y$, we let n_{xy} denote the number of transitions from state x to state y in $\widetilde{\omega}$. Furthermore, we let d_x denote the total duration spent in state x, that is, we let $d_x := \int_0^{t_{\max}} \mathbb{I}_x(\widetilde{\omega}(t)) \, dt$, where $\mathbb{I}_x$ is the indicator of $\{x\}$, defined by $\mathbb{I}_x(\widetilde{\omega}(t)) := 1$ if $\widetilde{\omega}(t) = x$ and $\mathbb{I}_x(\widetilde{\omega}(t)) := 0$ otherwise. We assume in the remainder that $d_x > 0$ for all $x \in \mathcal{X}$. Finally, for notational convenience, we define $q_{xy} := Q(x, y)$ for all $x, y \in \mathcal{X}$.

3.1 Precise Estimators

Under the assumption that the realisation ω was generated by a homogeneous continuous-time Markov chain with rate matrix Q, it is well known that the process dynamics can be modelled using exponentially distributed random variables whose parameters are given by the elements of Q. For various of such interpretations, we refer to e.g. [9]. What matters to us here is that, regardless of the interpretation, we can use this to obtain the following likelihood result (see e.g. [7]): for a given $\widetilde{\omega}$, the likelihood for a rate matrix Q is

$$L(\widetilde{\omega} \,|\, Q) = \prod_{\substack{x, y \in \mathcal{X} \\ x \neq y}} (q_{xy})^{n_{xy}} e^{-q_{xy} d_x}. \tag{1}$$

The corresponding maximum-likelihood estimator Q^{ML} is easily found [7]: $q_{xy}^{\mathrm{ML}} = n_{xy}/d_x$ if $x \neq y$ and $q_{xx}^{\mathrm{ML}} = -\sum_{y \in \mathcal{X} \setminus \{x\}} q_{xy}^{\mathrm{ML}}$, where the final expression follows from the (implicit) constraint that the rows of a rate matrix should sum to zero.

Inspection of the likelihood in (1) reveals that it belongs to an exponential family. This implies that there exists a conjugate prior for the rate matrix Q, such that its posterior distribution, given $\widetilde{\omega}$, belongs to the same family as this prior. This prior is given by a product of Gamma distributions, specifically on the off-diagonal elements q_{xy}, $x \neq y$, of the corresponding rate matrix [3]. We here use a slightly more general joint prior on Q whose "density" f is given by

$$f(Q \,|\, \boldsymbol{\alpha}, \boldsymbol{\beta}) := \prod_{\substack{x, y \in \mathcal{X} \\ x \neq y}} (q_{xy})^{\alpha_{xy} - 1} e^{-q_{xy} \beta_x} \propto \prod_{\substack{x, y \in \mathcal{X} \\ x \neq y}} \mathrm{Gamma}(q_{xy} \,|\, \alpha_{xy}, \beta_x), \tag{2}$$

with shapes α_{xy} and rates β_x in $\mathbb{R}_{\geq 0}$; we write $\boldsymbol{\alpha}, \boldsymbol{\beta}$ for the joint parameters.

Note that we have only defined the prior to equal a product of Gamma distributions up to normalisation, so that the prior $f(Q \,|\, \boldsymbol{\alpha}, \boldsymbol{\beta})$ may be improper. This has the advantage that it allows us to close the parameter domains and allow prior hyperparameters $\alpha_{xy} = 0$ and $\beta_x = 0$, for which the Gamma distribution is not properly defined. We acknowledge that the use of such improper priors is not entirely uncontroversial, and that their interpretation as a prior probability (which it indeed is not) leaves something to be desired. We will nevertheless, in this specific setting, be able to motivate their use here as a consequence of Theorem 5 further on.

Also, despite being improper, we can of course combine the prior (2) with the likelihood (1) and fix the normalisation in the posterior. As is well known, the means of the marginals of this posterior are then of the form[2]

$$\mathbb{E}\big[q_{xy}\,|\,\boldsymbol{\alpha},\boldsymbol{\beta},\widetilde{\omega}\big] = \frac{\alpha_{xy} + n_{xy}}{\beta_x + d_x} \qquad \forall x,y \in \mathcal{X}, x \neq y. \tag{3}$$

Furthermore, the (joint) posterior mean is well-known to be a Bayes estimator for Q under quadratic loss and given the prior $f(\cdot\,|\,\boldsymbol{\alpha},\boldsymbol{\beta})$ [2].

The question now remains of how to a priori settle on a "good" choice for these hyperparameters $\boldsymbol{\alpha},\boldsymbol{\beta}$, in the sense that they should adequately represent our prior beliefs. This is a non-trivial problem, and no general solution can be given. A popular (but not uncontroversial) attempt to characterise a non-informative prior consists in choosing the improper prior with $\boldsymbol{\alpha} = \boldsymbol{\beta} = 0$; the posterior mean (Bayes) estimator then equals Q^{ML}.

3.2   An Imprecise Probabilistic Estimator

Generalising the above Bayesian approach, we here suggest an imprecise probabilistic treatment. Following for example [11,14], this approach consists in using an entire *set* of prior distributions. Specifically, we consider a set of the form

$$\Big\{ f(\cdot\,|\,\boldsymbol{\alpha},\boldsymbol{\beta}) \,\Big|\, (\boldsymbol{\alpha},\boldsymbol{\beta}) \in C \Big\}, \tag{4}$$

with $f(\cdot\,|\,\boldsymbol{\alpha},\boldsymbol{\beta})$ as in (2), and where C is a set of possible prior parameters. In this way, we do not have to restrict our attention to one specific choice of the parameters $\boldsymbol{\alpha},\boldsymbol{\beta}$; rather, we can include all the parameter settings that we deem reasonable, by collecting them in C. Inference from $\widetilde{\omega}$ is then performed by point-wise updating each of these priors; we thereby obtain a set of posterior distributions on the space of all rate matrices. Each of these posteriors has a mean of the form (3), which is a Bayes estimator for Q under a specific prior in the set (4). This leads us to consider the imprecise, i.e., *set-valued*, estimator

$$\mathcal{Q}_C := \Big\{ Q \in \mathfrak{Q} \,\Big|\, \Big(\forall x,y \in \mathcal{X}, x \neq y : q_{xy} = \frac{\alpha_{xy} + n_{xy}}{\beta_x + d_x} \Big), (\boldsymbol{\alpha},\boldsymbol{\beta}) \in C \Big\}.$$

Note that even in this imprecise probabilistic approach, we still need to somehow specify the (now set-valued) prior model. That is, we need to be specific about the set C. Inspired by the well-known imprecise Dirichlet model [15], we may choose an "imprecision parameter" $s \in \mathbb{R}_{\geq 0}$, which can be interpreted as

---

[2] The assumption $d_x > 0$ prevents division by zero in (3). However, n_{xy} might be zero and, if then also $\alpha_{xy} = 0$, the posterior cannot be normalised and will still be improper. Nevertheless, using an intuitive (but formally cumbersome) argument we can still identify this posterior for q_{xy} with the (discrete) distribution putting all mass at zero. Alternatively, we can motivate (3) by continuous extension from the cases where $\alpha_{xy} > 0$, similarly yielding the estimate $\hat{q}_{xy} = 0$ at $\alpha_{xy} = n_{xy} = 0$.

a number of "pseudo-counts", to constrain $0 \leq \sum_{y \in \mathcal{X} \setminus \{x\}} \alpha_{xy} \leq s$ for all $x \in \mathcal{X}$, and to then vary all β_x over their domain $\mathbb{R}_{\geq 0}$. Unfortunately, similar to what is noted in [11], this leads to undesirable behaviour. For example, as is readily seen from e.g. (3), including unbounded β_x allows the off-diagonal elements q_{xy} to get arbitrarily close to zero, causing the model to a posteriori believe that transitions leaving x may be impossible, no matter the number of such transitions that we actually observed in $\widetilde{\omega}$! Hence, we prefer a different choice of C.

One way to circumvent this undesired behaviour is to constrain the range within which each β_x may be varied, to some interval $[0, \overline{\beta}_x]$, say. The downside is that this introduces a large number of additional hyperparameters; we then need to ("reasonably") choose a value $\overline{\beta}_x \in \mathbb{R}_{\geq 0}$ for each $x \in \mathcal{X}$. Fortunately, our main result – Theorem 5 further on – suggests that setting $\overline{\beta}_x = 0$ (and therefore $\beta_x = 0$) is in fact a very reasonable choice. This identification is obtained in the next section, using a limit result of discrete-time estimators, for which the hyperparameter settings follow entirely from first principles.

In summary, we keep the "imprecision parameter" $s \in \mathbb{R}_{\geq 0}$ and the constraint $0 \leq \sum_{y \in \mathcal{X} \setminus \{x\}} \alpha_{xy} \leq s$ for all $x \in \mathcal{X}$, and simply set $\beta_x = 0$ for all $x \in \mathcal{X}$. We then define C_s to be the largest set of parameters that satisfies these properties. Every $\boldsymbol{\alpha}$ in this set can be conveniently identified with the off-diagonal elements of a matrix sA, with $A \in \mathfrak{T}$ a transition matrix. Our set-valued estimator $\mathcal{Q}_s$ can thus be written as

$$\mathcal{Q}_s := \left\{ Q \in \mathfrak{Q} \,\middle|\, \left(\forall x, y \in \mathcal{X}, x \neq y : q_{xy} = \frac{sA(x,y) + n_{xy}}{d_x} \right), A \in \mathfrak{T} \right\}. \quad (5)$$

4   Discrete-Time Estimators and Limit Relations

A useful intuition is that we can consider a CTMC as a limit of DTMCs, where we assign increasingly shorter durations to the time steps at which the latter operate. In this section, we will use this connection to relate estimators for DTMCs to estimators for CTMCs. We start by discretising the observed path.

Because the realisation ω was only observed up to some time $t_{\max} \in \mathbb{R}_{\geq 0}$, we can discretise the (finite-duration) realisation $\widetilde{\omega}$ into a finite number of steps. For any $m \in \mathbb{N}$, we write $\delta^{(m)} := t_{\max}/m$, and we define the discretised path $w^{(m)} : \{0, \dots, m\} \to \mathcal{X}$ as $w^{(m)}(i) := \widetilde{\omega}(i\delta^{(m)})$ for all $i \in \{0, \dots, m\}$.

For any $m \in \mathbb{N}$ and $x, y \in \mathcal{X}$, we let $n_{xy}^{(m)} := \sum_{i=1}^{m} \mathbb{I}_x(w^{(m)}(i-1))\mathbb{I}_y(w^{(m)}(i))$ denote the number of transitions from state x to y in $w^{(m)}$, and we let $n_x^{(m)} := \sum_{y \in \mathcal{X}} n_{xy}^{(m)}$ denote the total number of time steps that started in state x.

4.1   Discrete-Time Estimators

For fixed $m \in \mathbb{N}$, we can interpret the discretised path $w^{(m)}$ as a finite-duration ($m+1$ steps long) realisation of a homogeneous discrete-time Markov chain with transition matrix $T^{(m)}$, with m keeping track of the discretisation level. Each transition along the path $w^{(m)}$, from state x to y, say, is then a realisation of a categorical distribution with parameters $T^{(m)}(x, \cdot)$. The likelihood for $T^{(m)}$, given $w^{(m)}$, is therefore proportional to a product of independent multinomial likelihoods. Hence, the maximum likelihood estimator follows straightforwardly and as expected: $T^{(m),\text{ML}}(x, y) = n^{(m)}_{xy}/n^{(m)}_x$ for all $x, y \in \mathcal{X}$; see [7] for details.

In a Bayesian analysis, and following e.g. [10], for fixed m we can model our uncertainty about the unknown $T^{(m)}$ by putting independent Dirichlet priors on the rows $T^{(m)}(x, \cdot)$. We write this prior as $g(\cdot \mid s, A)$, where $s \in \mathbb{R}_{\geq 0}$ is a "prior strength" parameter, and $A \in \text{int}(\mathfrak{T})$ is a prior location parameter. Note that we take A in the interior of $\mathfrak{T}$ – under the metric topology on $\mathfrak{T}$ – so that each row $A(x, \cdot)$ corresponds to a strictly positive probability mass function.

After updating with $w^{(m)}$, the posterior mean is an estimator for $T^{(m)}$ that is Bayes under quadratic loss and for the specific prior $g(\cdot \mid s, A)$; due to conjugacy, the posterior is again a product of independent Dirichlet distributions [10], whence the elements of the posterior mean are

$$\mathbb{E}\left[T(x, y) \,\middle|\, s, A, w^{(m)}\right] = \frac{sA(x, y) + n^{(m)}_{xy}}{s + n^{(m)}_x} \qquad \forall x, y \in \mathcal{X}.$$

What remains is again to determine a good choice for s and A. However, in an imprecise probabilistic context we do not have to commit to any such choice: the popular Imprecise Dirichlet Model generalises the above approach using a *set* of Dirichlet priors. This set is given by $\text{IDM}(\cdot \mid s) := \{g(\cdot \mid s, A) \mid A \in \text{int}(\mathfrak{T})\}$ and can be motivated from first principles [4,15]. Observe that only a parameter $s \in \mathbb{R}_{\geq 0}$ remains, which controls the "degree of imprecision". In particular, we no longer have to commit to a location parameter A; instead this parameter is freely varied over its entire domain $\text{int}(\mathfrak{T})$.

Element-wise updating with $w^{(m)}$ yields a set of posteriors which, due to conjugacy, are again independent products of Dirichlet distributions. The corresponding set $\mathcal{T}^{(m)}_s$ of posterior means thus contains estimators for $T^{(m)}$ that are Bayes for a specific prior from the IDM, and is easily verified to be

$$\mathcal{T}^{(m)}_s = \left\{ T \in \mathfrak{T} \,\middle|\, \left(\forall x, y \in \mathcal{X} : T(x, y) = \frac{sA(x, y) + n^{(m)}_{xy}}{s + n^{(m)}_x}\right), A \in \text{int}(\mathfrak{T}) \right\}.$$

4.2   Limits of Discrete-Time Estimators

As noted in Proposition 4, a rate matrix Q is connected to the transition probabilities $T_\Delta(x, y) := P(X_\Delta = y \mid X_0 = x)$ in the sense that $T_\Delta \approx (I + \Delta Q)$ for small Δ. Hence, for small Δ, we have that $Q \approx (T_\Delta - I)^1/\Delta$. This becomes exact in the limit for Δ going to zero.

This interpretation can also be used to connect discrete-time estimators for $T^{(m)}$ to estimators for Q. For example, if we let $Q^{(m)} := (T^{(m),\text{ML}} - I)^1/\delta^{(m)}$, then $Q^{\text{ML}} = \lim_{m \to +\infty} Q^{(m)}$. Similarly, we can connect our set-valued estimators for the discretised models to the set-valued continuous-time estimator in (5):

Theorem 5. *For all $m \in \mathbb{N}$, let $\mathcal{Q}^{(m)}_s := \left\{(T - I)^1/\delta^{(m)} \,\middle|\, T \in \mathcal{T}^{(m)}_s\right\}$. Then the Painlevé-Kuratowski [12] limit $\lim_{m \to +\infty} \mathcal{Q}^{(m)}_s$ exists, and equals $\mathcal{Q}_s$.*

5   Discussion

We have derived a set-valued estimator $\mathcal{Q}_s$ for the transition rate matrix of a homogeneous CTMC. It can be motivated both as a set of posterior means of a set of Bayesian models in continuous-time, and as a limit of set-valued discrete-time estimators based on the Imprecise Dirichlet Model. The only parameter of the estimator is a scalar $s \in \mathbb{R}_{\geq 0}$ that controls the degree of imprecision. In the special case where $s = 0$ there is no imprecision, and then $\mathcal{Q}_0 = \{Q^{\mathrm{ML}}\}$.

The set-valued representation $\mathcal{Q}_s$ is convenient when one is interested in the numerical values of the transition rates, e.g. for domain-analysis. If one aims to use the estimator to describe an *imprecise CTMC* [8,13], a representation using the *lower transition rate operator* $\underline{Q}$ is more convenient. This operator is the lower envelope of a set of rate matrices; for $\mathcal{Q}_s$ it is given, for all $h : \mathcal{X} \to \mathbb{R}$, by

$$\left[\underline{Q}h\right](x) := \inf_{Q \in \mathcal{Q}_s} \sum_{y \in \mathcal{X}} Q(x,y)h(y) = \frac{s}{d_x} \min_{y \in \mathcal{X}}\bigl(h(y)-h(x)\bigr) + \sum_{y \in \mathcal{X}\setminus\{x\}} \frac{n_{xy}}{d_x}\bigl(h(y)-h(x)\bigr),$$

for all $x \in \mathcal{X}$. Hence, $\underline{Q}h$ is straightforward to evaluate. This implies that when our estimator is used to learn an imprecise CTMC from data, the lower expectations of this imprecise CTMC can be computed efficiently [6].

Acknowledgements. The work in this paper was partially supported by H2020-MSCA-ITN-2016 UTOPIAE, grant agreement 722734. The authors wish to thank two anonymous reviewers for their helpful comments and suggestions.

References

1. Augustin, T., Coolen, F.P.A., De Cooman, G., Troffaes, M.C.M. (eds.): Introduction to Imprecise Probabilities. Wiley, Hoboken (2014)
2. Berger, J.O.: Statistical Decision Theory and Bayesian Analysis. Springer, Heidelberg (1985)
3. Bladt, M., Sørensen, M.: Statistical inference for discretely observed Markov jump processes. J. R. Stat. Soc.: Series B (Stat. Methodol.) **67**(3), 395–410 (2005)
4. De Cooman, G., De Bock, J., Diniz, M.A.: Coherent predictive inference under exchangeability with imprecise probabilities. J. Artif. Intell. Res. **52**, 1–95 (2015)
5. De Bock, J.: The limit behaviour of imprecise continuous-time Markov chains. J. Nonlinear Sci. **27**(1), 159–196 (2017)
6. Erreygers, A., De Bock, J.: Imprecise continuous-time Markov chains: efficient computational methods with guaranteed error bounds. In: Proceedings of ISIPTA 2017, pp. 145–156 (2017)
7. Inamura, Y.: Estimating continuous time transition matrices from discretely observed data. Bank of Japan (2006)
8. Krak, T., De Bock, J., Siebes, A.: Imprecise continuous-time Markov chains. Int. J. Approx. Reason. **88**, 452–528 (2017)
9. Norris, J.R.: Markov Chains. Cambridge University Press, Cambridge (1998)
10. Quaeghebeur, E.: Learning from samples using coherent lower previsions. Ph.D. thesis

11. Quaehebeur, E., De Cooman, G.: Imprecise probability models for inference in exponential families. In: Proceedings of ISIPTA 2005 (2005)
12. Rockafellar, T.R., Wets, R.J.B.: Variational Analysis. Springer, Heidelberg (1997)
13. Škulj, D.: Efficient computation of the bounds of continuous time imprecise Markov chains. Appl. Math. Comput. **250**(C), 165–180 (2015)
14. Walley, P.: Statistical Reasoning with Imprecise Probabilities. Chapman and Hall, London (1991)
15. Walley, P.: Inferences from multinomial data: learning about a bag of marbles. J. R. Stat. Soc. Ser. B **58**, 3–57 (1996)

Imprecise Probability Inference on Masked Multicomponent System

Daniel Krpelik[1,2($\boxtimes$)], Frank P. A. Coolen[1], and Louis J. M. Aslett[1]

[1] Department of Mathematical Sciences, University of Durham, South Road,
Durham DH1 3LE, UK
daniel.krpelik@durham.ac.uk
[2] FEECS, Department of Applied Mathematics,
VŠB - Technical University of Ostrava, 17. listopadu 15/2172,
708 33 Ostrava-Poruba, Czech Republic

Abstract. Outside of controlled experiment scope, we have only limited information available to carry out desired inferences. One such scenario is when we wish to infer the topology of a system given only data representing system lifetimes without information about states of components in time of system failure, and only limited information about lifetimes of the components of which the system is composed. This scenario, masked system inference, has been studied before for systems with only one component type, with interest of inferring both system topology and lifetime distribution of component composing it. In this paper we study similar scenario in which we consider systems consisting of multiple types of components. We assume that distribution of component lifetimes is known to belong to a prior-specified set of distributions and our intention is to reflect this information via a set of likelihood functions which will be used to obtain an imprecise posterior on the set of considered system topologies.

Keywords: System reliability · Masked system · Topology inference
Survival signatures · Imprecise likelihood

1 Introduction

Masked system inference concerns about carrying out inferences about the underlying system model from system failure time observations, rather than the more commonly studied situations where life test data is available on components. Our inference may concern lifetime distributions of system components and/or structure of the system. Also, the prior information may be available in various forms and sometimes prevents us from constructing suitable prior distributions for Bayesian inference.

We will study here a scenario in which we wish to infer unknown structure of the system from masked system lifetimes given prior distribution on system structure and a set of credible component lifetime distributions. System structures will be specified by survival signatures (introduced in [4]) and we will use

© Springer Nature Switzerland AG 2019
S. Destercke et al. (Eds.): SMPS 2018, AISC 832, pp. 133–140, 2019.
https://doi.org/10.1007/978-3-319-97547-4_18

theory of imprecise probabilities (IP; more in [3]) to describe and obtain inference results.

System reliability inferences with survival signatures based on component failure observations were described in [2] and further extended for IP framework in [5]. Masked system structural inference in Bayesian framework for single component type systems were studied by Aslett in [1], where further elaboration of the nature of inference on masked system with uncertain structure and its numerical solution by Monte Carlo algorithms is presented.

2   Problem Setting

2.1   Masked System Inference

Let Ω_S be a set of considered systems. We model underlying distribution of component lifetimes with a parametric model and we index collection of component lifetime distributions by multi-parameter $\theta \in \Omega_\Theta$. For each combination of system s and set of distributions indexed by θ we assume that we can calculate the system survival function $R(t|s, \theta) = Pr(T_{sys} > t)$.

We further assume that the observables, D, are distributed according to system lifetime distribution (elements d_i represent observations of system failure times, r.v. T_{sys}). With additional assumptions about dependency among observations (e.g. i.i.d.), we can construct the observation model $f(d|\theta, s) \triangleq \mathcal{L}(\theta, s; d)$, for inference purposes:

$$\mathcal{L}(s, \theta; D) = \prod_i f(d_i|s, \theta) = \prod_i \left(- \left[\frac{\partial}{\partial t} R_{sys}(t|s, \theta) \right]_{t=d_i} \right), \qquad (1)$$

where specific form of $f(d_i|\theta, s)$ depends on our system model and shall be given by Eq. 5.

The system design is considered unknown and is therefore included in the likelihood, which then enables joint inference about the reliability and the topology of the system.

2.2   Imprecise Probability Inference of Masked Systems

In IP inference we operate with set of models (set of priors, set of likelihoods). For each of singular model of this set, we can carry out standard inference and analyse the collection partial results. If our aim is to infer probability of some event of interest, in IP scenario we can calculate the bounds for coherent inferences - lower and upper probabilities, where lower probability is minimal inferred probability over the models in the set, and similarly for the upper probability.

In system inference with uncertainty about both component lifetime distributions and system structure, we can choose different uncertainty models for these respective variables. By imprecision we model situations in which we know only possible domain of random variable and are unable to specify prior distribution for Bayesian inference. Such case will lead to IP inference, where each particular value of imprecise random variable defines a stochastic model on which standard Bayesian inference can be performed and the results integrated.

In our case, we assume that we can construct prior distribution on system structures and know only set in which component lifetime distribution parameter θ lies. Lower bound on posterior predictive survival function can be obtained as:

$$\underline{Pr}(T_{sys} > t | D = d) = \min_{\theta \in \Omega_\Theta} \int_s R_{sys}(t|s,\theta) \frac{\mathcal{L}(s,\theta;d)}{Z(\theta,d)} f(s|\theta) ds, \qquad (2)$$

where R_{sys} is the system lifetime survival function, $\mathcal{L}$ is the likelihood function described in Eq. 1, f is prior density for Bayesian inference and factor Z is for posterior distribution normalization, i.e. $Z(\theta,d) = \int_s \mathcal{L}(s,\theta;d) f(s|\theta) ds$.

Upper bound is obtained via maximization of the same expression.

Similarly we can also introduce the lower posterior distribution on system structure as:

$$\underline{f}(S = s | D = d) = \min_{\theta \in \Omega_\Theta} \frac{\mathcal{L}(s,\theta;d)}{Z(\theta,d)} f(s|\theta), \qquad (3)$$

with respective maximizations in case of upper bound.

2.3 Survival Signatures for System State Modelling

Via component state space decomposition, we can express the system survival function for systems consisting of K distinct types of components, with M_k components of type k with i.i.d. lifetimes for each component type k, as:

$$R_{sys}(t|s,\theta) = \sum_l Pr(T_{sys} > t | \boldsymbol{C}(t) = \boldsymbol{l}, s, \theta) Pr(\boldsymbol{C}(t) = \boldsymbol{l}|s,\theta)$$

$$= \sum_l \phi_s(\boldsymbol{l}) \prod_{k=1}^{K} \binom{M_k}{l_k} R_{\theta,k}^{l_k}(t) F_{\theta,k}^{M_k - l_k}(t), \qquad (4)$$

where $\phi_s(\boldsymbol{l}) = Pr(T_{sys} > t | \boldsymbol{C}(t) = \boldsymbol{l}, s)$ is called survival signature of system s, random vector $\boldsymbol{C}(t)$ represents number of functioning components of each respective type at time t (i.e. C_i is number of functioning components of type i), summation is over all possible combinations $\boldsymbol{l}$ of numbers of functioning component of each type. Survival functions $R_{\theta,k}$ and cumulative distribution functions (CDFs) $F_{\theta,k}$ indexed by component type k and (multi-)parameter θ denote respective lifetime distribution characteristics for distinct component types.

The single observation density for systems described by survival signatures is therefore given by:

$$f(d_i|s,\theta) = -\left[\frac{\partial}{\partial t} R_{sys}(t|s,\theta)\right]_{t=d_i} =$$

$$-\sum_l \phi_s(\boldsymbol{l}) \sum_{k=1}^{K} \left\{ \binom{M_k}{l_k} \left[\frac{\partial}{\partial t}\left(R_{\theta,k}^{l_k}(t) F_{\theta,k}^{M_k - l_k}(t)\right)\right]_{t=d_i} \prod_{k \neq j=1}^{K} \binom{M_j}{l_j} R_{\theta,j}^{l_j}(d_i) F_{\theta,j}^{M_j - l_j}(d_i) \right\}$$

$$= \sum_l \left\{ \phi_s(\boldsymbol{l}) \prod_{k=1}^{K} \left[\binom{M_k}{l_k} F_{\theta,k}^{M_k - l_k}(d_i) R_{\theta,k}^{l_k}(d_i)\right] \sum_{k=1}^{K} \left[\left(\frac{l_k}{R_{\theta,k}(d_i)} - \frac{M_k - l_k}{F_{\theta,k}(d_i)}\right) f_{\theta,k}(d_i)\right] \right\}, \qquad (5)$$

where $f_{\theta,k}(.)$ is probability density function of kth component type lifetime.

We have derived everything necessary to be able to compute both the imprecise posterior and posterior predictive distributions in the setting where only masked system lifetime data are available and when the system design may be unknown. This allows us to perform joint inference on the component lifetime parameters and the topology of the system using imprecise probability.

In the remainder of the paper we will demonstrate the method for inference of system structure and predictive system lifetime. Inference on λ_2 is not further considered in this paper.

3   Examples

In the experiments, we shall assume that the real system structure is one of those described by survival signatures in Table 1 (those are all simply connected systems of order 4, as defined and listed in [1], each with a random component type assignment). These systems consists of $K = 2$ types components, 2 components of each type ($M_1 = M_2 = 2$). Underlying component type lifetime distributions are assumed to be exponential with rates $\lambda_1 = 0.45$ and $\lambda_2 \in [0.06, 1.12]$ ($\Omega_\Theta = \Omega_{\Lambda_1} \otimes \Omega_{\Lambda_2}$). Prior distribution on systems ($f(s|\theta)$ in Eqs. 2 and 3) is chosen to be uniform for all choices of λ_2.

The data, observed system failures, for experiments are simulated from system labeled as 6, which will be hereon referred to as the "ground truth" system. Ground truth hazard rate for components of type 2 is chosen to be $\lambda_2 = 0.32$.

Table 1. Survival signatures of systems in Ω_S. Zero row is being omitted ($\phi_s(\mathbf{0}) = 0$).

C1	C2	1	2	3	4	5	6	7	8	9	10	11
0	1	0.00	0.00	0.00	0.00	0.00	**0.00**	0.00	0.50	0.50	0.50	1.00
0	2	0.00	0.00	0.00	1.00	0.00	**0.00**	1.00	1.00	1.00	1.00	1.00
1	0	0.00	0.00	0.00	0.00	0.00	**0.00**	0.00	0.00	0.00	0.50	1.00
1	1	0.00	0.00	0.25	0.50	0.50	**0.75**	0.50	0.50	0.75	1.00	1.00
1	2	0.00	0.50	1.00	1.00	1.00	**1.00**	1.00	1.00	1.00	1.00	1.00
2	0	0.00	0.00	0.00	0.00	0.00	**0.00**	1.00	0.00	1.00	1.00	1.00
2	1	0.00	0.50	0.50	0.50	1.00	**1.00**	1.00	1.00	1.00	1.00	1.00
2	2	1.00	1.00	1.00	1.00	1.00	**1.00**	1.00	1.00	1.00	1.00	1.00

3.1 Survival Predictions are not Monotonic, nor Convex in λ_2

Since the predictions are defined by their bounds, it is necessary to acquire them by optimization. Optimization problems are greatly simplified for monotonic functions (we only need to investigate bounds of the set) and/or convex/concave functions (where efficient gradient based algorithms may be employed). Although the survival function predictions are monotone in case of know system structure ($|\Omega_s| = 1$), neither of these desired properties could be proven analytically in general for predictions with unknown system structure. Conducted experiment (see Fig. 1) provides a counterexample for monotonicity,

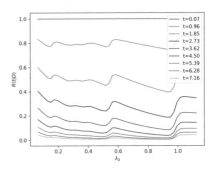

Fig. 1. Dependency of posterior survival function predictions for various selected times on imprecise λ_2 obtained by analysing 250 data samples

convexity and concavity of posterior predictions in case of unknown system structure. Furthermore, Fig. 3 provides a counterexample for the same in case of system structure posterior inference.

3.2 Imprecise Structure Posterior and System Identification

Two basic inferences of our interest are for the system lifetime survival function (via Eq. 2), and for posterior system distribution (via Eq. 3). An example of predictive and structure inferences are shown in Fig. 2. On the left side, the intervals for each system represents lower and upper bounds for posterior on the set Ω_Θ. On the right picture, one set of prediction bounds for prior distribution on system structures (before updating by observations) and another for posterior obtained via Bayesian updating are compared with the Kaplan-Meier estimate and the ground truth survival function.

The system identification, which would be done by comparing system posterior probabilities in Bayesian decision making, has to be done in IP setting. As can be seen in Fig. 2, left, upper probabilities for multiple systems approach 1 in this experiment whilst the lower remain near 0. Therefore, there are several systems for which we are indecisive. The explanation of this wide range is illuminated in Fig. 3, where we plot system posterior distributions obtained for various fixed λ_2 by standard Bayesian inference (i.e. inner function which is optimized in Eq. 3). In different regions of Ω_Θ, one system becomes dominant over others and this effect is further increased with increasing sample size.

We can observe that an useful informative inference, we might obtain in IP setting, is that of rejection of several system structures. As is apparent from Fig. 2 and also from Fig. 4, upper posterior probability for some of the systems tends to approach 0, which indicates their unfitness to observations. Although, further analyses have to be performed to investigate properties of these rejections in IP decision-making framework(s), which is out of the scope of this paper.

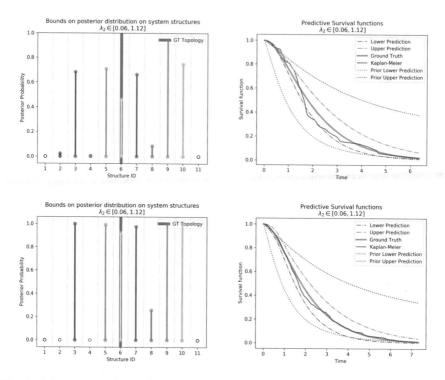

Fig. 2. Inference results with imprecise distribution parameter λ_2 for sample sizes 50 (top) and 250 (bottom). Left: imprecise posterior distribution on systems. Right: predictions of system lifetime survival function.

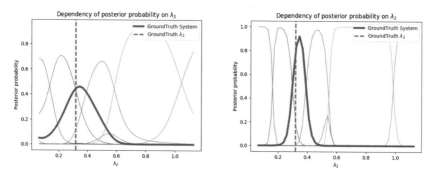

Fig. 3. Dependency of system structure posterior distributions on fixed λ_2. Each vertical slice at selected λ_2 represents system posterior distribution (i.e. sums to 1). Left image for 50 data samples, right for 250. Thick curve denotes the evolution of posterior distribution of the ground truth system.

3.3 Response to Varying the Support of λ_2

Next example investigates differences between disjoint choices of underlying support set Ω_Θ. We perform two imprecise inferences separately for λ_2 support divided by the value of (known) λ_1. The resulting imprecise system posteriors are shown in Fig. 4.

From Fig. 4 it is apparent, that some structures like 3 and 10, which were comparable by the means of inference in original support set (Fig. 2, left), exhibit significant differences in case when the support is focused because the likelihood of these systems is small in these regions (see Fig. 3). Similar behaviour was also observed in case of simply narrowing the λ_2 support where upper posterior probability of many systems approached 0. These results are being omitted here due to space limitations.

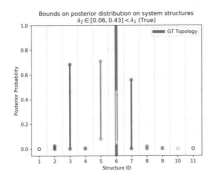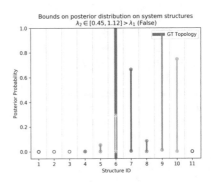

Fig. 4. Influence of choice of the support set for λ_2 on structure posterior distribution. In the left picture, the GT λ_2 lies in investigated set, in the right one it does not.

This scenario might be applicable for purposes of experimental design towards inference about adversarial systems. Proper choice of the support set Ω_Θ, and therefore the experimental settings, seems to influence identifiability of underlying unknown system structure.

4 Concluding Remarks

We have demonstrated a novel methodology for inference in limited prior knowledge scenario, which allows us to avoid introducing some redundant and possibly unjustified modelling assumptions.

For the described situation, we have shown that the optimized functions of interest are nor monotone nor convex and, so far, have to be solved by general optimization procedures (in Sect. 3.1). It has also been indicated in Sects. 3.2 and 3.3, that IP inference cannot generally serve for proper system identification, as IP reasoning allows for indecisiveness, but rather as a tool for system rejection in case of low upper posterior probability.

The behaviour which was presented was observed among multiple experiments that were conducted, although no analytical guarantees may be given at this stage of research. A follow-up generalizing study which would take into account even aspects which were only touched here (symmetrical properties of systems and rigorous IP decision theory) is necessary to further understand advantages and limitations of proposed methodology.

Acknowledgement. This work is funded by the European Commission's H2020 programme, through the UTOPIAE Marie Curie Innovative Training Network, H2020-MSCA-ITN-2016, Grant Agreement number 722734.

References

1. Aslett, L.J.M.: MCMC for inference on phase-type and masked system lifetime models. Ph.D. thesis, Trinity College Dublin (2012)
2. Aslett, L.J.M., Coolen, F.P.A., Wilson, S.P.: Bayesian inference for reliability of systems and networks using the survival signature. Risk Anal. **35**(9), 1640–1651 (2015). https://doi.org/10.1111/risa.12228
3. Augustin, T., Coolen, F.P., De Cooman, G., Troffaes, M.C. (eds.): Introduction to Imprecise Probabilities. Wiley Series in Probability and Statistics. Wiley, Hoboken (2014)
4. Coolen, F.P.A., Coolen-Maturi, T.: Generalizing the signature to systems with multiple types of components. In: Zamojski, W., Mazurkiewicz, J., Sugier, J., Walkowiak, T., Kacprzyk, J. (eds.) Complex Systems and Dependability, pp. 115–130. Springer, Heidelberg (2012)
5. Walter, G., Aslett, L.J., Coolen, F.P.: Bayesian nonparametric system reliability using sets of priors. Int. J. Approx. Reason. **80**, 67–88 (2017). https://doi.org/10.1016/j.ijar.2016.08.005

Regression Ensemble with Linguistic Descriptions

Jiří Kupka and Pavel Rusnok[✉]

Institute for Research and Applications of Fuzzy Modeling,
Center of Excellence IT4Innovations, University of Ostrava, Ostrava, Czech Republic
{jiri.kupka,pavel.rusnok}@osu.cz

Abstract. In this contribution, we present a brief presentation of a method which allows automatically to create an ensemble of regression techniques and compare this method to standard approaches. This is done with the help of mined linguistic rule base which is further used by advanced Perception-based Logical Deduction. As a possible side effect, we can obtain a linguistic description of the evaluative process.

Keywords: Association analysis · Linguistic description
Perception-based logical deduction · Regression · Ensemble technique

1 Introduction

Linguistic descriptions are considered as sets of fuzzy IF-THEN rules with linguistic semantics [1]. They can be comprised of evaluative linguistic expressions, which are one of the basic building blocks of natural language expressions and were mathematically formalized for instance in [1]. Formally, evaluative expressions were developed as a part of Fuzzy Natural Logic [2,3] and, in a form of linguistic descriptions, were successfully used in many applications.

For instance, mining of linguistic IF-THEN rules created with the help of evaluative linguistic expressions was firstly proposed in [4]. Then mined linguistic IF-THEN rules were studied and used in flood predictions [5], time series prediction [6] and even in ensembling time series prediction methods [7]. Inspired by these applications, linguistic descriptions can be applied in regression tasks as was demonstrated in [8]. It is worth mentioning that more general linguistic descriptions, extending those based on evaluative linguistic expressions [9], were used the latter application.

In this contribution, we continue in this direction, and we investigate the usability of linguistic rules for building an ensemble of regression models. We propose a procedure similar to the one in [7], but some differences due to the application domain appeared. The reason why to consider another approach was to correct some issues which might appear in the original model and might influence the prediction (for details see [10]). In the next section, we describe mainly the idea behind creating and mining linguistic fuzzy IF-THEN rules, we

© Springer Nature Switzerland AG 2019
S. Destercke et al. (Eds.): SMPS 2018, AISC 832, pp. 141–148, 2019.
https://doi.org/10.1007/978-3-319-97547-4_19

sketch the mining procedure, and then we also sketch their application of mined linguistic description in regression tasks. Unfortunately, due to the strong page limit, we cannot provide the reader with a fully detailed description. For this, motivation and some ideas behind this method, we refer to [8]. Then in Sect. 3, we introduce our ensemble modeling approach and in Sect. 4 we provide several results related to our proposed approach on various datasets from UCI machine learning repository [11]. We then conclude with a short description of possible future work.

2   Description of the Mining Procedure

2.1   The Original Model

In this section, we briefly introduce our method which was published in [8]. The origin of this method is in [4] allowing to mine purely linguistic expressions from datasets. This method was partially based on pioneering GUHA method [12] and model of evaluative linguistic expressions (see e.g. [9]). However, based on some observations about this method [10] we recently developed an extension of the original method in [8]. This extension to the novel model of evaluative linguistic expressions created with the help of standard notion of fuzzy partition requires another definition of specificity ordering. We are going to briefly introduce these notions. However, for a much more detailed description of this method we refer the reader to [8] and references therein (see also [4]).

Table 1. A table representing an initial dataset D.

	X_1	X_2	$\ldots$	X_m	Y
o_1	e_{11}	e_{21}	$\ldots$	e_{m1}	y_1
o_2	e_{12}	e_{22}	$\ldots$	e_{m2}	y_2
$\vdots$	$\vdots$	$\vdots$	$\ldots$	$\vdots$	$\vdots$
o_N	e_{1N}	e_{2N}	$\ldots$	e_{mN}	y_N

In our method, we start with an $(m+1)$-dimensional table D of the following form: As usually, rows o_i are *objects*, and columns X_i and Y are called *attributes* or *variables*. Our usual task is to mine some knowledge from the dataset D and then predict the value of the *dependent* variable Y for a new m-dimensional vector (Table 1).

We assume that $e_{ij} \in \mathbb{R}$, so, for each attribute X_i (resp. Y), their linguistic descriptions (see the text below) has to be given beforehand by the user of the mining procedure. Thus, for each $i = 1, 2, \ldots, m$, a linguistic evaluation $\mathcal{C}^{X_i}$ (i.e. a suitable family of fuzzy sets) of the variable X_i is given on the interval $[a^i, b^i]$ which represents a *context* of the variable X_i. Similarly, a linguistic evaluation $\mathcal{C}^Y$ of Y is denoted on $[a^Y, b^Y]$.

In [4] a standard model of evaluative linguistic descriptions is elaborated. Originally, three groups of linguistic predications in the original model from [4] are considered. They represent prototypical groups of *small, medium* and *big* predications. More detailed nuances within each such group are given with the help of so-called *linguistic modifiers*. We distinguish among those with narrowing and widening effects. *Narrowing* modifiers can be, for example, "*extremely (Ex), significantly (Si), very (Ve)*", and *widening* ones can be "*more or less (ML), roughly (Ro), quite roughly (QR), very roughly (VR)*" and so on. For future work with modifiers (and hence the predications) so-called *specificity ordering* is defined in the following way:

$$Ex \preceq_s Si \preceq_s Ve \preceq_s \text{empty hedge} \preceq_s ML \preceq_s Ro \preceq_s QR \preceq_s VR. \tag{1}$$

This mathematical model (e.g. model of linguistic predications), together with so-called *perception-based logical deduction* using specificity ordering, was successfully used in several applications, e.g. in time series prediction [6,13].

2.2 The New Model

Now, we briefly recall a generalization of this method suggested in [10] and fully elaborated in [8]. This new method used in this manuscript is based on a different, more general model of linguistic predications, which is based on fuzzy partitions and involves the previous model of linguistic predications. In [8] we also elaborated a generalized version of perception-based logical deduction which used different specificity ordering. In the rest of this section, we slightly recall some notation. Due to the strong page limit, we have to omit many details, motivations, and explanations. Every careful reader is referred to our previous publication [8] and references therein.

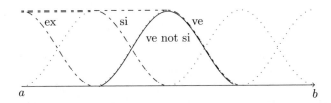

Fig. 1. Example of a fuzzy partition (dotted line) and $\mathcal{C}_1$ system of fuzzy sets (dashed and solid lines) modeling expressions "*extremely small*" (ex), "*significantly small*" (si), "*very small*" (ve), and "*very small but not significantly small*" (ve not si). More examples in [8].

The new mathematical model of linguistic predications is based on fuzzy partitions, fuzzy subpartitions and convex hulls based on them. For simplicity, we may assume that, for each variable X having values in an interval $[a, b]$, we consider a system $\mathcal{C}^X$ (being the union of systems $\mathcal{C}_1, \mathcal{C}_2, \ldots, \mathcal{C}_r$ of fuzzy sets

- see Fig. 1) of fuzzy sets on $[a,b]$, each of them having some suitable linguistic representation. Thus, for each $A \in \mathcal{C}^X$ one can have a sentence of the form

$$X \text{ is } A. \tag{2}$$

Naturally (and as we mentioned above), this new model of fuzzy sets requires a specificity ordering. In the simplest case, so-called *specificity ordering* $\leq_s$ on $\mathcal{C}_j$, $j = 1, \ldots, r$, is defined in the following way - for $E, F \in \mathcal{C}_j$, $E \leq_s F$ if and only if $E(x) \leq F(x)$ for each $x \in X$. In this case, one can say $\leq_s$ is an extension of $\preceq_s$ and this ordering is naturally extended to higher dimensions in [8].

The new model of evaluative predications can be used to mine so-called *fuzzy/linguistic IF-THEN rules* or simply IF-THEN rules [8]. They can be of the form

$$\mathcal{R} := \textbf{IF } X \text{ is } A \textbf{ THEN } Y \text{ is } B, \quad A \in \mathcal{C}^X, \ B \in \mathcal{C}^Y. \tag{3}$$

The expression X is A is called the *antecedent*, and Y is B is called the *consequent*. In the case of datasets of higher dimensions the antecedent "X is A" represents the conjunction of linguistic expressions of some variables, where the conjunction is computed with the help of a suitable t-norm. As a shortage for an rule $\mathcal{R}$ of the form (3), we use simply $A \Rightarrow B$. The goal of the mining step of the method is to find a *linguistic description* $LD = \{\mathcal{R}_1, \ldots, \mathcal{R}_m\}$, i.e. a finite set of *suitable* IF-THEN rules.

The suitability of a given IF-THEN rule $\mathcal{R}$ of the form (3) is given by two user-defined thresholds. Namely, if (*support* and *confidence*) measures

$$\text{supp}_\otimes(A \Rightarrow B) = \frac{\sum_{o \in D} A(o) \otimes B(o)}{N}, \quad \text{conf}_\otimes(A \Rightarrow B) = \frac{\sum_{o \in D} A(o) \otimes B(o)}{\sum_{o \in D} A(o)},$$

are higher than those user-defined thresholds, then $\mathcal{R} \in LD$.

As soon as the linguistic description LD of the dataset is created, it can be used in the deductive process. As we mentioned above, due to using the generalized *model of evaluative expressions*, the user needs to use a generalized perception-based logical deduction (PbLD). Briefly said, for a new instance u representing values from $X_1, X_2, \ldots, X_m$ we want to predict a value y_u from Y. The PbLD method works in the following way. First, having u fixed, the method chooses a *topic* $T^{LD}(u)$ of LD having non-zero evaluation of respective antecedent. Then another subset P^{LD} of T^{LD} is taken by considering the most specific antecedents. All this allows us to locally apply an *inference rule* and aggregate the predicted values into a single one. As we mentioned above, for further details we refer to [8].

3   Ensemble with Linguistic Rules

In general, we consider an $(m+1)$-dimensional table of $X_1, \ldots, X_m, Y$ variables, which is the first part of Table 2. We are going to build an ensemble model from given regression models. Let there be r regression models $f_1, \ldots, f_r$ with errors $E_1, \ldots, E_r$:

$$Y = f_1(X_1, \ldots, X_m) + E_1,$$
$$Y = f_2(X_1, \ldots, X_m) + E_2,$$
$$\vdots$$
$$Y = f_r(X_1, \ldots, X_m) + E_r, \tag{4}$$

The models' values for objects $o_1, \ldots, o_N$ are stored as variables $F_1, \ldots, F_r$. The regression models are obtained by a usual process of cross-validation [14], while the values of the best model are stored. We treat the models as new features in our dataset. Based on the regression models, we define weights variables as

$$W_i = \frac{1}{|Y - f_i(X_1, \ldots, X_m)|},$$

that are also added to the dataset. Then we get an extended version of dataset in Table 2. Similarly, to the construction of r regression models, we divide the dataset into learning and testing parts. We then mine rules from the learning part of dataset of the following form

Table 2. A table representing an extended dataset D with regression models with their weights.

	X_1	X_2	$\ldots$	X_m	Y	F_1	F_2	$\cdots$	F_r	W_1	W_2	$\cdots$	W_r
o_1	e_{11}	e_{21}	$\ldots$	e_{m1}	y_1	f_{11}	f_{21}	$\cdots$	f_{r1}	w_{11}	w_{21}	$\cdots$	w_{r1}
o_2	e_{12}	e_{22}	$\ldots$	e_{m2}	y_2	f_{12}	f_{22}	$\cdots$	f_{r2}	w_{12}	w_{22}	$\cdots$	w_{r2}
$\vdots$	$\vdots$	$\vdots$	$\ldots$	$\vdots$	$\vdots$	$\vdots$	$\vdots$	$\vdots$	$\vdots$	$\vdots$	$\vdots$	$\vdots$	$\vdots$
o_N	e_{1N}	e_{2N}	$\ldots$	e_{mN}	y_N	f_{1N}	f_{2N}	$\cdots$	f_{rN}	w_{1N}	w_{2N}	$\cdots$	w_{rN}

$$\textbf{IF } (X_i \text{ is } A_i \textbf{ AND } F_j \text{ is } A_j \ldots) \textbf{ THEN } W_i \text{ is } B, \quad \text{for } A_i \in \mathcal{C}^{X_i}, F_j \in \mathcal{C}^{F_j}, B \in \mathcal{C}^{W_i}. \tag{5}$$

We mine set of rules $\mathcal{R}_i$ for every weight variable W_i. We are then given testing part of dataset $X_1, X_2, \ldots, X_m$ and regression models variables $F_1, F_2, \ldots, F_r$ and with the rules $\mathcal{R}_i$ we predict weights W_i in testing part of a dataset using the PbLD inference method described in Sect. 2. From the weights and regression models we build two ensemble models ens and $ensMax$ as, $i = 1, 2, \ldots, N$,

$$ens_i = \frac{\sum_j w_{ji} \cdot f_{ji}}{\sum_j w_{ji}},$$

and

$$ensMax_i = f_{ji}, \quad \text{where } j = \arg\max_{j \in \{1, \ldots, r\}} \{w_{ji}\}.$$

The first ensemble model ens is a weighted average, but its weights are dynamically changing based on input dataset and mined rules. The ensemble model $ensMax$ dynamically chooses the best model f_j (i.e., with the highest weight) again based on input dataset and mined rules.

4   Experiments

For testing the proposed ensemble methods from the previous section, we have downloaded several datasets from UCI machine learning repository [11], of which some had multiple independent variables. Overall, we performed twenty-nine regression tasks. The methods we used for our ensemble learning were Linear Regression (LR), Multivariate Adaptive Regression Splines (MARS), Support Vector Machines with Polynomial Kernel (SVM), k-nearest neighbor (KNN), regression trees (RT) via caret package [15]. With these methods we obtained five models $f_1, \ldots, f_5$. For mining of rules and inference with the rules, we used the implementation in Linguistic Fuzzy Logic package [16].

For every dataset, we mined five sets of rules $\mathcal{R}_1, \ldots, \mathcal{R}_5$, which formed the linguistic model of weights $W_1, \ldots, W_5$. We predicted the weights on testing parts of datasets and created the ensemble models described in the previous section. We compared our ensemble methods with a simple Mean of models $f_1, \ldots, f_5$ as a baseline.

We computed root mean squared errors (RMSE) on every dataset for all five models $f_1, \ldots, f_5$ and three ensemble models (Mean, ens, and $ensMax$). Based on the RMSE we calculated the ranking $(1, \ldots, 8)$ of all eight models based on RMSE, i.e., the model with the lowest RMSE was assigned rank 1 and the model with the highest RMSE was assigned the lowest rank 8. The target variables were standardized and therefore the averaging RMSE through all data sets is reasonable.

We calculated the average ranking and average RMSE of every method through all the datasets and results are shown in Table 3. Both our ensemble

Table 3. Average ranking and root mean square errors for multiple regressions.

	LR	SVM	MARS	KNN	RT	Mean	$ensMax$	ens
Average ranking	6.310	4.621	3.621	6.310	5.828	2.724	3.138	3.448
Average RMSE	12.151	9.626	9.637	10.373	10.476	9.341	9.247	9.249

methods $ensMax$ and ens have the best average RMSE, 9.247 and 9.249 respectively. They are not always the best as the average ranking of simple mean (2.724) is below the average ranking of our ensemble methods.

The models $f_1, \ldots, f_5$ were quite correlated. The median of average correlation of models on all datasets was 0.67. So we divided the datasets into two groups. The first one was with the average correlation between models above 0.7 and the second was with the average correlation between models below 0.7. In the higher correlation scenario (see Table 4) our proposed ensemble method ens was the best with respect to other models in both ranking and RMSE. In the scenario of less correlated models the performance of our ensemble methods is slightly worse, see Table 5. It seems that in lower correlation scenario it is better to simply average the models.

Table 4. Average ranking and root mean square errors for multiple regressions for datasets with average correlation between regression models *above* 0.7.

	LR	SVM	MARS	KNN	RT	Mean	*ensMax*	*ens*
Average ranking	7.600	4.333	3.200	5.667	5.600	3.400	3.267	2.933
Average RMSE	13.989	10.136	10.027	11.326	11.462	9.722	9.447	9.427

Table 5. Average ranking and root mean square errors for multiple regressions for datasets with average correlation between regression models *below* 0.7.

	LR	SVM	MARS	KNN	RT	Mean	*ensMax*	*ens*
Average ranking	4.929	4.929	4.071	7.000	6.071	2.000	3.000	4.000
Average RMSE	10.182	9.080	9.219	9.352	9.420	8.933	9.033	9.058

Our ensemble approach, through mining linguistic descriptions, has an advantage, that we can study the relationship between the models (LR, SVM, MARS, KNN, and RT) more deeply. In the remaining part, we present an example of a mined rule of the form (5) stated in natural language, which we used for prediction of model weights in our ensemble models.

A rule

IF X_1 is *small* AND MARS is *medium* THEN W_1 is *significantly small*,

states that when the first variable in dataset has *small* values and MARS model is modeling *medium* values then the weight W_1 (LR weight) is supposed to be *significantly small*. In other words, the Linear Regression behaves poorly when X_1 has *small* values, and Multivariate Adaptive Regression Splines have *medium* values. Another rule

IF X_1 is *medium* AND X_2 is *small* THEN W_5 is *big*,

states that if the first variable in dataset has *medium* values and the second variable has *small* values then the weight W_5 (RT weight) is *big*.

Beware, that such rules are not valid for all datasets, but only for one particular dataset. Notice that different linguistic descriptions are usually obtained for the different datasets.

5   Conclusion

We proposed an approach to ensemble regression models with linguistic rules. We tested our proposed approach on twenty-nine different regression tasks, and we observed that our proposed method behaves very well in situations when the models in the ensemble are highly correlated. However, the higher (resp. lower) correlation scenario was based only on 14 (resp. 15) datasets and more detailed investigation of this phenomenon has to be done in the future. It is not completely obvious what other (not only statistical characteristics) may influence our prediction.

It is worth mentioning that our ensembling approach strongly depends on the choice of initial fuzzy partitions and the mining procedure itself. In our future work, we plan to propose new more sophisticated methods for mining such linguistic descriptions - possible approaches are the following. For instance, the use of some other measures (besides support and confidence measures) can be investigated and used. We already partially incorporated so-called lift measure in [8].

References

1. Novák, V.: A comprehensive theory of trichotomous evaluative linguistic expressions. Fuzzy Sets Syst. **159**(22), 2939–2969 (2008)
2. Novák, V.: Towards formalized integrated theory of fuzzy logic. In: Fuzzy Logic and Its Applications to Engineering, Information Sciences, and Intelligent Systems, pp. 353–363. Springer (1995)
3. Novák, V., Perfilieva, I., Dvořák, A.: Insight Into Fuzzy Modeling. Wiley, Hoboken (2016)
4. Novák, V., Perfilieva, I., Dvořák, A., Chen, G., Wei, Q., Yan, P.: Mining pure linguistic associations from numerical data. Int. J. Approx. Reason. **48**(1), 4–22 (2008)
5. Burda, M., Rusnok, P., Štěpnička, M.: Mining linguistic associations for emergent flood prediction adjustment. Adv. Fuzzy Syst. 4 (2013)
6. Štěpnička, M., Dvořák, A., Pavliska, V., Vavříčková, L.: A linguistic approach to time series modeling with the help of f-transform. Fuzzy Sets Syst. **180**(1), 164–184 (2011)
7. Štěpnička, M., Burda, M., Štěpničková, L.: Fuzzy rule base ensemble generated from data by linguistic associations mining. Fuzzy Sets Syst. **285**, 141–161 (2015)
8. Kupka, J., Rusnok, P.: Regression analysis based on linguistic associations and perception-based logical deduction. Expert Syst. Appl. **67**, 107–114 (2017)
9. Novák, V., Perfilieva, I.: On the semantics of perception-based fuzzy logic deduction. Int. J. Intell. Syst. **19**(11), 1007–1031 (2004)
10. Kupka, J., Tomanová, I.: Some extensions of mining of linguistic associations. Neural Netw. World **20**(1), 27–44 (2010)
11. Bache, K., Lichman, M.: UCI machine learning repository (2013)
12. Hájek, P., Havránek, T.: Mechanizing Hypothesis Formation: Mathematical Foundations for a General Theory. Springer, Heidelberg (1978)
13. Štěpnička, M., Cortez, P., Donate, J.P., Štěpničková, L.: Forecasting seasonal time series with computational intelligence: on recent methods and the potential of their combinations. Expert Syst. Appl. **40**(6), 1981–1992 (2013)
14. Han, J.: Data Mining: Concepts and Techniques. Morgan Kaufmann Publishers Inc., San Francisco (2005)
15. Kuhn, M., Wing, J., Weston, S., Williams, A., Keefer, C., Engelhardt, A., Cooper, T., Mayer, Z., Kenkel, B., The R Core Team, Benesty, M., Lescarbeau, R., Ziem, A., Scrucca, L., Tang, Y., Candan, C., Hunt., T.: Caret: Classification and Regression Training. R package version 6.0-77 (2017)
16. Burda, M.: Linguistic fuzzy logic in R. In: IEEE International Conference on Fuzzy Systems (FUZZ-IEEE), Istanbul, pp. 1–7. IEEE (2015)

Dynamic Classifier Selection Based on Imprecise Probabilities: A Case Study for the Naive Bayes Classifier

Meizhu Li$^{(\boxtimes)}$, Jasper De Bock, and Gert de Cooman

ELIS, SYSTeMS, Ghent University, Ghent, Belgium
{meizhu.li,jasper.debock,gert.decooman}@ugent.be

Abstract. Dynamic classifier selection is a classification technique that, for every new instance to be classified, selects and uses the most competent classifier among a set of available ones. In this way, a new classifier is obtained, whose accuracy often outperforms that of the individual classifiers it is based on. We here present a version of this technique where, for a given instance, the competency of a classifier is based on the robustness of its prediction: the extent to which the classifier can be altered without changing its prediction. In order to define and compute this robustness, we adopt methods from the theory of imprecise probabilities. As a proof of concept, we here apply this idea to the simple case of naive Bayes classifiers. Based on our preliminary experiments, we find that the resulting classifier outperforms the individual classifiers it is based on.

1 Introduction

In machine learning and statistics, classification is the problem of predicting the class of a new instance, using only its features and a given data set with labelled instances. A model for performing this task is called a classifier.

The usual approach to solving such problems is to consider some tractable family of possible classifiers and to try to select from among them the classifier that achieves the highest classification accuracy on all instances. The method of Dynamic Classifier Selection [2] switches this around: instead of selecting a single classifier to be uson all instances, it selects a (possibly) different classifier for each and every instance, and then uses the result of this classifier to predict the class of that instance. If all goes well, the resulting combined classifier outperforms each of the individual classifiers it is based on.

The key to the success of this Dynamic Classifier Selection method is to find a way to assess, for a given instance, which classifier is most likely to classify it correctly. We here propose to base this assessment on imprecise-probabilistic measures for the robustness of a prediction: the extent to which a classifier can be altered without changing its prediction. Our main reason for this proposal is the recent discovery that these measures of robustness correlate surprisingly well

© Springer Nature Switzerland AG 2019
S. Destercke et al. (Eds.): SMPS 2018, AISC 832, pp. 149–156, 2019.
https://doi.org/10.1007/978-3-319-97547-4_20

with the accuracy of their classifier (when evaluated on instances with similar robustness) [3].

To the best of our knowledge, this idea has never been tried before. For that reason, we here start by applying it to the case of the Naive Bayes Classifier, as it is a simple but well-known and surprisingly performant classifier. Furthermore, since the Naive Bayes topology is a special case of the imprecise graphical models that are considered in Ref. [3], this will allow us to compute the required robustness measures with existing algorithms.

The rest of this paper is organized as follows. After a brief introduction to the Naive Bayes Classifier in Sects. 2 and 3 goes on to introduce its imprecise-probabilistic extension, called the Naive Credal Classifier, and the critical perturbation threshold that can be derived from it. Next, in Sect. 4, we explain how these thresholds can be used as a tool to choose, for a given instance, between two different classifiers. Experiments on real data sets from the UCI Machine Learning Repository are reported on in Sect. 5. We conclude the paper in Sect. 6, where we also hint at possible avenues for future research.

2   Naive Bayes Classifiers

A Naive Bayes Classifier (NBC) is a simple probabilistic model that is used to estimate the class of an instance based on the value of its features. The class variable is denoted by C and takes values c in a finite set $\mathcal{C}$. If C is binary, as will be the case in our experiments, we denote its elements by c and $\bar{c}$. The number of features is denoted by m. For every $i \in \{1, \ldots, m\}$, the i-th feature variable is denoted by F_i and takes values f_i in a finite set $\mathcal{F}_i$. For notational convenience, we gather all feature variables in a single vector $\mathbf{F} = (F_1, \ldots, F_m)$ that takes values $\mathbf{f} = (f_1, \ldots, f_m)$ in $\mathcal{F}_1 \times \cdots \times \mathcal{F}_m$.

For any given feature vector $\mathbf{f}$, a Naive Bayes Classifier will return as its estimate the class $\hat{c}$ that has the highest probability $P(\hat{c}|\mathbf{f})$ given the features $\mathbf{f}$. In fact, it has this in common with every other probabilistic classifier. In the case of a Naive Bayes Classifier, the computation of these probabilities is facilitated by the (naive) assumption that the features are independent given the class; see Fig. 1.

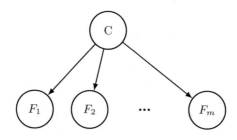

Fig. 1. Example of a Naive Bayes Classifier

Due to this assumption, the conditional probability that is to be maximised is given by

$$P(\hat{c}|\mathbf{f}) = \frac{P(\hat{c}) \prod_{i=1}^{m} P(f_i|\hat{c})}{\sum_{c \in \mathcal{C}} P(c) \prod_{i=1}^{m} P(f_i|c)}.$$

In this expression, the (conditional) probabilities that appear on the right hand side are typically learned from data. To avoid probability zero, we adopt Laplace smoothing, meaning that for all $i \in \{1, \ldots, m\}$, $c \in \mathcal{C}$ and $f_i \in \mathcal{F}_i$:

$$P(c) = \frac{n(c) + 1}{n + |\mathcal{C}|} \text{ and } P(f_i|c) = \frac{n(c, f_i) + 1}{n(c) + |\mathcal{F}_i|},$$

with n the total number of data points, $n(c)$ the number of data points with class c and $n(c, f_i)$ the number of data points with class c and i-th feature f_i.

Example 1. Most of our experiments further on will be conducted on the Balloons data set of the UCI Machine Learning Repository [4]. It has two possible class values, which we denote by c and $\bar{c}$ and 4 features, which we denote by F_1, F_2, F_3 and F_4. For the purposes of this contribution, the meaning of these features is irrelevant.

The data set consists of 76 instances, which we randomly split in a training set (70%) and a testing set (30%). The training set is used to learn the local probabilities of a Naive Bayes Classifier (using Laplace smoothing), and this classifier is then used to predict the class of the instances in the testing set. This process is repeated a hundred times, and the average accuracy over these hundred runs is reported.

For an NBC that uses all four features, the obtained (average) accuracy is 0.776502. We refer to this classifier as 'Classifier 1'. We also consider a second classifier, called 'Classifier 2', that only uses F_1, F_3 and F_4. For that classifier, the obtained accuracy is 0.746948. After performing feature selection using the Sequential Forward Selection (SFS) method, these two classifiers came out first and second, respectively.

For a given instance, these two classifiers may of course yield different predictions. For didactic purposes and future reference, we here mention one particular unspecified instance where, in the first of our hundred runs, Classifier 1 predicted c while Classifier 2 predicted $\bar{c}$.

3   Naive Credal Classifiers and Their Thresholds

The Naive Credal Classifier (NCC) [5] is an extension of the Naive Bayes Classifier to the framework of imprecise probabilities that can be used to robustify the inferences of an NBC. Basically, the idea is to consider an NBC whose local probabilities are only partially specified.

In particular, instead of considering a probability mass function $P(C)$ that contains the probabilities $P(c)$ of each of the classes $c \in \mathcal{C}$, an NCC considers a set of such probability mass functions, which we denote by $\mathcal{P}(C)$. Similarly, for every class $c \in \mathcal{C}$ and every $i \in \{1, \ldots, m\}$, it considers a set $\mathcal{P}(F_i|c)$ of conditional probability mass functions. In general, these local sets can be learned from data, elicited from experts, or obtained by considering neighbourhoods around the local models of an NBC. We here consider the first option. In particular, we use a version of the Imprecise Dirichlet Model (IDM) [1], suitably adapted such that it is guaranteed to contain the result of Laplace smoothing. In particular, $P(C)$ is taken to belong to $\mathcal{P}(C)$ if and only if there is a probability mass function t on $\mathcal{C}$ such that

$$P(c) = \frac{n(c) + 1 + st(c)}{n + |\mathcal{C}| + s} \text{ for all } c \in \mathcal{C},$$

where s is a fixed hyperparameter that determines the degree of imprecision. For every $i \in \{1, \ldots, m\}$ and $c \in \mathcal{C}$, the local set $\mathcal{P}(F_i|c)$ is defined similarly.

If we now choose a single probability mass function $P(C)$ in $\mathcal{P}(C)$ and, for every $c \in \mathcal{C}$ and $i \in \{1, \ldots, m\}$, a single conditional probability mass function $P(F_i|c)$ in $\mathcal{P}(F_i|c)$, we obtain a single NBC. By doing this in every possible way, we obtain a set of NBCs. This set is a Naive Credal Classifier (NCC) [5].

Classification for such an NCC is done by performing classification with each of the NBCs it consists of separately. If all these NBCs agree on which class to return, then the output of the NCC will be that class. If they do not agree, the result of the NCC is indeterminate and consists of a set of possible classes, amongst which it is unable to choose.

Example 2. For the particular instance at the end of Example 1, for the same first run, using $s = 0.6$ makes Classifier 2 indeterminate, in the sense that it then returns the uninformative set $\{c, \bar{c}\}$, while for that value of s, the NCC that corresponds to Classifier 1 continuous to predict class c.

For our present purposes, however, we are not interested in the indeterminate predictions of an NCC. Instead, we are interested in the maximum value of s for which it still remains determinate. This value is a particular case of the critical perturbation threshold in Ref. [3]. Quite remarkably, it has been observed that for any given instance, the corresponding critical perturbation threshold serves as a good indicator for the performance of the original NBC: instances with higher thresholds are classified correctly more often [3]. Furthermore, this threshold can be computed efficiently using the algorithms in that same reference.

Example 3. Continuing with Example 2, we find that for that same instance and that same run, the critical perturbation threshold for Classifier 1 is 0.63, while that for Classifier 2 is 0.52. This explains why, for $s = 0.6$, Classifier 1 continued to be informative whereas Classifier 2 was indeterminate.

4   A Dynamic Classifier Selection Method

The main takeway message of the previous section was that, for a given instance and classifier, the corresponding critical perturbation threshold serves as a good indicator for the performance of that classifier on this instance, in the sense that instances with higher thresholds have a higher chance of being classified correctly. Inspired by this observation, we will now try to use this threshold to perform dynamic classifier selection, that is, to derive a method that, for every new instance to be classified, is able to select the most competent classifier among a set of available ones.

As a first idea, one could consider to simply choose the classifier with the highest threshold. The problem with that approach, however, is that it does not make sense to directly compare the thresholds of different classifiers, because the empirical relation between threshold and performance differs from classifier to classifier. Classifiers with more features, for example, tend to have lower thresholds for all instances, but this does not mean that they achieve lower accuracies. In general, the only thing that we know is that for a given classifier, instances with higher thresholds are classified correctly more often.

What we need, therefore, is a method for determining the empirical relation between the thresholds of different classifiers and their probabilities of correctly classifying the instance that is considered. In particular, for the case of two classifiers that we will here consider, we need a method for determining the empirical relation between a pair of thresholds—one for each of the two classifiers—and the corresponding probabilities of successful classification. Given such a relation, for any given instance, we can use its pair of thresholds to estimate which of the two considered classifiers has the highest probability of successful classification, and then use that classifier to predict the instance's class. We propose to do this in the following way.

For every new test instance that is to be classified, we start by searching the training set for instances that have a similar pair of thresholds. In particular, we choose those k training instances whose pair of thresholds is closest to that of the test instance, according to some given distance measure. In our experiments, we consider two different distance measures: the Euclidean distance and the Chebyshev distance. On the chosen subset of training instances, we then compare the success rate of each of the two classifiers, select the classifier with the highest success rate, and use that classifier to perform the classification. The resulting classifier is a combination of the two classifiers it is based on: for every test instance, it uses the one that is expected to perform best.

Figure 2 illustrates this idea with a fictitious example that is imagined to have fifty training instances, whose pairs of thresholds are depicted on the plane. The threshold value of Classifier 1 and 2 are the x- and y-coordinate, respectively. Every instance in the training set corresponds to a black point. Consider now a test instance whose pair of thresholds corresponds to the red dot and let $k = 10$. Our method then starts by considering the ten points that are closest to the red dot, according to the chosen distance measure. In Fig. 2, the green diamonds correspond to the Euclidean distance, while the purple stars are for Chebyshev distance. Next, we compare the accuracy of both classifiers on this set of points. Whichever classifier performs best on them is the one that we will use to classify this particular test instance.

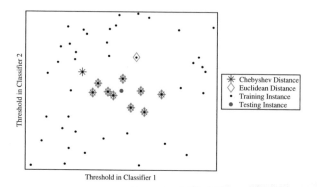

Fig. 2. Illustration of the chosen k-nearest instances, using a fictituous data set with fifty training points, and for $k = 10$ and two different distance measures

Our next example tests this method on the Balloons data set that we considered before.

Example 4. We consider the two classifiers that were discussed in Example 1 and compare them with two new classifiers that are obtained by the method that we introduced above, one for each of the two considered distance measures. As before, we randomly split the data into training and testing instances and report average accuracies over hundred such runs. In order to study the effect of the parameter k, we vary it from 2 to 20. The results are depicted in Fig. 3. The pink line shows the accuracy of the combined classifier with the Chebyshev distance (AC_{ch}), while the green one depicts the accuracy for the Euclidean distance (AC_{eu}). The purple line and the orange line correspond to the constant accuracy of Classifier 1 (AC_{C1}) and Classifier 2 (AC_{C2}), respectively.

A first important observation is that our new classifiers outperform the individual classifiers on which they are based, regardless of the value of k. We also see that the difference between using the Euclidean or Chebychev distance seems negligible. For $k = 3$, both of our new classifiers reach their peak accuracies, which is given by $AC_{ch} = 0.788485$ and $AC_{eu} = 0.788961$, respectively. The accuracy of the original classifiers are $AC_{C1} = 0.776502$ and $AC_{C2} = 0.746948$.

5   Additional Experiments

As a final test, we apply our method to four more data sets, again from the UCI Machine Learning Repository [4]. The Breast Cancer Wisconsin (original) data set (BCW), the Statlog (Australian Credit Approval) data set (ACA), the Auto MPG data set (MPG) and the Tic-Tac-Toe Endgame Data Set (TIC). A brief description of these data sets is given in Table 1, along with the Balloons data set that we considered before. For the sake of simplicity, instances with missing values and features with string values were ignored and continuous variables

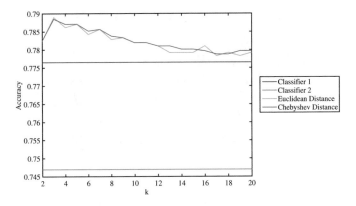

Fig. 3. The achieved accuracy as a function of the parameter k, for four different classifiers: the two original ones (which do not depend on k) and two combined classifiers (one for each of the considered distance measures)

were discretized by their median. For the remaining features, feature selection was performed to select the best set of features. Here too, the corresponding classifier is called 'Classifier 1' (C1). 'Classifier 2' (C2) is a suboptimal classifier with slightly different features. The features of these two classifiers are reported in Table 1.

Table 1. Description of data sets

Name	# Data	# Class values	# Features	Features C1	Features C2
Balloons	76	2	4	$(1, 2, 3, 4)$	$(1, 3, 4)$
BCW	699	2	9	$(1, 2, 5, 6, 8)$	$(1, 2, 6, 8)$
ACA	690	2	14	$(3, 4, 6, 13)$	$(1, 3, 4, 6, 7)$
MPG	398	2	8	$(1, 5, 6)$	$(1, 4, 5, 6)$
TIC	958	2	9	$(1, 5, 7, 8)$	$(1, 2, 5, 7, 8)$

The results of our experiments are given in Table 2. In contrast with Example 4, the parameter k was not kept constant over all runs. Rather, for each run, an optimal value of k was determined through cross-validation on the training set. Once more, our combined classifiers consistently outperform the individual ones on which they are based. Here too, the choice of distance measure seems to have very little effect.

Table 2. A comparison of all four classifiers

Data Set	AC_{C1}	AC_{C2}	AC_{eu}	AC_{ch}
Balloons	0.776502	0.746948	0.781039	0.781970
BCW	0.974221	0.972496	0.974710	0.974710
ACA	0.723675	0.723190	0.724884	0.724884
MPG	0.920696	0.917610	0.921039	0.920697
TIC	0.724366	0.724363	0.733661	0.732731

6   Conclusions and Future Work

The main conclusion of this contribution is that imprecise-probabilistic robustness measures, such as the critical perturbation threshold that we here considered, can be used to develop dynamic classifier selection methods that outperform the individual classifiers they select from. Given the restricted scope of our experiments, this conclusion is of course preliminary. We regard our results as a proof of concept, and hope that they will inspire some of you to apply similar techniques to other—perhaps more advanced—types of classifiers as well.

In our own future work, we intend to start by deepening our study of the case of the Naive Bayes Classifier. In particular, we will study the performance of our method on classification problems with more than two classes, and will extend it to allow for dynamic classifier selection among more than two classifiers. Finally, we would like to explore the effect of the specific set of classifiers among which our method chooses. In our current experiments, the classifiers that we started from had similar features and accuracies; it might very well be beneficial to start from classifiers that are more diverse.

Acknowledgements. We acknowledge support for this project from the China Scholarship Council. We also thank two anonymous reviewers for their generous constructive feedback.

References

1. Bernard, J.M.: An introduction to the imprecise Dirichlet model for multinomial data. Int. J. Approx. Reason. **39**(2–3), 123–150 (2005)
2. Cruz, R.M.O., Sabour, R., Cavalcanti, G.D.C.: Dynamic classifier selection: recent advances and perspectives. Inf. Fusion **41**, 195–216 (2018)
3. De Bock, J., De Campos, C.P., Antonucci, A.: Global sensitivity analysis for MAP inference in graphical models. In: Advances in Neural Information Processing Systems 27, Proceedings of NIPS 2014, pp. 2690–2698 (2014)
4. UCI Machine Learning Repository. http://mlr.cs.umass.edu/ml/index.html
5. Zaffalon, M.: The naive credal classifier. J. Stat. Plan. Inference **105**(1), 5–21 (2002)

Case Study-Based Sensitivity Analysis of Scale Estimates w.r.t. the Shape of Fuzzy Data

María Asunción Lubiano[(✉)], Carlos Carleos, Manuel Montenegro, and María Ángeles Gil

Departamento de Estadística, I.O. y D.M., Universidad de Oviedo,
C/ Federico García Lorca 18, 33007 Oviedo, Spain
{lubiano,carleos,mmontenegro,magil}@uniovi.es

Abstract. For practical purposes, and to ease both the drawing and the computing processes, the fuzzy rating scale was originally introduced assuming values based on such a scale to be modeled by means of trapezoidal fuzzy numbers. In this paper, to know whether or not such an assumption is too restrictive, we are going to examine on the basis of a real-life example how statistical conclusions concerning location-based scale estimates are affected by the shape chosen to model imprecise data with fuzzy numbers. The discussion will be descriptive for the considered scale estimates, but for the Fréchet-type variance it will be also inferential. The study will lead us to conclude that statistical conclusions are scarcely influenced by data shape.

Keywords: Fuzzy data · Fuzzy shape · Location-based scale estimates

1 Introduction

In previous papers we have discussed the influence of the shape of fuzzy data coming from a random process in some statistical conclusions about this process. Although the assumption of the trapezoidal shape is not at all mandatory to develop statistics with fuzzy data, such an assumption substantially eases computations. Moreover, several authors have provided with different arguments either to employ trapezoidal fuzzy numbers or to employ trapezoidal approximations of fuzzy numbers preserving some key features (like ambiguity, expected interval, etc.).

The already developed discussions concern location of the random processes generating fuzzy data (see Lubiano *et al.* [7,9]), and a few ones regard the Fréchet-type variance of these processes (see De la Rosa de Sáa *et al.* [2,3]).

This paper presents a discussion involving some scale estimates for fuzzy data sets that have been recently introduced (see [3]). The discussion is to be based on a case study and will include both, descriptive and inferential conclusions.

© Springer Nature Switzerland AG 2019
S. Destercke et al. (Eds.): SMPS 2018, AISC 832, pp. 157–165, 2019.
https://doi.org/10.1007/978-3-319-97547-4_21

2 Preliminaries

By a (bounded) **fuzzy number** we mean a mapping $\widetilde{U} : \mathbb{R} \to [0,1]$ such that for all $\alpha \in [0,1]$, the α-level set $\widetilde{U}_\alpha = \{x \in \mathbb{R} : \widetilde{U}(x) \geq \alpha\}$ if $\alpha \in (0,1]$, $= \mathrm{cl}\{x \in \mathbb{R} : \widetilde{U}(x) > 0\}$ if $\alpha = 0$ (with 'cl' denoting the closure of the set) is a nonempty compact interval.

As frequently used examples of fuzzy numbers we can consider those in Fig. 1, which are instances of the so-called LU-fuzzy numbers (see Stefanini *et al.* [13]).

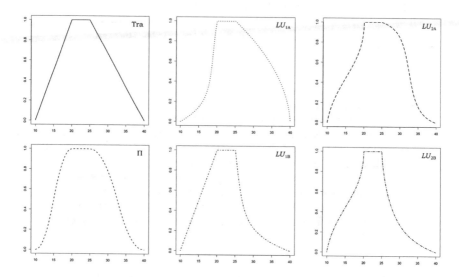

Fig. 1. Six types of fuzzy numbers sharing core $[20, 25]$ and support $(10, 40)$ and differing in shape. On the left, trapezoidal (top) and Π-curve (bottom), along with four different LU fuzzy numbers on the middle and the right

Random processes generating (intrinsically-valued) fuzzy data can be soundly formalized by means of **random fuzzy numbers** (for short RFN's), the one-dimensional convex version of fuzzy random variables, as defined by Puri and Ralescu [10] (i.e., a random fuzzy number is a fuzzy number-valued mapping $\mathcal{X}$ associated with a probability space and such that, for each α, the α-level interval-valued mapping is a random interval associated with the probability space).

Let $\mathcal{X}$ be an RFN associated with a probability space, and let $\widetilde{\mathbf{x}}_n = (\widetilde{x}_1, \ldots, \widetilde{x}_n)$ be a sample of observations from $\mathcal{X}$. The **sample Aumann-type mean** is the fuzzy number such that for each α

$$(\overline{\widetilde{\mathbf{x}}}_n)_\alpha = \left[\sum_{i=1}^{n} \inf(\widetilde{x}_i)_\alpha / n, \sum_{i=1}^{n} \sup(\widetilde{x}_i)_\alpha / n \right],$$

and the **sample 1-norm median** (Sinova *et al.* [12]) is the fuzzy number such that for each α

$$(\widehat{\mathrm{Me}}(\widetilde{\mathbf{x}}_n))_\alpha = [\mathrm{Me}_i \inf(\widetilde{x}_i)_\alpha, \mathrm{Me}_i \sup(\widetilde{x}_i)_\alpha].$$

3   Case Study

The discussions in this paper will be based on the following case study.

Example 1 (Gil *et al.* [4]). This example is related to the well-known questionnaire TIMSS-PIRLS 2011 which is conducted on the **population** of Grade 4 students (i.e., nine to ten years old) and concerns their opinion and feeling on aspects regarding reading, math, and science. This questionnaire is rather standard and most of the involved questions have to be answered according to a 4-point Likert scale, responses being DISAGREE A LOT, DISAGREE A LITTLE, AGREE A LITTLE, and AGREE A LOT.

To get more expressive responses and informative conclusions, some items selected from the original **questionnaire form** (see Table 1) have been adapted to allow a double-type response: the original Likert and a fuzzy rating scale-based one with reference interval [0, 10] (see Fig. 2 for one of the items, and http:// bellman.ciencias.uniovi.es/smire/Archivos/FormandDatasetFRS-TP.pdf for the full paper-and-pencil form, and Hesketh *et al.* [5] and Lubiano *et al.* [6,8]).

The questionnaire involving these double response questions has been conducted in 2014 on a **sample** of 69 fourth grade students from Colegio San Ignacio (Oviedo-Asturias, Spain). These students have been distributed in accordance with (their usual) three groups, so that the teachers have decided that the students in one of the three classrooms have to fill out the paper-and-pencil format and the students from the other two groups have to complete the computerized version.

The **training** of the students to let them know about the meaning and purpose of the case study, as well as the aim of the double response, has been carried out in up to 15 min, and three researchers from the Department of Statistics, OR and Math Teaching have been in charge of the explanation and conduction of the survey. At this point, it should be remarked that the students had no idea on the concept of real-valued functions and they have just learned that of a trapezium. With the guidelines enclosed in the form, the students have not had understanding problems, they have catched the philosophy behind and they have been able to provide us with quite coherent responses in most of the cases. Actually, for all the questions, the number of 'no response"s has been very small and smaller for the fuzzy rating than for the Likert scale. In summary, the training has been surprisingly much easier and more effective than we had expected.

Datasets associated with responses to this questionnaire can be found in http://bellman.ciencias.uniovi.es/smire/Archivos/FormandDatasetFRS-TP. pdf.

Table 1. Questions selected from the student questionnaire in Example 1

READING IN SCHOOL	
R.1	I like to read things that make me think
R.2	I learn a lot from reading
R.3	Reading is harder for me than any other subject
MATHEMATICS IN SCHOOL	
M.1	I like mathematics
M.2	My math teacher is easy to understand
M.3	Mathematics is harder for me than any other subject
SCIENCE IN SCHOOL	
S.1	My teacher taught me to discover science in daily life
S.2	I read about science in my spare time
S.3	Science is harder for me than any other subject

Mathematics in school

Mathematics

How much do you agree with these statements about learning mathematics?

M.2 . My teacher is easy to understand

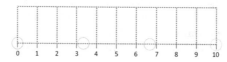

Fig. 2. Example of the response (paper-and-pencil) form to an item in Example 1

4   Metrics and Scale Measures for Fuzzy Data

Distances have been computed by considering two different metrics: the L^2 metric ρ_2 and the L^1 metric ρ_1 (see Diamond and Kloeden [1]), where for fuzzy numbers $\widetilde{U}, \widetilde{V}$ they are given by

$$\rho_2(\widetilde{U}, \widetilde{V}) = \sqrt{\frac{1}{2} \int_{[0,1]} \left[(\inf \widetilde{U}_\alpha - \inf \widetilde{V}_\alpha)^2 + (\sup \widetilde{U}_\alpha - \sup \widetilde{V}_\alpha)^2 \right] d\alpha},$$

$$\rho_1(\widetilde{U}, \widetilde{V}) = \frac{1}{2} \int_{[0,1]} \left[|\inf \widetilde{U}_\alpha - \inf \widetilde{V}_\alpha| + |\sup \widetilde{U}_\alpha - \sup \widetilde{V}_\alpha| \right] d\alpha.$$

Let $\mathcal{X}$ be an RFN associated with the probability space $(\Omega, \mathcal{A}, P)$, $\widetilde{\mathbf{x}}_n = (\widetilde{x}_1, \ldots, \widetilde{x}_n)$ a sample of observations from $\mathcal{X}$.

Then, the *(sample) Fréchet-type ρ_2-Standard Deviation, ρ_1-Average Distance Deviation, ρ_2-Average Distance Deviation, ρ_2-Median Distance Deviation, ρ_1-Median Distance Deviation* are given by

$$\rho_2\text{-SD}(\widetilde{\mathbf{x}}_n) = \sqrt{\frac{1}{n}\sum_{i=1}^{n}\left[\rho_2(\widetilde{x}_i, \overline{\overline{\mathbf{x}}}_n)\right]^2},$$

$$\rho_2\text{-}\widehat{\text{ADD}}(\widetilde{\mathbf{x}}_n) = \frac{1}{n}\sum_{i=1}^{n}\rho_2(\widetilde{x}_i, \overline{\overline{\mathbf{x}}}_n), \quad \rho_1\text{-}\widehat{\text{ADD}}(\widetilde{\mathbf{x}}_n) = \frac{1}{n}\sum_{i=1}^{n}\rho_1\left(\widetilde{x}_i, \widehat{\widetilde{\text{Me}}}(\widetilde{\mathbf{x}}_n)\right),$$

$$\rho_2\text{-}\widehat{\text{MDD}}(\widetilde{\mathbf{x}}_n) = \text{Me}_i\left\{\rho_2(\widetilde{x}_i, \overline{\overline{\mathbf{x}}}_n)\right\}, \quad \rho_1\text{-}\widehat{\text{MDD}}(\widetilde{\mathbf{x}}_n) = \text{Me}_i\left\{\rho_1\left(\widetilde{x}_i, \widehat{\widetilde{\text{Me}}}(\widetilde{\mathbf{x}}_n)\right)\right\}.$$

5   Case Study-Based Descriptive Discussion

A descriptive comparative study has been developed by computing the scale estimators in the last section over the samples of fuzzy-valued responses to Items in Table 1.

By considering the 4-tuples characterizing the fuzzy responses, we have built the six LU-fuzzy numbers in Fig. 1, along with the triangular ones $\text{Tri}(a, b, c, d)$

Table 2. Scale estimates values for the responses to Items $R.1$–$R.3$, concerning READING in Example 1, depending on the considered shape

$R.1$	Tra	Π	LU_{1A}	LU_{1B}	LU_{2A}	LU_{2B}	Tri	TriS
ρ_2-SD	2.2609	2.2577	2.2578	2.2581	2.2573	2.2573	2.2329	2.2447
ρ_2-ADD	1.4413	1.4390	1.4689	1.3751	1.4284	1.3800	1.4330	1.5640
ρ_1-ADD	1.3683	1.3647	1.2640	1.3104	1.3658	1.3101	1.3364	1.4309
ρ_2-MDD	1.8205	1.8130	1.8176	1.8105	1.8131	1.8104	1.7645	1.7852
ρ_1-MDD	1.7189	1.7142	1.7113	1.7201	1.7121	1.7178	1.6944	1.7203
$R.2$	Tra	Π	LU_{1A}	LU_{1B}	LU_{2A}	LU_{2B}	Tri	TriS
ρ_2-SD	1.8780	1.8733	1.8819	1.8957	1.8794	1.8983	1.8107	1.7554
ρ_2-ADD	1.4514	1.4476	1.4228	1.4082	1.4552	1.4291	1.3200	1.3448
ρ_1-ADD	1.3875	1.3835	1.3339	1.4234	1.3649	1.4037	1.3423	1.3351
ρ_2-MDD	1.6332	1.6237	1.6327	1.6543	1.6312	1.6570	1.5390	1.4835
ρ_1-MDD	1.4959	1.4932	1.4862	1.5323	1.4958	1.5317	1.4498	1.3834
$R.3$	Tra	Π	LU_{1A}	LU_{1B}	LU_{2A}	LU_{2B}	Tri	TriS
ρ_2-SD	2.8987	2.8919	2.9193	2.8947	2.8950	2.8978	2.8535	2.8303
ρ_2-ADD	2.2743	2.2713	2.2857	2.2421	2.2667	2.2353	2.2446	2.0791
ρ_1-ADD	1.5844	1.6013	1.6006	1.5571	1.5652	1.5542	1.6123	1.6688
ρ_2-MDD	2.4180	2.4073	2.4457	2.4028	2.4124	2.4081	2.3459	2.3357
ρ_1-MDD	2.2435	2.2395	2.2807	2.2301	2.2458	2.2368	2.2198	2.1938

Table 3. Scale estimates values for the responses to Items $M.1$–$M.3$, concerning MATHEMATICS in Example 1, depending on the considered shape

$M.1$	Tra	Π	LU_{1A}	LU_{1B}	LU_{2A}	LU_{2B}	Tri	TriS
ρ_2-SD	2.7000	2.6972	2.6866	2.7058	2.6961	2.7035	2.6378	2.6355
ρ_2-ADD	2.3100	2.2998	2.3368	2.4626	2.3166	2.4672	2.2321	2.0475
ρ_1-ADD	2.1719	2.1754	2.1970	2.3441	2.1809	2.3529	2.0469	2.0123
ρ_2-MDD	2.3777	2.3732	2.3642	2.3781	2.3720	2.3761	2.2803	2.2659
ρ_1-MDD	2.2741	2.2734	2.2565	2.2835	2.2706	2.2801	2.2163	2.2048
$M.2$	Tra	Π	LU_{1A}	LU_{1B}	LU_{2A}	LU_{2B}	Tri	TriS
ρ_2-SD	2.3419	2.3380	2.3012	2.3785	2.3357	2.3722	2.2803	2.2165
ρ_2-ADD	1.7976	1.7954	1.7880	1.8233	1.7842	1.8017	1.7443	1.6821
ρ_1-ADD	1.0571	1.0564	0.9956	1.1065	1.0558	1.1030	1.1344	1.1188
ρ_2-MDD	1.9837	1.9788	1.9465	2.0164	1.9781	2.0116	1.8949	1.8272
ρ_1-MDD	1.7374	1.7371	1.6887	1.7951	1.7332	1.7870	1.7159	1.6497
$M.3$	Tra	Π	LU_{1A}	LU_{1B}	LU_{2A}	LU_{2B}	Tri	TriS
ρ_2-SD	3.4820	3.4811	3.4686	3.4973	3.4803	3.4951	3.4744	3.4573
ρ_2-ADD	3.4066	3.4061	3.4307	3.3773	3.4142	3.3874	3.3268	3.3392
ρ_1-ADD	3.0357	3.0445	3.0127	3.0160	3.0447	3.0191	3.0068	2.9780
ρ_2-MDD	3.0942	3.0927	3.0844	3.1052	3.0928	3.1041	3.0829	3.0658
ρ_1-MDD	3.0276	3.0273	3.0169	3.0410	3.0268	3.0394	3.0278	3.0097

Table 4. Scale estimates values for the responses to Items $S.1$–$S.3$, concerning SCIENCE in Example 1, depending on the considered shape

$S.1$	Tra	Π	LU_{1A}	LU_{1B}	LU_{2A}	LU_{2B}	Tri	TriS
ρ_2-SD	2.5932	2.5880	2.5909	2.5907	2.5928	2.5924	2.5419	2.5203
ρ_2-ADD	1.7080	1.7053	1.7131	1.7086	1.6889	1.6895	1.6893	1.7510
ρ_1-ADD	1.6899	1.6926	1.6245	1.7540	1.7098	1.7496	1.5653	1.6506
ρ_2-MDD	2.1580	2.1483	2.1427	2.1572	2.1564	2.1590	2.0765	2.0665
ρ_1-MDD	2.0268	2.0208	2.0209	2.0337	2.0233	2.0337	1.9951	1.9901
$S.2$	Tra	Π	LU_{1A}	LU_{1B}	LU_{2A}	LU_{2B}	Tri	TriS
ρ_2-SD	2.3401	2.3316	2.3792	2.3127	2.3368	2.3192	2.2958	2.3077
ρ_2-ADD	1.6748	1.6716	1.7131	1.6096	1.6840	1.6268	1.5829	1.6988
ρ_1-ADD	1.6022	1.5897	1.6734	1.6384	1.5795	1.6343	1.5716	1.6289
ρ_2-MDD	1.9297	1.9182	1.9659	1.8889	1.9239	1.8960	1.8726	1.9007
ρ_1-MDD	1.8317	1.8267	1.8667	1.7972	1.8299	1.8024	1.8150	1.8452
$S.3$	Tra	Π	LU_{1A}	LU_{1B}	LU_{2A}	LU_{2B}	Tri	TriS
ρ_2-SD	2.9307	2.9247	2.9530	2.9302	2.9286	2.9339	2.8818	2.8394
ρ_2-ADD	2.4098	2.4072	2.4812	2.3751	2.4132	2.3808	2.3283	2.2254
ρ_1-ADD	2.2448	2.2007	2.1720	2.2917	2.1704	2.2904	2.2457	2.1063
ρ_2-MDD	2.5827	2.5742	2.6057	2.5717	2.5790	2.5768	2.5068	2.4671
ρ_1-MDD	2.4685	2.4614	2.4941	2.4676	2.4666	2.4722	2.4379	2.3951

$= \mathrm{Tra}(a, (b+c)/2), d)$, $\mathrm{TriS}(a, b, c, d) = \mathrm{Tra}(a, (a+d)/2), d)$. After computing the scale estimates we have obtained the outputs in Tables 2, 3, and 4. For each of the Items and scale estimates, one can conclude that the outputs scarcely differ.

6   Case Study-Based Inferential Discussion

In this section, we are going to examine, by means of an inferential analysis of the case study in Example 1, the influence of the shape of fuzzy data on the statistical conclusions.

The discussion is carried out on the basis of the test about the equality of variances with fuzzy data, the **bootstrapped homoscedasticity test of** k **independent RFN's**, developed by Ramos-Guajardo and Lubiano [11], which is now algorithmically summarized for the two-sample case. If $\mathcal{X}_1, \mathcal{X}_2$ are independent RFN's, consider a sample of independent observations $\widetilde{\boldsymbol{x}}_i = (\widetilde{x}_{i1}, \dots, \widetilde{x}_{in_i})$ from $\mathcal{X}_i$, $i = 1, 2$, the two samples being also independent, with $n = n_1 + n_2$. Denote $\overline{\widetilde{\boldsymbol{x}}_i} = \frac{1}{n_i} \cdot (\widetilde{x}_{i1} + \dots + \widetilde{x}_{in_i})$ the sample Aumann-type mean for $\widetilde{\boldsymbol{x}}_i$, $S^2_{\widetilde{\boldsymbol{x}}_i} = \sum_{j=1}^{n_i} \left[\rho_2(\widetilde{x}_{ij}, \overline{\widetilde{\boldsymbol{x}}_i})\right]^2 / n_i$ the sample Fréchet-type variance for $\widetilde{\boldsymbol{x}}_i$, and $\overline{S^2_{\widetilde{\boldsymbol{x}}}} = \sum_{i=1}^{2} n_i \cdot S^2_{\widetilde{\boldsymbol{x}}_i} / n$.

Then, the bootstrapped algorithm to test the null hypothesis $H_0 : \sigma^2_{\mathcal{X}_1} = \sigma^2_{\mathcal{X}_2}$ (equality of the population Fréchet-type variances) proceeds as follows:

Step 1. Compute the value of the statistic

$$T_{n_1, n_2} = \frac{\displaystyle\sum_{i=1}^{2} n_i \left(S^2_{\widetilde{\boldsymbol{x}}_i} - \overline{S^2_{\widetilde{\boldsymbol{x}}}}\right)^2}{\displaystyle\sum_{i=1}^{2} \frac{1}{n_i} \sum_{j=1}^{n_i} \left(\left[\rho_2(\widetilde{x}_{ij}, \overline{\widetilde{\boldsymbol{x}}_i})\right]^2 - S^2_{\widetilde{\boldsymbol{x}}_i}\right)^2}$$

Step 2. For each $i \in \{1, 2\}$, obtain a bootstrap sample from $\left(\widetilde{x}_{i1} \cdot \sqrt{\overline{S^2_{\widetilde{\boldsymbol{x}}}}/S^2_{\widetilde{\boldsymbol{x}}_i}}\right.$, $\dots, \widetilde{x}_{in_i} \cdot \sqrt{\overline{S^2_{\widetilde{\boldsymbol{x}}}}/S^2_{\widetilde{\boldsymbol{x}}_i}}\Big)$, $\widetilde{\boldsymbol{x}}_i^* = (\widetilde{x}_{i1}^*, \dots, \widetilde{x}_{in_i}^*)$, and compute the value of the bootstrap statistic

$$T^*_{n_1, n_2} = \frac{\displaystyle\sum_{i=1}^{2} n_i \left(S^2_{\widetilde{\boldsymbol{x}}_i^*} - \overline{S^2_{\widetilde{\boldsymbol{x}}^*}}\right)^2}{\displaystyle\sum_{i=1}^{2} \frac{1}{n_i} \sum_{j=1}^{n_i} \left(\left[\rho_2(\widetilde{x}_{ij}^*, \overline{\widetilde{\boldsymbol{x}}_i^*})\right]^2 - S^2_{\widetilde{\boldsymbol{x}}_i^*}\right)^2}$$

Step 3. *Step 2* should be repeated a large number B of times to get a set of estimates, denoted by $\{t_1^*, \dots, t_B^*\}$.

Step 4. Compute the bootstrap p-value as the proportion of values in $\{t_1^*, \dots, t_B^*\}$ being greater than T_{n_1, n_2}.

Table 5 collects the corresponding p-values (with $B = 10000$) for testing the equality of the population Fréchet-type variances for the trapezoidal RFN vs the other seven LU-valued RFN's.

Table 5. p-Values for the equality of population Fréchet's variances of trapezoidal vs other LU's responses for to Items $R.1$ to $S.3$ in Example 1

p-values Tra vs ...	Π	LU_{1A}	LU_{1B}	LU_{2A}	LU_{2B}	Tri	TriS
$R.1$	0.992	0.993	0.993	0.990	0.990	0.917	0.957
$R.2$	0.981	0.991	0.940	0.999	0.908	0.740	0.537
$R.3$	0.984	0.954	0.991	0.985	0.998	0.896	0.842
$M.1$	0.992	0.966	0.988	0.990	0.985	0.814	0.814
$M.2$	0.992	0.913	0.925	0.988	0.903	0.851	0.704
$M.3$	0.996	0.956	0.955	0.995	0.949	0.976	0.913
$S.1$	0.987	0.994	0.992	0.997	0.996	0.861	0.793
$S.2$	0.977	0.900	0.931	0.984	0.937	0.886	0.919
$S.3$	0.979	0.933	0.995	0.984	0.989	0.839	0.705

On the basis of the obtained p-values, one can immediately conclude that in computing Fréchet-type variance, data shape seems not to be significantly influential.

For all the usual significance levels one can consider, there are no significant differences between population Fréchet's variances for the seven developed comparisons, even for the triangular shaped data.

Ackowledgements. The research is this paper has been partially supported by the Spanish Ministry of Economy, Industry and Competitiveness Grant MTM2015-63971-P. Its support is gratefully acknowledged.

References

1. Diamond, P., Kloeden, P.: Metric spaces of fuzzy sets. Fuzzy Sets Syst. **35**, 241–249 (1990)
2. De la Rosa de Sáa, S., Carleos, C., López, M.T., Montenegro, M.: A case study-based analysis of the influence of the fuzzy data shape in quantifying their Fréchet's variance. In: Gil, E., et al. (eds.) The Mathematics of the Uncertain. A Tribute to Pedro Gil. Studies in Systems, Decision and Control, vol. 142, pp. 709–720. Springer, Cham (2018)
3. De la Rosa de Sáa, S., Lubiano, S., Sinova, S., Filzmoser, P.: Robust scale estimators for fuzzy data. Adv. Data Anal. Classif. **11**, 731–758 (2017)
4. Gil, M.A., Lubiano, M.A., De la Rosa de Sáa, S., Sinova, B.: Analyzing data from a fuzzy rating scale-based questionnaire: a case study. Psicothema **27**, 182–191 (2015)

5. Hesketh, T., Pryor, R., Hesketh, B.: An application of a computerized fuzzy graphic rating scale to the psychological measurement of individual differences. Int. J. Man-Mach. Stud. **29**, 21–35 (1988)

6. Lubiano, M.A., De la Rosa de Sáa, S., Montenegro, M., Sinova, B., Gil, M.A.: Descriptive analysis of responses to items in questionnaires. Why not using a fuzzy rating scale? Inf. Sci. **360**, 131–148 (2016)

7. Lubiano, M.A., De la Rosa de Sáa, S., Sinova, B., Gil, M.A.: Empirical sensitivity analysis on the influence of the shape of fuzzy data on the estimation of some statistical measures. In: Grzegorzewski, P., et al. (eds.) Strengthening Links Between Data Analysis and Soft Computing. Advances in Intelligent Systems and Computing, vol. 315, pp. 123–131. Springer, Heidelberg (2015)

8. Lubiano, M.A., Montenegro, M., Sinova, B., De la Rosa de Sáa, S., Gil, M.A.: Hypothesis testing for means in connection with fuzzy rating scale-based data: algorithms and applications. Eur. J. Oper. Res. **251**, 918–929 (2016)

9. Lubiano, M.A., Salas, A., Gil, M.A.: A hypothesis testing-based discussion on the sensitivity of means of fuzzy data with respect to data shape. Fuzzy Sets Syst. **328**, 54–69 (2017)

10. Puri, M.L., Ralescu, D.A.: Fuzzy random variables. J. Math. Anal. Appl. **114**, 409–422 (1986)

11. Ramos-Guajardo, A.B., Lubiano, M.A.: K-sample tests for equality of variances of random fuzzy sets. Comp. Stat. Data. Anal. **56**(4), 956–966 (2012)

12. Sinova, B., Gil, M.A., Colubi, A., Van Aelst, S.: The median of a random fuzzy number. The 1-norm distance approach. Fuzzy Sets Syst. **200**, 99–115 (2012)

13. Stefanini, L., Sorini, L., Guerra, M.L.: Parametric representation of fuzzy numbers and applications to fuzzy calculus. Fuzzy Sets Syst. **157**, 2423–2455 (2006)

Compatibility, Coherence and the RIP

Enrique Miranda[1(✉)] and Marco Zaffalon[2]

[1] Department of Statistics and O.R., University of Oviedo, Oviedo, Spain
mirandaenrique@uniovi.es
[2] IDSIA, Lugano, Switzerland
zaffalon@idsia.ch

Abstract. We generalise the classical result on the compatibility of marginal, possible non-disjoint, assessments in terms of the running intersection property to the imprecise case, where our beliefs are modelled in terms of sets of desirable gambles. We consider the case where we have unconditional and conditional assessments, and show that the problem can be simplified via a tree decomposition.

1 Introduction

This paper deals with the *marginal problem*, that of the compatibility of a number of marginal assessments with a global model. This problem is trivial when the marginal models are defined on disjoint sets of variables: in that case, we could for instance determine a compatible joint model by considering the independent product of the marginal distributions. However, when the sets of variables where our marginal assessments are defined are not disjoint then the problem is not immediate anymore, and it has been received quite some attention in the literature [1–3].

A necessary condition for the compatibility of a number of marginal assessments is their pairwise compatibility. Using the theory of hypergraphs, Beeri et al. [4] established a necessary and sufficient condition for pairwise compatibility to imply global compatibility: the *running intersection property*, that means that the sets of indices $S_1, \ldots, S_r$ satisfy that $S_i \cap (\cup_{j<i} S_j)$ is included in some S_{j^*} for $j^* < i$, and this for every i.

Another way to find a compatible joint is the *Iterative Proportional Fitting Procedure* (IPFP) [5]; when there is a compatible joint, this procedure determines a sequence of probability measures that converges to the compatible joint that maximizes the Kullback-Leibler information [6].

The works above investigate the compatibility of probability measures; if the possibility spaces are infinite, they are assumed to be countably additive on a suitable σ-field. Nevertheless, there are situations where the available information does not allow us to model our knowledge by means of a precise probability measure. In those cases, we may consider a number of alternative models, which are sometimes gathered under the term *imprecise probabilities* [7]. The marginal problem has been investigated for some of these models by Studeny [8], Vejnarová [9] and Jirousek [10], using the IPFP.

© Springer Nature Switzerland AG 2019
S. Destercke et al. (Eds.): SMPS 2018, AISC 832, pp. 166–174, 2019.
https://doi.org/10.1007/978-3-319-97547-4_22

In this paper, we study the problem of the compatibility of some partial assessments with a global one in as great a generality as possible: one the one hand, we assume that the assessments are expressed by means of an imprecise probability model that includes as particular cases all the models considered so far in the literature: *sets of desirable gambles*. In addition, we also investigate the case where the marginal assessments may be of a conditional nature.

After recalling some preliminary concepts in Sect. 2, in Sect. 3 we extend the classical result on the compatibility of marginal probability measures to the imprecise case. In Sect. 4 we deal with the case of conditional and unconditional information, and show that the compatibility problem can be simplified by means of a graphical procedure. Some additional comments are given in Sect. 5. Due to the space limitations, proofs have been omitted.

2   Preliminary Concepts

2.1   Sets of Desirable Gambles and Coherent Lower Previsions

Consider a possibility space $\mathcal{X}$. A *gamble* on $\mathcal{X}$ is a bounded real-valued function $f : \mathcal{X} \to \mathbb{R}$. We denote the set of all gambles on $\mathcal{X}$ by $\mathcal{L}(\mathcal{X})$, and denote by $\mathcal{L}^+(\mathcal{X}) := \{f \geq 0 : f \neq 0\}$, or simply $\mathcal{L}^+$ when no confusion is possible, the set of positive gambles. A subset $\mathcal{D} \subseteq \mathcal{L}(\mathcal{X})$ is called *coherent* when $0 \notin \mathcal{D}$ and moreover $\mathcal{D} = \mathrm{posi}(\mathcal{D} \cup \mathcal{L}^+)$, where posi denotes the set of positive linear combinations. One trivial example is the *vacuous* (least-informative) set of gambles $\mathcal{L}^+$.

We say that $\mathcal{D}$ *avoids partial loss* when it is included in some coherent set of gambles. In that case, the smallest such set is called its *natural extension*, and it is given by $\mathcal{E} = \mathrm{posi}(\mathcal{L}^+ \cup \mathcal{D})$. Moreover, $\mathcal{D}$ avoids partial loss if and only if $0 \notin \mathcal{E}$.

Given possibility spaces $\mathcal{X}_1, \ldots, \mathcal{X}_n$ and a subset S of $\{1, \ldots, n\}$, we shall let $\mathcal{X}_S := \bigtimes_{j \in S} \mathcal{X}_j$. In order to simplify the notation, we shall use $\mathcal{X}^n := \mathcal{X}_{\{1,\ldots,n\}}$. Let π_S be the *projection operator*, given by

$$\pi_S : \mathcal{X}^n \to \mathcal{X}_S$$
$$x \hookrightarrow (x_j)_{j \in S}.$$

We shall say that a gamble f on $\mathcal{X}^n$ is S-measurable if and only if $f(x) = f(y)$ for every $x, y \in \mathcal{X}^n$ such that $\pi_S(x) = \pi_S(y)$, and we shall denote by $\mathcal{K}_S$ the subset of $\mathcal{L}(\mathcal{X}^n)$ given by the $\mathcal{X}_S$-measurable gambles. There exists a one-to-one correspondence between $\mathcal{L}(\mathcal{X}_S)$ and $\mathcal{K}_S$, and we will sometimes abuse the notation by writing $\mathcal{D} \cap \mathcal{L}(\mathcal{X}_S)$ when we mean $\mathcal{D} \cap \mathcal{K}_S$ for a given set of gambles $\mathcal{D} \subseteq \mathcal{L}(\mathcal{X}^n)$. In this sense, we shall say that a set $\mathcal{D} \subseteq \mathcal{K}_S$ is *coherent relative* to $\mathcal{K}_S$ when the set $\mathcal{D}' \subseteq \mathcal{L}(\mathcal{X}_S)$ that we can make a one-to-one correspondence with, is coherent.

In addition, we shall also consider in this paper *separately coherent* sets of desirable gambles. If we consider two disjoint subsets S_1, S_2 of $\{1, \ldots, n\}$, a separately coherent set of desirable gambles $\mathcal{D}_{S_1|S_2}$ will be given by

$$\mathcal{D}_{S_1|S_2} := \cup_{x \in \mathcal{X}_{S_2}} \mathcal{D}_{\cdot|x},$$

where $\mathcal{D}_{|x}$ is a coherent set of desirable gambles relative to $\mathcal{K}_{S_1}$ for every $x \in \mathcal{X}_{S_2}$. Formally, $\mathcal{D}_{S_1|S_2}$ is a subset of $\mathcal{K}_{S_1 \cup S_2}$, but it need not be coherent relative to it: it is only coherent once we focus on each particular element $x \in \mathcal{X}_{S_2}$.

A slightly more restrictive imprecise probability model is that of coherent lower previsions. A functional $\underline{P} : \mathcal{L}(\mathcal{X}) \to \mathbb{R}$ is called a *coherent lower prevision* when it satisfies $\underline{P}(f) \geq \inf f$, $\underline{P}(\lambda f) = \lambda \underline{P}(f)$ and $\underline{P}(f + g) \geq \underline{P}(f) + \underline{P}(g)$ for every $f, g \in \mathcal{L}(\mathcal{X})$ and every $\lambda > 0$. When the third of these conditions is satisfied with equality for every f, g, then $\underline{P}$, denoted by P in the special case, is called a *coherent (linear) prevision*, and it corresponds to the expectation operator with respect to the finitely additive probability on $\mathcal{P}(\mathcal{X})$ that is its restriction to events.

A coherent set of desirable gambles $\mathcal{D}$ induces a coherent lower prevision $\underline{P}$ on $\mathcal{L}(\mathcal{X})$ by means of the formula

$$\underline{P}(f) = \sup\{\mu : f - \mu \in \mathcal{D}\}; \tag{1}$$

however, there may be different coherent sets of desirable gambles $\mathcal{D}_1, \mathcal{D}_2$ that induce the same coherent lower prevision $\underline{P}$ by means of Eq. (1), so in this sense coherent sets of desirable gambles are more general than coherent lower previsions.

2.2 Compatibility, Coherence and RIP

Consider now subsets $S_1, \ldots, S_r$ of $\{1, \ldots, n\}$, and let $P_1, \ldots, P_r$ be (finitely additive) probability measures, so that P_i is defined on the power set $\mathcal{P}(\mathcal{X}_{S_i})$. Then each P_i has a unique extension as a coherent (linear) prevision P_i' on $\mathcal{L}(\mathcal{X}_{S_i})$ or, using the one-to-one correspondence mentioned above, to $\mathcal{K}_{S_i}$. The compatibility problem studies if it is possible to find a joint probability measure on $\mathcal{P}(\mathcal{X}^n)$ with marginals $P_1, \ldots, P_r$. Taking into account the one-to-one correspondence between previsions and probability measures in the precise case, this is equivalent to the existence of a coherent prevision P' on $\mathcal{L}(\mathcal{X}^n)$ such that $P'(f) = P_i'(f)$ for every $i = 1, \ldots, r$ and every $f \in \mathcal{K}_{S_i}$.

Of course, one necessary condition for the existence of P' is the *pairwise compatibility* of $P_1', \ldots, P_r'$, that means that for every $i \neq j$ it holds that $P_i'(f) = P_j'(f)$ for every $f \in \mathcal{K}_{S_i} \cap \mathcal{K}_{S_j}$. If that is the case, then $P_1', \ldots, P_r'$ allow us to define a lower prevision Q' on $\mathcal{K} := \cup_{i=1}^r \mathcal{K}_{S_i}$ by $Q'(f) = P_i'(f)$ if $f \in \mathcal{K}_{S_i}$. Pairwise compatibility simply means that Q' is well-defined. Compatibility then means that there exists a coherent prevision on $\mathcal{L}(\mathcal{X}^n)$ (or equivalently, a finitely additive probability on $\mathcal{P}(\mathcal{X}^n)$) that coincides with Q' on $\mathcal{K}$, and this in turn is equivalent to the coherence of Q' on its domain, in the sense considered in [11].

What the classical result tells us then is that, given $P_1, \ldots, P_r$ on $\mathcal{K}_{S_1}, \ldots, \mathcal{K}_{S_r}$, their pairwise compatibility guarantees the coherence of Q' if and only if the sets of variables $S_1, \ldots, S_r$ satisfy the *running intersection property (RIP)*: for every $i = 2, \ldots, r$ it holds that

$$S_i \cap (\cup_{j<i} S_j) = S_i \cap S_{j^*} \text{ for some } j^* < i.$$

Now, the compatibility problem can be expressed in terms of coherence in the imprecise case: if we have marginal assessments given by coherent lower previsions $\underline{P}_1, \ldots, \underline{P}_r$ on $\mathcal{K}_{S_1}, \ldots, \mathcal{K}_{S_r}$, then the existence of a coherent lower prevision $\underline{P}$ on $\mathcal{L}(\mathcal{X}^n)$ with these marginals is equivalent to the coherence of the lower prevision $\underline{Q}'$ that we can define on $\mathcal{K} = \cup_{i=1}^r \mathcal{K}_{S_i}$ by means of $\underline{P}_1, \ldots, \underline{P}_r$, provided these are pairwise compatible. In Sect. 3 we shall prove that the classical result can be extended to the imprecise case.

3   Compatibility of Sets of Desirable Gambles

In this section, we shall study the compatibility problem of a number of partial assessments, when these assessments are modelled by coherent sets of desirable gambles; this includes as particular cases those of coherent lower previsions and finitely additive probability measures.

We consider therefore subsets $S_1, \ldots, S_r$ of $\{1, \ldots, n\}$, and for every $j = 1, \ldots, r$ let $\mathcal{D}_j$ be a subset of $\mathcal{L}(\mathcal{X}^n)$ that is coherent with respect to the set $\mathcal{K}_{S_j}$ of $\mathcal{X}_{S_j}$-measurable gambles. Our goal is to find conditions that guarantee the existence of a coherent set of desirable gambles $\mathcal{D}$ that is compatible with $\mathcal{D}_1, \ldots, \mathcal{D}_r$. In order to alleviate the notation, we shall use $\mathcal{K}_j := \mathcal{K}_{S_j}$.

Let us clarify what we mean by pairwise and global compatibility, in terms of sets of desirable gambles. On the one hand, if we consider $i \neq j$ in $\{1, \ldots, r\}$, we say that $\mathcal{D}_i, \mathcal{D}_j$ are pairwise compatible if and only if

$$\mathcal{D}_i \cap \mathcal{K}_j = \mathcal{D}_j \cap \mathcal{K}_i.$$

In other words, those gambles on $\mathcal{D}_i$ that are S_j-measurable belong to $\mathcal{D}_j$, and viceversa. On the other hand, we shall say that a set of gambles $\mathcal{D}$ is (globally) compatible with $\mathcal{D}_1, \ldots, \mathcal{D}_r$ if and only if it is pairwise compatible with each of them, in the sense that $\mathcal{D} \cap \mathcal{K}_j = \mathcal{D}_j$ for every $j = 1, \ldots, r$.

As mentioned in Sect. 2.2, the running intersection property is the key for pairwise compatibility to imply global compatibility in the precise case. We next extend this result to the imprecise case:

Proposition 1. *If $S_1, \ldots, S_r$ satisfy RIP and the sets $\mathcal{D}_1, \ldots, \mathcal{D}_r$ are pairwise compatible, then there exists a coherent set of desirable gambles $\mathcal{D} \subseteq \mathcal{L}(\mathcal{X}^n)$ that is globally compatible with $\mathcal{D}_1, \ldots, \mathcal{D}_r$.*

As a corollary, we can establish a similar result in terms of coherent lower previsions:

Corollary 1. *Consider subsets $S_1, \ldots, S_r$ of $\{1, \ldots, r\}$ satisfying RIP and for every j let $\underline{P}_j$ be a coherent lower prevision on $\mathcal{X}_{S_j}$. Then there exists a coherent lower prevision $\underline{P}$ on $\mathcal{X}^n$ such that $\underline{P}(f) = \underline{P}_j(f)$ for every $f \in \mathcal{K}_j$, $j = 1, \ldots, r$ if and only if $\underline{P}_i(f) = \underline{P}_j(f) \ \forall f \in \mathcal{K}_i \cap \mathcal{K}_j$, and for every $i \neq j \in \{1, \ldots, r\}$.*

As a particular case of Corollary 1, we obtain the result for linear previsions, that is, expectation operators with respect to finitely additive probabilities, and as a consequence also for countably additive probabilities.

It is not difficult to see that without RIP, pairwise compatibility does not imply global compatibility. This is actually not surprising: we deduce from [4, Theorem 3.4] that, if the sets of variables $S_1, \ldots, S_r$ do not satisfy RIP, then it is possible to find marginal probability measures $P_1, \ldots, P_r$ that are pairwise compatible but not globally compatible. Using the correspondence with sets of desirable gambles, it is possible to express the result in terms of sets of desirable gambles, too.

More generally, we may have unconditional and conditional information, and we should study whether they can be encompassed into a joint model. However, the meaning of compatibility is not as clear as in our previous results, in the sense that such a joint may necessarily induce additional assessments that are not immediately present in the original ones.

Taking this into account, given a number of sets of desirable gambles $\mathcal{D}_1, \ldots, \mathcal{D}_r$ that gather the information on different sets of variables $S_1, \ldots, S_r$, we shall investigate to which extent these sets avoid partial loss, meaning that they have a joint coherent superset; but we are not requiring anymore that $\mathcal{D} \cap \mathcal{K}_j = \mathcal{D}_j$ for every $j = 1, \ldots, r$.

Our first result tells us that if a variable appears only in one of these sets, then our assessments on this variable are not relevant for the problem:

Proposition 2. *Consider subsets $S_1, \ldots, S_r$ of $\{1, \ldots, n\}$ and coherent sets of desirable gambles $\mathcal{D}_1, \ldots, \mathcal{D}_r$, where $\mathcal{D}_j$ is coherent relative to the set $\mathcal{K}_j$ of $\mathcal{X}_{S_j}$-measurable gambles. For every $i = 1, \ldots, r$, let $\mathcal{D}_i^*$ be the restriction of $\mathcal{D}_i$ to the class of $\mathcal{X}_{S_i \cap (\cup_{j \neq i} S_j)}$-measurable gambles. Then $\cup_{i=1}^r \mathcal{D}_i$ avoids partial loss if and only if $\cup_{i=1}^r \mathcal{D}_i^*$ avoids partial loss.*

We may think that the problem considered above can be simplified into the study of pairwise compatibility, in the sense that, given $i \neq j$, we can let $\mathcal{D}_i^j$ be the restriction of $\mathcal{D}_i$ to the class of $\mathcal{X}_{S_i \cap S_j}$-measurable gambles. Then $\cup_{i \neq j} \mathcal{D}_i^j \subseteq \cup_i \mathcal{D}_i^*$, and as a consequence we have that

$$\cup_{i=1}^r \mathcal{D}_i \text{ avoid partial loss } \Rightarrow \cup_{i \neq j} \mathcal{D}_i^j \text{ avoid partial loss.}$$

However, the converse is not true, as the following example shows:

Example 1. Consider $\mathcal{X}_1 = \cdots = \mathcal{X}_6 = \{0,1\}$, and the sets $S_1 = \{1,2,3\}$, $S_2 = \{1,2,4\}, S_3 = \{1,3,5\}, S_4 = \{2,3,6\}$. Take the following sets of desirable gambles:

$$\mathcal{D}_1 := \{f : fI_A \geq 0, \text{ where } A = \{(x_1,x_2,x_3) : x_1+x_2+x_3 \in \{1,2\}\} \setminus \{0\}$$
$$\mathcal{D}_2 := \{f : fI_B \geq 0, \text{ where } B = \{(x_1,x_2,x_4) : x_1+x_2 \in \{0,2\}\} \setminus \{0\}$$
$$\mathcal{D}_3 := \{f : fI_C \geq 0, \text{ where } C = \{(x_1,x_3,x_5) : x_1+x_3 \in \{0,2\}\} \setminus \{0\}$$
$$\mathcal{D}_4 := \{f : fI_D \geq 0, \text{ where } D = \{(x_2,x_3,x_6) : x_2+x_3 \in \{0,2\}\} \setminus \{0\}.$$

To see that $\cup_{i \neq j} \mathcal{D}_i^j$ avoids partial loss, note that $\mathcal{D}_i^j$ is vacuous for every $i \neq j$. However, $\mathcal{D}_1, \ldots, \mathcal{D}_4$ do not avoid partial loss, as we can see by considering the gambles $f_1, \ldots, f_4$ given in the following table:

(X_1, X_2, X_3)	(1,1,1)	(1,1,0)	(1,0,1)	(1,0,0)	(0,1,1)	(0,1,0)	(0,0,1)	(0,0,0)
f_1	-4	1	1	1	1	1	1	-4
f_2	1	1	-4	-4	-4	-4	1	1
f_3	1	-4	1	-4	-4	1	-4	1
f_4	1	-4	-4	1	1	-4	-4	1

Then $f_i \in \mathcal{D}_i$ for $i = 1, \ldots, 4$, but $f_1 + \cdots + f_4 < 0$. ◆

4 Graphical Representation and Compatibility

When some of the assessments are of a conditional type, the notion of coherence of lower previsions can be extended in two different manners: *weak* and *(strong) coherence*. In [12], we showed that the verification of these conditions can be simplified by means of a graphical representation known as *coherence graphs*, which partition the set of assessments by means of the so-called *superblocks*. We proved that it suffices to verify the (weak or strong) coherence of those assessments that belong to the same superblock to automatically deduce the global coherence of all of them together. This allows to make a first simplification of the compatibility problem.

Consider thus a number of conditional templates $O_1|I_1, \ldots, O_r|I_r$, and assume that we have a belief assessment for the variable X_{O_j} conditional on X_{I_j}, and this for $j = 1, \ldots, r$. We represent these templates in a coherence graph. Taking into account the results from [12] mentioned above, we shall assume that this coherence graph consists of only one superblock; otherwise we treat each superblock separately.

Next we make a graphical representation of these templates so that we put the variables $O_j \cup I_j$ in one node, for $j = 1, \ldots, r$, and connect two nodes when their associated sets of variables have non-empty intersection. From this graphical representation, it is always possible to make a tree of cliques called *join tree*, so that the sets of variables present in the different cliques satisfy the running intersection property condition (see [15] for more details). RIP guarantees that if a variable k is in two cliques in the same path, then it also belongs to all the other cliques in the same path.

We assume that on each of the cliques of the join tree we have a coherent set of desirable gambles $\mathcal{D}_j$ on the corresponding set of variables. This set can be obtained by aggregating the information of the different nodes from the initial graph, that in turn is modelled by means of separately coherent sets of desirable gambles. The set $\mathcal{D}_j$ is coherent relative to the set $\mathcal{K}_j$ of $\mathcal{X}_{S_j}$-measurable gambles, and we assume moreover that the sets are pairwise compatible. The RIP condition guarantees that the intersection $S_i \cap (\cup_{j \neq i} S_j)$ can be obtained as the union of the intersections $S_i \cap S_j$, where the node j is adjacent to i. Since the different sets of variables satisfy the running intersection property, we can deduce from Proposition 1 that there exists a coherent set of desirable gambles that includes all these assessments as soon as the sets of desirable gambles we are considering avoid partial loss. The smallest such set can be obtained by means of the procedure of natural extension.

We next give an iterative procedure for determining this natural extension (it is junction tree propagation adapted to dealing with sets of desirable gambles):

- We pick any node as a root. Since the tree is undirected and totally connected, we can make a partition of its set of nodes $\{1, \ldots, r\}$ into sets $A_0, A_1, \ldots, A_k$, $k < r$, where A_i includes those nodes that are at a distance i from the root. Thus, A_0 includes only the root.
- Step 1. We consider the nodes in A_k. For each of them, we take its associated set of desirable gambles. Note that no pair of nodes in A_k can be adjacent, because of the tree structure.
- Step 2. We consider the nodes in A_{k-1}. For each node j of them, we have two possibilities:
  - If it has no adjacent nodes in A_k, we define $\mathcal{D}'_j$ as its set $\mathcal{D}_j$ of desirable gambles.
  - If it has adjacent nodes in A_k, we take the set A of adjacent nodes, and define $\mathcal{D}'_j$ as the natural extension of $\mathcal{D}_j \cup \bigcup_{l \in A} \mathcal{D}'_{l|S_j \cap S_l}$.
- We proceed in this manner until step $k + 1$, that produces a set of desirable gambles $\mathcal{D}'_0$ on the root node.

We have proven that the procedure above provides us with the restriction of the natural extension of $\mathcal{D}_1, \ldots, \mathcal{D}_r$ to those gambles that depend on the variables in the root node:

Proposition 3. *The set $\mathcal{D}'_0$ constructed in the manner described above is the restriction of the natural extension $\mathcal{E}$ of $\mathcal{D}_1, \ldots, \mathcal{D}_r$ to $\mathcal{K}_{S_0}$. It follows that $\mathcal{D}_1, \ldots, \mathcal{D}_r$ avoid partial loss if and only if $\mathcal{D}'_0$ is coherent.*

In order to obtain the natural extension of $\mathcal{D}_1, \ldots, \mathcal{D}_r$ on $\mathcal{L}(\mathcal{X}^n)$, we can consider the reverse procedure: given the same root node as before and the sets of desirable gambles $\mathcal{D}'_0, \ldots, \mathcal{D}'_{r-1}$ we generated above, we define iteratively sets $\mathcal{D}''_0, \ldots, \mathcal{D}''_{r-1}$ as follows:

- We make $\mathcal{D}''_0 := \mathcal{D}'_0$.
- Step 1: if a node i belongs to A_1, we define $\mathcal{D}''_i := \mathrm{posi}(\mathcal{D}'_i \cup \mathcal{D}'_{0|S_i \cap S_0} \cup \mathcal{L}^+(\mathcal{X}_{S_i}))$.
- Step 2: for any $i \in A_2$, we let B_i denote its neighbours in A_1, and let $\mathcal{D}''_i := \mathrm{posi}(\mathcal{D}'_i \cup \bigcup_{j \in B_i} \mathcal{D}''_{j|S_j \cap S_i} \cup \mathcal{L}^+(\mathcal{X}_{S_i}))$.

We proceed iteratively in this manner until we get to the nodes in A_k. Then the set $\mathcal{D}''_i$ we obtain with this procedure is the natural extension of the natural extension of $\mathcal{D}_1, \ldots, \mathcal{D}_r$ to the class of $\mathcal{X}_{S_i}$-measurable gambles. Note that this holds in particular for the root node, taking into account Proposition 3.

Proposition 4. *Let $\mathcal{E}$ be the natural extension of $\mathcal{D}_1, \ldots, \mathcal{D}_r$. If we follow the procedure above, then $\mathcal{D}''_i = \mathcal{E} \cap \mathcal{K}_i$ for every $i = 1, \ldots, r$.*

It is worth noting that the procedure above cannot be simplified, in the sense that, for a given index $i = 1, \ldots, r$, it does not hold that

$$\text{posi}(\cup_{j=1}^{r}\mathcal{D}_j \cup \mathcal{L}^+) \cap \mathcal{K}_i = \text{posi}(\cup_{S_j \cap S_i \neq \emptyset}\mathcal{D}_j \cup \mathcal{L}^+) \cap \mathcal{K}_i;$$

that is, even if a set of desirable gambles does not involve any variable in the set S_i, it could be that it has behavioural implications on S_i when we propagate through the tree:

Example 2. Let $\mathcal{X}_1, \mathcal{X}_2, \mathcal{X}_3$ be binary variables, and consider the conditional assessments $X_2|X_1$ and $X_3|X_2$ given by $X_1 = 0 \Rightarrow X_2 = 1; X_1 = 1 \Rightarrow X_2 = 1; X_2 = 0 \Rightarrow X_3 = 0; X_2 = 1 \Rightarrow X_3 = 1$. These produce the sets

$$\mathcal{D}_{12} = \{f \in \mathcal{L}(\mathcal{X}_1 \times \mathcal{X}_2) : f(0,1) \geq 0, f(1,1) \geq 0, \max\{f(0,1), f(1,1)\} > 0\};$$
$$\mathcal{D}_{23} = \{f \in \mathcal{L}(\mathcal{X}_2 \times \mathcal{X}_3) : f(0,0) \geq 0, f(1,1) \geq 0, \max\{f(0,0), f(1,1)\} > 0\}.$$

Their natural extension includes the gamble $g = I_{X_3=1} - 2I_{X_3=0}$: to see this, note that $g \geq f_1 + f_2$, for $f_1 = 0.5I_{X_2=1} - 3I_{X_2=0} \in \mathcal{D}_{12}$ and $f_2 = 0.5I_{X_2=X_3} - 3I_{X_2 \neq X_3} \in \mathcal{D}_{23}$. However, g does not belong to the restriction of $\mathcal{D}_{23}$ to $\mathcal{L}(\mathcal{X}_3)$. ◆

5   Conclusions

We have showed that the classical compatibility result based on the running intersection property can be generalised in a number of ways: first, to finitely additive probabilities, getting rid of measurability constraints; secondly, to coherent lower previsions, which are equivalent to sets of probability measures; and finally to sets of desirable gambles, which include coherent lower previsions as a particular case and are more suited for dealing with the problem of conditioning on sets of probability zero.

In addition, we have combined this result with our earlier work in [12] to simplify the study of the coherence of unconditional and conditional assessments, going beyond coherence graphs and using a tree decomposition to give a necessary and sufficient condition for avoiding partial loss. A study of the computational complexity, the expression of our desirability results in terms of conditional lower previsions [13], the relation to conglomerable extensions [14] and to other works are future lines of research.

Acknowledgements. We acknowledge the financial support by project TIN2014-59543-P.

References

1. Kellerer, H.: Verteilungsfunktionen mit gegebenen marginalverteilungen. Z. Wahrscheinlichkeitstheorie **3**, 247–270 (1964)
2. Skala, H.J.: The existence of probability measures with given marginals. Ann. Probab. **21**(1), 136–142 (1993)
3. Fritz, T., Chaves, R.: Entropic inequalities and marginal problems. IEEE Trans. Inf. Theory **59**(3), 803–817 (2013)
4. Beeri, C., Fagin, R., Maier, D., Yannakis, M.: On the desirability of acyclic database schemes. J. ACM **30**, 479–513 (1983)

5. Deming, W., Stephan, F.: On a least square adjustment of a sampled frequency table when the expected marginal totals are known. Ann. Math. Stat. **11**, 427–444 (1940)
6. Csiszár, I.: I-divergence geometry of probability distributions and minimization problems. Ann. Probab. **3**(1), 146–158 (1975)
7. Augustin, T., Coolen, F., de Cooman, G., Troffaes, M. (eds.): Introduction to Imprecise Probabilities. Wiley, Hoboken (2014)
8. Studeny, M.: Marginal problem in different calculi of AI. In: Advances in Intelligent Computing - IPMU 1994, pp. 348–359. Springer, Heidelberg (1994)
9. Vejnarová, J.: A note on the interval-valued marginal problem and its maximum entropy solution. Kybernetika **34**(1), 19–26 (1998)
10. Jirousek, R.: Solution of the marginal problem and decomposable distributions. Kybernetika **27**(5), 403–412 (1991)
11. Walley, P.: Statistical Reasoning with Imprecise Probabilities. Chapman and Hall, London (1991)
12. Miranda, E., Zaffalon, M.: Coherence graphs. Artif. Intell. **173**(1), 104–144 (2009)
13. Miranda, E., Zaffalon, M.: Notes on desirability and conditional lower previsions. Ann. Math. Artif. Intell. **60**(3–4), 251–309 (2010)
14. Miranda, E., Zaffalon, M.: Conglomerable natural extension. Int. J. Approx. Reason. **53**(8), 1200–1227 (2012)
15. Jensen, F., Nielsen, T.: Bayesian Networks and Decision Graphs. Springer, Heidelberg (2007)

Estimation of Classification Probabilities in Small Domains Accounting for Nonresponse Relying on Imprecise Probability

Aziz Omar[1,2(⊠)] and Thomas Augustin[1]

[1] Department of Statistics, LMU Munich, Munich, Germany
[2] Department of Mathematics, Insurance and Applied Statistics,
Helwan University, Cairo, Egypt
aziz.omar@stat.uni-muenchen.de

Abstract. Nonresponse treatment is usually carried out through imposing strong assumptions regarding the response process in order to achieve point identifiability of the parameters of interest. Problematically, such assumptions are usually not readily testable and fallaciously imposing them may lead to severely biased estimates. In this paper we develop generalized Bayesian imprecise probability methods for estimation of proportions under potentially nonignorable nonresponse using data from small domains. Namely, we generalize the imprecise Beta model to this setting, treating missing values in a cautious way. Additionally, we extend the empirical Bayes model introduced by Stasny (1991, JASA) by considering a set of priors arising, for instance, from neighborhoods of maximum likelihood estimates of the hyper parameters. We reanalyze data from the American National Crime Survey to estimate the probability of victimization in domains formed by cross-classification of certain characteristics.

Keywords: Small area estimation · Missing data
Imprecise Beta model · Generalized empirical Bayes estimation
National Crime Survey

1 Introduction

Censuses and sample surveys are the main sources of data for both official authorities and private sector entities. Since censuses are very time and resources consuming, sample surveys are the practical choice to provide estimates for the parameters of interest.

Initially, these estimates are sought for planned domains defined geographically and/or demographically. It is common that different data users demand estimates for unplanned domains such as low level administrative units, certain demographic groups or cross classifications of both. A problem arises when

© Springer Nature Switzerland AG 2019
S. Destercke et al. (Eds.): SMPS 2018, AISC 832, pp. 175–182, 2019.
https://doi.org/10.1007/978-3-319-97547-4_23

the samples corresponding to such domains are too small to permit reliable estimates. A set of statistical methods has been developed to estimate domain-specific parameters in such situations. These methods are known as Small Area Estimation (SAE) with the terms area and domain being used exchangeably. The main approach of SAE is to compensate insufficient domain samples through the use of data from other domains in a practice known as borrowing strength carried out under the assumption of domain similarity. Rao and Molina (2015) gave a comprehensive account of theoretical and practical aspects of SAE.

In an SAE setting, nonresponse imposes a further issue, as it may seriously reduce the amount of information in the – already small – domain sample and cause a severe bias. Remedies for nonresponse depend mainly on the implied missingness mechanism defined in the sense of Little and Rubin (2014). For example, application of weighting and imputations techniques, which are among the common practices, assumes nonresponse to occur (conditionally) randomly, i.e. independent of the process generating the data (given observed covariates). Modeling the response process along with the target variable through a precise model, however, is the usual approach under the assumption of nonrandom nonresponse. These assumptions regarding the missingness mechanism and other – usually accompanied – distributional assumptions are required, and even enforced, to achieve point identifiability of the parameters aimed to be estimated. Problematically, such assumptions are usually not testable and fallaciously imposing them may lead to severely biased estimates.

Manski (2003), among others, argued in favor of an assumption free framework that is capable to produce credible results. This is achieved by moving from point-identifiability mindset to partial identifiability mindset, where the sought true value is located within an interval whose limits are attainable without imposing unjustified assumptions. This (cautious) approach has already been applied to produce credible estimates in the field of official statistics, for example by Manski (2016). In an SAE setting, Plass et al. (2017) followed a similar approach to produce cautious versions of some prominent small area estimators based on a likelihood approach.

In this article we estimate classification probability of a binary variable in small domains using data that has nonresponse. We refrain from imposing strong assumptions regarding the missingness process and propose two distinctive frameworks to do so. First, we communicate the uncertainty implied by nonresponse through utilizing an appropriately generalized version of the imprecise Beta model of Walley (1991). Second, we extend the empirical Bayes approach suggested by Stasny (1991) by considering priors in the neighborhood of maximum likelihood estimators of the hyper parameters. We use data from the National Crime Survey regarding the victimization status of the residents in certain domains within the United States. This data has been previously analyzed by Stasny (1991) and Nandarm and Choi (2002).

The rest of this article is organized as follows. Section 2 presents the data and outlines previous analyses run on it. Section 3 describes the imprecise Beta model and expresses its usage to convey the uncertainty ensuing due to nonresponse.

Section 4 introduces an extension to the empirical Bayes estimation framework. Section 5 gives some concluding remarks.

2   Previous Analyses of the National Crime Survey Data

The National Crime Survey (NCS) is a large-scale household survey conducted by the U.S. Census Bureau. In the NCS, members of households are asked about crimes committed against them or against their properties in the previous six months. Stasny (1991) created a subset of the NCS data pertaining the first half of 1975. The data are post-stratified into 10 domains according to three domain characteristics: (1) urban (U) and rural (R); (2) central city (C), other incorporated place (I), and unincorporated or not a place (N); and (3) low poverty level (L) (less than 10% of the families are below the poverty level) and high poverty level (H) (at least 10% of the families are below the poverty level). The data is presented in Table 1. Due to the sensitive nature of certain crimes, it is suspected that not all victims of such crimes would openly report them during the survey. Hence, nonresponse should not treated as random.

Let the binary variables Y_{ij} and R_{ij} be, respectively, victimization status and response indicator of household j in domain i, $j = 1, \cdots, n_i$, $i = 1, \cdots, M$, where n_i is the ith domain sample size and M is the number of domains. It is of interest to estimate $\theta_i = \mathrm{P}(Y_{ij} = 1)$, the victimization probability in domain i, that is the probability that at least one victimization occurred for a household member in domain i during the previous six months. Additionally, let r_i and $\mathfrak{z}_i$ represent, respectively, number of respondents and number of observed victimizations (successes) in the ith domain sample. Finally, $\boldsymbol{n}$, $\boldsymbol{r}$ and $\boldsymbol{\mathfrak{z}}$ represent vectors of their counterparts n_i, r_i and $\mathfrak{z}_i$, respectively.

Stasny (1991) assumed Y_{ij} to be independently distributed Bernoulli(θ_i) random variables and modeled nonresponse to be nonrandom by defining the response probabilities $\pi_{it} = \mathrm{P}(R_{ij}|Y_{ij} = t)$, $t = 0, 1$. She also assumed a prior similarity among domains regarding both the victimization process and the response process, and borrowed strength across different domains by using the following common priors: $\theta_i \overset{iid}{\sim} \text{Beta}(a, b)$ and $\pi_{it} \overset{iid}{\sim} \text{Beta}(\alpha_t, \beta_t)$. Stasny (1991, p. 299) then derived the following posterior expectation

$$\mathrm{E}(\theta_i|r_i, \mathfrak{z}_i) = \left(\sum_{k=0}^{n_i - r_i} T_{ik} \left(\mathfrak{z}_i + a + k/n_i + a + b \right) \right) \left(\sum_{k=0}^{n_i - r_i} T_{ik} \right)^{-1} \tag{1}$$

where $T_{ik} = \binom{n_i - r_i}{k} \cdot B(\mathfrak{z}_i + a + k, n_i - \mathfrak{z}_i + b - k) \cdot B(\mathfrak{z}_i + \alpha_1, \beta_1 + k) \cdot B(r_i - \mathfrak{z}_i + \alpha_0, n_i - r_i + \beta_0 - k)$. She then followed an empirical Bayes approach by estimating the vector of hyper parameters $\boldsymbol{\phi} = (a, b, \alpha_0, \beta_0, \alpha_1, \beta_1)^{\mathsf{T}}$ from the marginal likelihood and treated it as fixed thereafter receiving the empirical Bayes estimator $\hat{\boldsymbol{\theta}}^{(E)}$.

Nandram and Choi (2002) analyzed the same data using similar priors, but they followed a full Bayesian approach by imposing further prior distributions on the hyper parameters ϕ with parameters that are fixed on a higher level. They used an MCMC algorithm to sample from the posterior distribution, hence obtaining the full Bayes estimator $\hat{\boldsymbol{\theta}}^{(F)}$. Table 1 shows a comparison between the naive estimator of $\hat{\boldsymbol{\theta}}^{(N)}$ (obtained using only complete observed data) and the posterior estimators $\hat{\boldsymbol{\theta}}^{(E)}$ and $\hat{\boldsymbol{\theta}}^{(F)}$.

Table 1. National Crime Survey Data (Stasny 1991) and a comparison of θ estimates under the naive, empirical Bayes (Stasny 1991) and full Bayes (Nandram and Choi 2002) estimation schemes

Domain	$\mathfrak{z}$	r	n	$\hat{\theta}^{(N)}$	$\hat{\theta}^{(E)}$	$\hat{\theta}^{(F)}$
UCL	156	711	815	0.219	0.272	0.270
UCH	95	459	532	0.207	0.265	0.261
UIL	162	719	820	0.225	0.276	0.263
UIH	72	334	370	0.216	0.254	0.244
UNL	92	389	468	0.237	0.305	0.300
UNH	15	55	64	0.273	0.287	0.270
RIL	11	47	54	0.234	0.265	0.256
RIH	10	115	135	0.087	0.185	0.210
RNL	35	309	341	0.113	0.166	0.184
RNH	79	492	556	0.161	0.213	0.217

In contrast to these analyses that followed a traditional Bayesian framework treating the respective hyper parameters of the final models as fixed, and additionally imposing strict distributional assumptions while modeling the nonresponse, we argue in favor of a less restrictive and cautious treatment of nonresponse. The proposed treatment results in set-valued estimates instead of the usual single-valued estimates. Nevertheless, these set-valued estimates entertain a high degree of credibility as they do not depend on untestable assumptions. In the following two sections we propose two different frameworks reflecting the suggested treatment.

3 Direct Cautious Modeling Based on the Imprecise Beta Model

In this section we discuss how to express the uncertainty associated with nonresponse using the imprecise Beta model. Let us first consider the ideal situation where no nonresponse occurs in the ith domain sample. In this case, $\mathfrak{z}_i$ is equivalent to z_i, the "true" number of successes. To estimate θ_i under the traditional Bayesian framework, it is natural to consider the conjugate Beta-Binomial model, where $\theta_i \sim \text{Beta}(a, b)$ (allowing, again, for borrowing strength across domains in the SAE setting). The following conjugate posterior mean estimator is then straightforwardly received

$$\hat{\theta}_i(z_i) = \frac{z_i + a}{n_i + a + b} \tag{2}$$

To expresses uncertainty associated with prior ignorance about the probability of success in a Binomial experiment, Walley (1991, Chap. 5) introduced the imprecise Beta model (IBM). Unlike the traditional Bayes approach, where a single (vague) prior is utilized, the IBM expresses prior uncertainty by defining a set of Beta prior distributions, connected through a common hyper parameter. This set of priors are to be updated into an equivalent set of Beta posteriors after obtaining the sample.

Adopting the IBM to the SAE setting, we can express uncertainty about each θ_i by defining $\mathcal{M}_0$, the set of all Beta distributions that have a fixed value, ν, as the summation of their parameters. After observing z_i successes in the ith domain sample, each prior in $\mathcal{M}_0$ is updated into a corresponding posterior that is in $\mathcal{M}_{z_i}$, the set of all Beta distributions that have $\nu + n_i$ as the summation of their parameters.

Inference about certain events and parameters defined on the convex space of θ_i is attainable in the form of set-valued estimates whose minimum and maximum values are found by optimizing (2) w.r.t. a constraining for $\nu = a + b$. It is straightforward to deduce that

$$\hat{\theta}_i(z_i) \in \left(\frac{z_i}{n_i + \nu}, \frac{z_i + \nu}{n_i + \nu} \right) \tag{3}$$

Now, consider the case of nonresponse, where z_i is no longer available. Without forcing any assumptions pertaining the nonresponse process, it is certain that $z_i \in [\mathfrak{z}_i, \mathfrak{z}_i + n_i - r_i]$. Hence, $\hat{\theta}_i(r_i, \mathfrak{z}_i)$, the conjugate posterior estimator that is equivalent to (2) in case of nonresponse, is certain to satisfy

$$\hat{\theta}_i(r_i, \mathfrak{z}_i) \in \left[\frac{\mathfrak{z}_i + a}{n_i + \nu}, \frac{\mathfrak{z}_i + n_i - r_i + a}{n_i + \nu} \right] \tag{4}$$

The same way (3) is obtained from (2), the set-valued estimator $\hat{\Theta}_i$ can be obtained from (4) by optimizing the limits of $\hat{\theta}_i(r_i, \mathfrak{z}_i)$ w.r.t. a constraining for ν, where

$$\hat{\Theta}_i := \left(\underline{\hat{\Theta}}_i, \overline{\hat{\Theta}}_i \right) := \left(\frac{\mathfrak{z}_i}{n_i + \nu}, \frac{\mathfrak{z}_i + n_i - r_i + \nu}{n_i + \nu} \right), i = 1, \ldots, M \tag{5}$$

The borrowing of strength, expressed by the common parametrization of the Beta distribution with parameters (a, b), confines the vector $\boldsymbol{\theta} = (\theta_1, \cdots, \theta_M)^\mathsf{T}$ to be within $\hat{\boldsymbol{\Theta}}$, a proper subset of the cube $\times_{i=1}^M \hat{\Theta}_i$ produced from (5), where

$$\hat{\boldsymbol{\Theta}} := \left\{ \left(\frac{z_1 + a}{n_1 + \nu}, \ldots, \frac{z_M + a}{n_M + \nu} \right)^\mathsf{T} \, \middle| \, a \in (0, \nu), \boldsymbol{z} \in [\mathfrak{z}, \mathfrak{z} + \boldsymbol{n} - \boldsymbol{r}] \right\} \tag{6}$$

Since, however, $\hat{\Theta}$ cannot be fully described component-wise, we show in the left panel of Table 2 just the area specific set-valued estimates $\hat{\Theta}_i$ for the NCS data under certain values of ν.

Our procedure resembles the approaches of Ramoni (2001), de Cooman and Zaffalon (2004) and Utkin and Augustin (2007). It can also be seen as a natural application of the cautious data completion discussed in Augustin et al. (2014, Sect. 7.8.2) to adjust for missingness, refraining from any distributional assumptions on the missingness process. It may be noted that the set-valued estimates $\hat{\Theta}_i$ contain, for each domain, the naive estimates, the empirical Bayes estimates and the full Bayes estimates shown in Table 1. This confirms the cautious nature of our approach depicted in its similarity to the aforementioned approaches, where cautious set-valued estimates contain single-valued estimates produced by traditional precise approaches.

Table 2. Interval limits of the posterior expectations $\hat{\Theta}(\phi)$ for the NCS data under the extension of the IBM and the empirical Bayes approach

Domain	$\nu = 1$		$\nu = 2$		$\epsilon = 0.9 * \hat{\phi}$		$\hat{\Phi}$ [a]	
	$\underline{\hat{\Theta}}_i$	$\overline{\hat{\Theta}}_i$	$\underline{\hat{\Theta}}_i$	$\overline{\hat{\Theta}}_i$	$\underline{\hat{\Theta}}_i(\hat{\Phi})$	$\overline{\hat{\Theta}}_i(\hat{\Phi})$	$\underline{\hat{\Theta}}_i(\hat{\Phi})$	$\overline{\hat{\Theta}}_i(\hat{\Phi})$
UCL	0.191	0.320	0.191	0.321	0.269	0.275	0.232	0.317
UCH	0.178	0.317	0.178	0.318	0.261	0.269	0.210	0.331
UIL	0.197	0.322	0.197	0.322	0.273	0.279	0.236	0.321
UIH	0.194	0.294	0.194	0.296	0.248	0.260	0.182	0.345
UNL	0.196	0.367	0.196	0.368	0.300	0.310	0.240	0.380
UNH	0.231	0.385	0.227	0.394	0.268	0.308	0.102	0.626
RIL	0.200	0.345	0.196	0.357	0.244	0.287	0.078	0.633
RIH	0.074	0.228	0.073	0.234	0.174	0.197	0.073	0.386
RNL	0.102	0.199	0.102	0.201	0.160	0.172	0.103	0.261
RNH	0.142	0.259	0.142	0.260	0.209	0.217	0.165	0.275

[a] $\hat{\Phi} := \left\{ \phi \mid 0 < \phi \leq 3 * \hat{\phi} \right\}$

4   Imprecise Probability Extension of the Empirical Bayes Approach

As shown in Sect. 2, traditional Bayes inference founds the analysis on a single fixed value $\hat{\phi}$ of the hyper-parameter ϕ. We regard this to be over-optimistic, and argue that considering a class of prior distributions, instead, is more realistic. Therefore, we propose in this section another framework that can be seen as an extension to the usual empirical Bayes approach.

Recalling the nonrandom nonresponse model of Stasny (1991) introduced in Sect. 2, let the prior distributions of θ_i and π_{it} be characterized by having a compact and convex set of hyper parameters $\boldsymbol{\Phi}$. Let $\hat{\theta}_i(\phi)$ be the domain estimator resulting from the direct application of the empirical Bayes approach, where $\hat{\theta}_i(\phi)$ corresponds to $\mathrm{E}(\theta_i|r_i, \mathfrak{z}_i)$ in (1) for a specific value of ϕ.

Instead of depending solely on $\hat{\theta}_i(\hat{\phi})$, where $\hat{\phi}$ is the maximum likelihood estimator of ϕ, it is more credible to use the set-valued estimator $\hat{\Theta}_i(\hat{\boldsymbol{\Phi}})$, where

$$\hat{\Theta}_i(\hat{\boldsymbol{\Phi}}) := \left[\underline{\hat{\Theta}}_i(\hat{\boldsymbol{\Phi}}), \overline{\hat{\Theta}}_i(\hat{\boldsymbol{\Phi}}) \right] := \left[\inf_{\phi \in \hat{\boldsymbol{\Phi}}} \hat{\theta}_i(\phi), \sup_{\phi \in \hat{\boldsymbol{\Phi}}} \hat{\theta}_i(\phi) \right], i = 1, \ldots, M \qquad (7)$$

and $\hat{\boldsymbol{\Phi}}$ is an appropriately specified subset of $\boldsymbol{\Phi}$. To preserve the principle of borrowing strength, a simultaneous optimization of $\hat{\theta}_i(\phi)$ for all domains w.r.t. ϕ results in the set-valued estimator $\hat{\boldsymbol{\Theta}}(\hat{\boldsymbol{\Phi}})$, where

$$\hat{\boldsymbol{\Theta}}(\hat{\boldsymbol{\Phi}}) := \left\{ \left(\hat{\theta}_1(\phi), \ldots, \hat{\theta}_M(\phi) \right)^{\mathsf{T}} \middle| \phi \in \hat{\boldsymbol{\Phi}} \right\} \qquad (8)$$

Fundamentally, there are several ways to specify $\hat{\boldsymbol{\Phi}}$. We suggest, as a kind of a generalized empirical Bayes approach, to let $\hat{\boldsymbol{\Phi}}$ be data-dependent by taking $\hat{\boldsymbol{\Phi}}$ to be a defined neighborhood of $\hat{\phi}$. A natural definition of such neighborhood is to rely on a certain threshold δ of the relative marginal likelihood. That is, let $\boldsymbol{Y}, \boldsymbol{R}$ represent the data from all domains and take $\hat{\boldsymbol{\Phi}}$ as the convex hull of all values ϕ with $\mathrm{f}(\boldsymbol{Y}, \boldsymbol{R}|\phi)/\mathrm{f}(\boldsymbol{Y}, \boldsymbol{R}|\hat{\phi}) \geq \delta$. This idea is related to the likelihood-based imprecise probability framework developed by Cattaneo (2013).

Defining a neighborhood as such requires complete knowledge of the behavior of $\mathrm{f}(\boldsymbol{Y}, \boldsymbol{R}|\phi)$ over all possible ϕ. Such knowledge is not straightforwardly attainable, though, specially for complex likelihood functions as the one considered here. For illustrative proposes, we use two neighborhoods with radii in relative terms of the components of $\hat{\phi}$: the first is defined with radius $\epsilon = 0.9 * \hat{\phi}$ and the second is defined such that $\hat{\boldsymbol{\Phi}} := \left\{ \phi \middle| 0 < \phi \leq 3 * \hat{\phi} \right\}$. As a surrogate for (8), the right panel of Table 2 shows $\hat{\Theta}_i(\hat{\boldsymbol{\Phi}})$ for each neighborhood for the NCS data.

It worth noting that $\hat{\Theta}_i(\hat{\boldsymbol{\Phi}})$ contains the empirical Bayes estimates $\hat{\theta}_i(\hat{\phi})$ since $\hat{\boldsymbol{\Phi}}$ is defined so as to contain $\hat{\phi}$. Observing that domains with small samples (e.g. RIL, UNH and RIH) have set-valued estimates with large interval widths may suggest an effect of the domain sample size. Such effect suggests limitation of borrowing strength using only response variable data, therefore motivates inclusion of auxiliary covariates through an appropriate model. This approach is known as model based small area estimation (cf. e.g. Datta 2009).

5 Concluding Remarks

We discussed cautious treatment of nonresponse using data pertaining small domains with the aim of estimating proportions utilizing an imprecise probability perspective. First we extended the IBM through a generalization that allows to handle missing data by cautious data completion and allowing for borrowing strength across domains. Then, we proposed an imprecise probability extension of the empirical Bayes estimation of the nonrandom nonresponse model of Stasny (1991) through considering a neighborhood of the hyper parameters instead of depending on the traditional maximum likelihood single-valued estimate.

Both our extensions can readily form bases of further development to nonresponse treatments refraining from strong assumptions that are hardly testable. Additionally, our second extension demonstrated the attractiveness of looking at a combination of traditional empirical Bayes estimation and imprecise probability concepts, aiming at a sort of imprecise empirical Bayes methods. Far beyond the ideas sketched here, the different ways to construct the neighborhood of empirical Bayes estimates still have to be investigated further.

Acknowledgements. The authors are thankful to the two referees for their valuable comments. The first author acknowledges the support provided by both the Egyptian government and the German Academic Exchange Service (DAAD).

References

Augustin, T., Walter, G., Coolen, F.: Statistical inference. In: Augustin, T., Coolen, F., de Cooman, G., Troffaes, M. (eds.) Introduction to Imprecise Probabilities, pp. 135–189. Wiley (2014)

Cattaneo, M.: Likelihood decision functions. Electron. J. Stat. **7**, 2924–2946 (2013)

Datta, G.S.: Model-based approach to small area estimation. In: Rao, C.R. (eds.) Sample Surveys: Inference and Analysis, Handbook of Statistics, vol. 29, no. B, pp. 251–288. Elsevier (2009)

de Cooman, G., Zaffalon, M.: Updating beliefs with incomplete observations. Artif. Intell. **159**, 75–125 (2004)

Little, R., Rubin, D.: Statistical Analysis with Missing Data, 2nd edn. Wiley, Hoboken (2014)

Manski, C.: Partial Identification of Probability Distributions. Springer, New York (2003)

Manski, C.: Credible interval estimates for official statistics with survey nonresponse. J. Econom. **191**(2), 293–301 (2016)

Nandram, B., Choi, J.: Hierarchical Bayesian nonresponse models for binary data from small areas with uncertainty about ignorability. J. Am. Stat. Assoc. **97**, 381–388 (2002)

Plass, J., Omar, A., Augustin, T.: Towards a cautious modeling of missing data in small area estimation. In: Antonucci, A., Corani, G., Couso, I., Destercke, S. (eds.) Proceedings of the Tenth International Symposium on Imprecise Probability: Theories and Applications. PMLR, vol. 62, pp. 253–264 (2017)

Ramoni, M.: Robust learning with missing data. J. Mach. Learn. **45**(2), 147–170 (2001)

Rao, J., Molina, I.: Small Area Estimation, 2nd edn. Wiley, Hoboken (2015)

Stasny, E.: Hierarchical models for the probabilities of a survey classification and nonresponse: an example from the National Crime Survey. J. Am. Stat. Assoc. **86**, 296–303 (1991)

Utkin, L., Augustin, T.: Decision making under imperfect measurement using the imprecise Dirichlet model. Int. J. Approx. Reason. **44**, 322–338 (2007)

Walley, P.: Statistical Reasoning with Imprecise Probabilities. Chapman and Hall, London (1991)

Beyond Doss and Fréchet Expectation Sets

Juan Jesus Salamanca$^{(\boxtimes)}$

Escuela Politécnica de Ingeniería, Departamento de Estadística e I.O. y D.M.,
Universidad de Oviedo, 33071 Gijón, Spain
salamancajuan@uniovi.es

Abstract. In this work, we study a family of sets which generalizes the notions of Fréchet and Doss expectation for a random variable and a random set. They appear as a natural generalization of the Fréchet functional. We study their main properties, paying special attention to the cases of finite and finite-dimensional metric spaces.

1 Introduction

Given a real random variable, its expectation and its moments are basic characteristics used in probability theory. Since we are interested in random variables (and random sets) of an arbitrary metric space (M, d), we need to consider analogous parameters. Given a random variable X of a metric space (M, d), the function $p \mapsto \mathbb{E}\left[d(X, p)^2\right]$, $p \in M$ is essential to get an important statistical set, the Fréchet expectation. This set is defined as the global minimizers of previous function [4, Sect. 3.2]. The infimum value of previous function is defined as the Fréchet variance of X. Note that previous definition of Fréchet expectation requires that the Fréchet variance is finite. Without that hypothesis, we have the Doss expectation, defined by [4, Sect. 3.3],

$$\mathbb{E}_D[X] = \bigcap_{p \in M} \mathbb{B}_p\left(\mathbb{E}\left[d(X, p)\right]\right).$$

We remark that, in any Euclidean space, both expectation sets coincide with the usual one [4, Theorem 3.6].

In order to get more information, we consider the family of functions $\mathbb{E}\left[d(X, p)^r\right]$, $r \geq 1$, which generalizes the original one. Then, we define the Fréchet r-moment as the infimum of $\mathbb{E}\left[d(X, p)^r\right]$, and the r-Fréchet expectation as the set of such global minimizers (it may not exist) (see, for instance, [3] or [6] and references therein).

In another setting, given an integrable closed random set, $\widehat{X}$, it is possible to define a set playing the role of Doss expectation. Recall that, in any metric space (M, d), the family of non-empty bounded closet sets forms a metric under the Hausdorff metric d_H. Making use of this metric, the Herer expectation of $\widehat{X}$ is defined by [4]

© Springer Nature Switzerland AG 2019
S. Destercke et al. (Eds.): SMPS 2018, AISC 832, pp. 183–190, 2019.
https://doi.org/10.1007/978-3-319-97547-4_24

$$\mathbb{E}_H\left[\widehat{X}\right] = \bigcap_{p\in M} \mathbb{B}_p\left(\mathbb{E}\left[d_H(\widehat{X}, \{p\})\right]\right).$$

The Hausdorff metric has the disadvantage that it may be very difficult to compute. Moreover, several assumptions need to be satisfied before computing the Herer expectation. We will find a suitable function which simplifies computations and it may serve to build nice sets which we interpret them as a kind of r-moment of a random set. Moreover, we will see that it satisfies suitable properties to play that role.

Summing up, we will study sets which generalize the notion of Fréchet expectation set. We interpret them as moments (of a random variable or of a random set). Note that they provide important statistical information.

The study of Fréchet expectation sets has important applications (see [1,2] or [7]). Then, our results can be used in further developments of those applications.

Next section is aimed to study statistical sets related to a random variable of a metric space. Section 3 is aimed to get analogous notions related to a random set. Finally, some final comments on the potential implications and applications are given the last section.

2   The Case of a Random Variable

In the literature there exist several notions for the expectation of a random variable X of a metric space (M, d). We remark that all of them are related to expected values of the distance of X to a point. The first one we expose here is related to the following natural function,

$$D^2(X) : M \to [0, \infty]$$
$$p \mapsto \mathbb{E}\left[d(p, X)^2\right].$$

The Fréchet expectation is the set of the global minimizers of previous functional [4]. That is,

$$\mathbb{E}_F[X] := \operatorname*{argmin}_{p\in M} D^2(X).$$

The Fréchet variance of X is the infimum of $D^2(X)$ [4].

On the other hand, we define the Doss expectation as the set

$$\mathbb{E}_{D^2}[X] := \bigcap_{p\in M} \mathbb{B}_p\left(\mathbb{E}\left[d(p, X)\right]\right),$$

where $\mathbb{B}_p(r)$, $p \in M$, $r \geq 0$, denotes the ball centered at p with radius r.

In general, the existence of any of previous expectation is not guaranteed [4]. However, in a Euclidean space, both notions coincide with the usual one.

Now, we study further some interesting statistical parameters.

In order to obtain more information about the behavior of a random variable, we desire to get a generalization of moments in an arbitrary metric space. We can consider the following family of functions, which generalize the starting one,

$$D^r(X) : M \to [0, \infty]$$
$$p \mapsto \mathbb{E}\left[d(p, X)^r\right],$$

where $r \in \mathbb{N}$. We will call r-moment the infimum of this function; and we will call the r-Fréchet expectation the absolute minimizers; that is,

$$\mathbb{E}_F[X] := \underset{p \in M}{\operatorname{argmin}} \, D^r(X).$$

Likewise, we will call the r-Doss expectation the set

$$\mathbb{E}_{D^r}[X] := \bigcap_{p \in M} \mathbb{B}_p \left(\sqrt[r]{\mathbb{E}\left[d(p, X)^r\right]} \right).$$

The following example shows a random variable whose r-Doss expectation sets are all different, illustrating that previous definitions are meaningful.

Example 1. Consider the metric space $[0, 1]$ endowed with the Euclidean distance. Let X be a continuous random variable of that space with uniform distribution. It is easy to check that, for any $r \in \mathbb{N}$, $x \in [0, 1]$,

$$\mathbb{E}\left[d(X, x)^r\right] = \frac{1}{r + 1}\left[x^{r+1} + (1 - x)^{r+1}\right].$$

Hence, the r-moment takes the value $2^{-r}/(r + 1)$. The r-Doss expectation can be calculated making use of the symmetry of the distribution. We have,

$$\mathbb{E}_{D^r}[X] = \left[1 - \frac{1}{\sqrt[r]{r + 1}}, \frac{1}{\sqrt[r]{r + 1}}\right].$$

The following example shows a random variable whose r-Fréchet mean depends on r,

Example 2. Consider $\mathbb{R}$ endowed with the Euclidean distance. Let X be a random variable with two atoms, one at 0 with probability 0.2 and another at 1. The r-Fréchet mean is attained at $2^{(1-r)/(1+r)}$.

Moreover, we are able to show a metric space where there exists no Doss expectation but an r-Doss expectation,

Example 3. Let $M = \{a, b, c\}$, endowed with the following distance: $d(a, b) = 1$, $d(a, c) = 2$ and $d(b, c) = 3$. Consider a random variable X characterized by $P\{X = a\} = 0.1$, $P\{X = b\} = 0.5$ and $P\{X = c\} = 0.4$. The table below summarizes some expectation values.

$\mathbb{E}[d(X, a)] = 1.3$	$\sqrt{\mathbb{E}[d(X, a)^2]} \cong 1.4$
$\mathbb{E}[d(X, b)] = 1.3$	$\sqrt{\mathbb{E}[d(X, b)^2]} \cong 1.9$
$\mathbb{E}[d(X, c)] = 1.7$	$\sqrt{\mathbb{E}[d(X, c)^2]} \cong 2.2$

Since $\mathbb{B}_a(1.3) = \{a, b\}$ and $\mathbb{B}_c(1.7) = \{c\}$, we have that the Doss-expectation of X is the empty set. However, we can show that the 2-Doss expectation is non-empty -it consists of the point a. In fact, $\mathbb{B}_a(1.4) = \{a, b\}$, $\mathbb{B}_b(1.9) = \{a, b\}$ and $\mathbb{B}_c(2.2) = \{a\}$. Hence,

$$\mathbb{B}_a\left(\sqrt{\mathbb{E}\left[d(X, a)^2\right]}\right) \cap \mathbb{B}_b\left(\sqrt{\mathbb{E}\left[d(X, b)^2\right]}\right) \cap \mathbb{B}_c\left(\sqrt{\mathbb{E}\left[d(X, c)^2\right]}\right) = \{a\}.$$

We have a nice relationship between the different r-Doss expectations. The following result follows from the nesting of the L^p spaces and we omit the proof,

Proposition 1. *Let X be a random variable. For any $r < s$, $\mathbb{E}_{D^r}[X] \subseteq \mathbb{E}_{D^s}[X]$.*

Clearly, an isometry preserves r-Doss expectation,

Proposition 2. *Let X be a random variable of (M, d), and let $\phi : M \to M$ be an isometry. For the random variable $\phi \circ X$, it is satisfied,*

$$\phi\left(\mathbb{E}_{D^r}[X]\right) = \mathbb{E}_{D^r}[\phi(X)]$$

Now, we are interested in obtaining global behaviors when the metric space admits a splitting, in terms of a Cartesian product. Let (M, d_0) and (N, d_1) be two metric spaces. Then, on their Cartesian product, $M \times N$, we can define a product distance d by Euclidean analogy: given (m, n) and (m', n') points of $M \times N$, the product distance obeys the law $d\left((m, n), (m', n')\right)^2 = d_0(m, m')^2 + d_1(n, n')^2$. Let us see the properties related to the r-Doss expectation,

Proposition 3. *Let X be a random variable of (M, d_0) with r-Doss expectation $\mathbb{E}_{D^r}[X]$, and let Y be a random variable of (N, d_1) with r-Doss expectation $\mathbb{E}_{D^r}[Y]$. Assume X and Y are independent. Then, the random variable (X, Y) on $M \times N$ endowed with the product distance satisfies*

$$\mathbb{E}_{D^r}[X] \times \mathbb{E}_{D^r}[Y] \subseteq \mathbb{E}_{D^r}[(X, Y)].$$

Proof. Let $p \in \mathbb{E}_{D^r}[X]$ and $l \in \mathbb{E}_{D^r}[Y]$. We have to prove that $(p, l) \in \mathbb{E}_{D^r}[(X, Y)]$. Equivalently, for any $(x, y) \in M \times N$, we have to show that

$$\mathbb{E}\left[d\left((X, Y), (x, y)\right)^r\right] \geq d\left((p, l), (x, y)\right)^r.$$

It is easy to show, making use of Hölder's inequality,

$$\mathbb{E}\left[d\left((X, Y), (x, y)\right)^r\right] \geq \left[\mathbb{E}\left(d_0(X, x)^2 + d_1(Y, y)^2\right)\right]^{r/2}.$$

From Proposition 1, we have $\mathbb{E}\left[d_0(X, x)^2\right] \geq d_0(p, x)^2$ and $\mathbb{E}\left[d_1(Y, y)^2\right] \geq d_1(l, y)^2$. These equations lead to

$$\mathbb{E}\left[d\left((X, Y), (x, y)\right)^r\right] \geq \left[d_0(p, x)^2 + d_1(l, y)^2\right]^{r/2} = d\left((p, l), (x, y)\right)^r,$$

which concludes the proof.

Slight modifications in the proof of the last result provide,

Proposition 4. *Let X be a random variable of (M, d_0) with r-Doss expectation $\mathbb{E}_{D^r}[X]$, and let Y be a random variable of (N, d_1) with r-Doss expectation $\mathbb{E}_{D^r}[Y]$. Assume X and Y are statistically independent. Then, the random variable (X, Y) on $M \times N$ endowed with distance d obeying $d\left((x, y), (x', y')\right) \geq \sqrt{d_0(x, x')^2 + d_1(y, y')^2}$, $(x, y), (x', y') \in M \times N$, satisfies*

$$\mathbb{E}_{D^r}[X] \times \mathbb{E}_{D^r}[Y] \subseteq \mathbb{E}_{D^r}[(X, Y)].$$

For the special case of the Euclidean space, we have,

Proposition 5. *In the Euclidean space, the r-Doss expectation coincides with the usual mean if the r-th moment is finite.*

Proof. Let us reason by contradiction. Assume there exists $p \in \mathbb{E}_{D^r}[X]$, $p \neq \mathbb{E}[X]$ -observe that $\mathbb{E}[X]$ exists since we are assuming that the r-th moment is finite. On the Euclidean space $(\mathbb{R}^m, d)$, take a coordinate system at $\mathbb{E}[X]$, $\{x, y_1, \ldots, y_{m-1}\}$, such that $p \equiv (-p_x, 0, \ldots, 0)$, with $p_x > 0$. Considering Proposition 1, there is no loss of generality assuming r is even. We will show that for any sequence $p_n := (n, 0, \ldots, 0)$, $n \to \infty$,

$$\lim_{n \to \infty} \frac{\mathbb{E}\left[d(X, p_n)^r\right]}{d\left(p, p_n\right)^r} = \lim_{n \to \infty} \frac{\mathbb{E}\left[d(X, p_n)^r\right]}{(n + p_x)^r} \to 1.$$

Denote by $\widehat{X}, \widehat{Y}_1, \ldots, \widehat{Y}_{m-1}$ the projections onto the coordinates of X; that is, $X = \left(\widehat{X}, \widehat{Y}_1, \ldots, \widehat{Y}_{m-1}\right)$. Then, we have

$$\lim_{n \to \infty} \frac{\mathbb{E}\left[d(X, p_n)^r\right]}{(n + p_x)^r} = \lim_{n \to \infty} \frac{\mathbb{E}\left[d(X, p_n)^r\right]}{n^r} = \lim_{n \to \infty} \frac{\mathbb{E}\left[\left((\widehat{X} - n)^2 + \sum_{j=1}^{m-1} \widehat{Y}_j^2\right)^{r/2}\right]}{n^r}$$

$$= \lim_{n \to \infty} \mathbb{E}\left[\left(\left(\frac{1}{n^2}\widehat{X} - 1\right)^2 + \sum_{j=1}^{m-1} \frac{1}{n^2}\widehat{Y}^2\right)^{r/2}\right].$$

This limit shows, that for an $\epsilon > 0$, there exists n^* such that

$$p \notin \mathbb{B}_{p_{n^*}}\left(\sqrt[r]{d\left(X, p_{n^*}\right)}\right).$$

Previous result could be seen a little disappointing for reaching non-trivial r-Doss convexity. However, there exist interesting metric spaces where r-Doss expectation behaves in a very different way (see [4,8] for further related questions),

Proposition 6. *Let $M = \{p_i\}_{i=1}^{n}$ be a finite space endowed with a distance function d. Let $p_j \in M$ satisfying: $d(p_j, p_i) < \max_l d(p_i, p_l)$, for any $j \in 1, \ldots, n$. Then, for any random variable X with support M, there exists $r \in \mathbb{N}$ such that $p_j \in \mathbb{E}_{D^r}[X]$.*

Proof. Fix $p_k \in M$. Clearly, $\sqrt[r]{\mathbb{E}\left[d(X, p_k)\right]} = \sqrt[r]{\sum_l d(p_l, p_k)\, \pi_l}$, where $\pi_l = P\{X = p_l\}$. Note that $\pi_l > 0$ by assumptions. Taking $r \to \infty$, we have

$$\sqrt[r]{\mathbb{E}\left[d(X, p_k)\right]} \to \max_i d(p_k, p_i) > d(p_k, p_j).$$

Hence, we can get an integer r' (depending on p_k) such that $\sqrt[r]{\mathbb{E}\left[d(X, p_k)\right]} \geq d(p_k, p_j)$. The same procedure can be done for any point of the space: then it is enough to consider the maximum integer of all of them.

The main hypothesis in previous result admits a nice geometrical interpretation: the point p_j of M must not be the furthest of any other one -let us say p_f. The underlying reason to impose this assumption is clear: otherwise, the ball at p_f with radius provided by the definition cannot reach p_j. The same idea can be considered when the metric space is not finite. However, it is not enough to claim a point not to be the furthest of other one; Proposition 5 shows it cannot be possible in general. Requiring that the metric space has finite diameter allows us to get such an extension. In order to clarify some ideas let us define here a set which will be fundamental in the proposition below. If (M, d) is a metric space with finite diameter, we define the function Di by $Di(p) := \mathrm{Sup}_{q \in M} d(p, q)$, for any $p \in M$. Now, we define the set $\overline{\partial} M$ by those points q for which given $\epsilon > 0$ there exists a point p_ϵ such that $d(q, p_\epsilon) + \epsilon > Di(q)$ and $d(q, p_\epsilon) + \epsilon > Di(p_\epsilon)$. Observe that, by definition, the set $\overline{\partial} M$ contains the points which are the furthest from (at least) another. The last ingredient for the following result is that X must have support M, as in previous proposition. Recall that we consider here the support of X as the set $\mathrm{supp}\, X = \{p \in M : \forall \epsilon > 0, P\{X \in \mathbb{B}_p(\epsilon)\} > 0\}$. Now, we are able to announce,

Proposition 7. *Let (M, d) be an n-dimensional metric space with finite diameter* $\mathrm{diam}(M)$. *Let $p \in M$ satisfying:* $\exists \delta > 0$, $\mathbb{B}_p(\delta) \cap \overline{\partial} M = \emptyset$. *If X is a random variable with support M, there exists $r^* \in \mathbb{N}$ such that $p \in \mathbb{E}_{D^{r^*}}[X]$.*

Proof. From the fact that M is finite-dimensional, we have that $\mathbb{B}_p(\delta)$ is compact. Moreover, since $\mathrm{diam}(M)$ is finite, the map Di is upper bounded by this quantity. Define the following function on $\mathbb{B}_p(\delta)$:

$$\pi(q) = \mathrm{Sup}_{q' \in M} P\{X \in \mathbb{B}_{q'}(d(q', q))\}.$$

We claim that $\pi(q) < 1$. Otherwise, there exists a sequence of points of M, $(q'_n)_n$ such that $P\{X \in \mathbb{B}_{q'_n}(d(q'_n, q))\} \to 1$. That is, $M \setminus \mathbb{B}_{q'_n}(d(q'_n, q)) \to \emptyset$. This implies that, given $\epsilon > 0$, there exists q'_n from which $M = \mathbb{B}_{q'_n}(d(q'_n, q) + \epsilon)$. That is, $d(q, q'_n) + \epsilon \geq \mathrm{diam}(M) > Di(q)$ and $d(q, q'_n) + \epsilon \geq \mathrm{diam}(M) > Di(p_\epsilon)$. Contradiction with the fact that $q \notin \overline{\partial} M$. Now, we can get $\pi_0 := \max \pi$. Clearly, $\pi_0 < 1$. It is easy to see now that, for any $q \in M$, $P\{X \in M \setminus \mathbb{B}_q(d(q, p) + \delta)\} \geq 1 - \pi_0$. This implies that $\sqrt[r]{\mathbb{E}\left[d(X, q)^r\right]} \geq (1 - \pi_0)^{1/r}\, (d(q, p) + \delta)$. When taking r^* obeying,

$$(1 - \pi_0)^{1/r^*}\, (\mathrm{diam}(M) + \delta) = \mathrm{diam}(M),$$

we have $\sqrt[r]{\mathbb{E}\left[d(X, q)^r\right]} \geq d(q, p)$. That is, $p \in \sqrt[r]{\mathbb{E}\left[d(X, q)^r\right]}$. Since our reasoning is valid for any q, we have that $p \in \mathbb{E}_{D^{r^*}}[X]$.

Corollary 1. *Let M be a subset of an Euclidean space $(\mathbb{R}^n, d_0)$ whose closure is compact. Consider M endowed with the induced distance, $d := d_0|_M$. Let X be a random variable of (M, d). Then, for any point p of the interior of M, there exists $r^* \in \mathbb{N}$ such that $p \in \mathbb{E}_{D^{r^*}}[X]$.*

Note that previous corollary agrees with the behavior of the r-Doss expectation found in Example 1.

3   The Case of a Random Set

In this section we ask whether an analogous r-Doss expectation for random sets can be considered. The main idea is to get intersections of balls. The radii are characterized by a function playing the role of mean of the r-th power of the distance to the random set. The question is to find a suitable function which represents correctly that role. The Choquet integral (see [9] for instance) provides a nice context where expectations have been studied. Given a random set $\widehat{X}$ of a metric space (M, d), and $r \in \mathbb{N}$, we define the function $E^r[\widehat{X}] : M \to [0, \infty]$ by

$$E^r[\widehat{X}](p) = \left[\int_0^\infty \left(1 - P\left\{ \widehat{X} \subseteq \mathbb{B}_p(\sqrt[r]{\alpha}) \right\} \right) d\alpha \right]^{1/r}.$$

The reason of this function will be seen explicitly in the examples below studied. We define the r-Doss expectation of the random set $\widehat{X}$ as the set

$$\mathbb{E}_{D^r}\left[\widehat{X} \right] = \bigcap_{p \in M} \mathbb{B}_p(E^r[\widehat{X}](p)). \tag{1}$$

Let us see in simple spaces that this function is the answer to previous question.

Let $M = \bigcup_{i=1}^n p_i$ be a finite metric space. Let $\widehat{X}$ be the random set described by $P(\{p_1\}) = 1$. For any p_i, let us study the function $P\left\{ \widehat{X} \subseteq \mathbb{B}_{p_i}(\sqrt[r]{\alpha}) \right\}(\alpha)$. This function identically vanishes up $\sqrt[r]{\alpha} = d(p_i, p_1)$; from that point, this function is 1. In other words, this function is the unit-step function at $d(p_i, p_1)^r$. Hence, $E^r[\widehat{X}](p_i) = \left[\int_0^{\alpha : \alpha = d(p_i, p_1)^r} d\alpha \right]^{1/r} = d(p_i, p_1)$. Clearly, $E^r[\widehat{X}](p_1) = 0$. This means that $\{p_1\} = \bigcap_{i=1}^n \mathbb{B}_{p_i}(E^r[\widehat{X}](p_i))$.

Now, let us consider another deterministic random set $\widehat{Y}$ of the same space described by $P(\{p_1, p_2, \ldots, p_m\}) = 1$. For any p_i, the function $P\left\{ \widehat{Y} \subseteq \mathbb{B}_{p_i}(\alpha^{1/r}) \right\}$ is the unit-step function at $\max_{1 \leq j \leq m} d(p_i, p_j)^r$. This means that for any p_i, p_j, $j \leq m$, $p_j \in \mathbb{B}_{p_i}\left(E^r[\widehat{X}](p_i) \right)$. In other words, $\{p_1, \ldots, p_m\}$ is contained in the r-Doss expectation of $\widehat{Y}$. Observe that we cannot state, in general, that $\mathbb{E}_{D^r}\left[\widehat{X} \right] = \{p_1, \ldots, p_m\}$.

To get a suitable interpretation of previous definition, let us study carefully the case where $\widehat{X}$ admits a simple description. We will say that a random set is finite-supportly if there exists a finite collection of sets, A_i, $i = 1, \ldots, n$ such that $P\left(\widehat{X} = A_i\right) > 0$ and $\sum_{i=1}^{n} P\left(\widehat{X} = A_i\right) = 1$. For a random set satisfying this assumption, $E^r[\widehat{X}](p) = \sqrt[r]{\sum_{i=1}^{n} \left(\mathrm{Sup}_{q \in A_i} d(q,p)\right)^r}$. That is, for fixed $p \in M$, $E^r[\widehat{X}](p)$ absorbs the probability of A_i in the highest distance from p to A_i.

Note that there exist random sets whose r-Doss expectation is the whole space. For instance, in $\mathbb{R}$, if a non-bounded interval appears with a non-null probability.

Now, let us study some properties of this definition, following [4, p. 190]. We omit the proof due to space limit.

Proposition 8. *Let $\widehat{X}$ be a random set of a metric space (M, d). We have,*

(i) $\mathbb{E}_{D^r}\left[\widehat{X}\right] \subseteq \mathbb{E}_{D^s}\left[\widehat{X}\right]$ *if $r < s$.*

(ii) $\mathbb{E}_{D^r}\left[\widehat{X}\right] \subseteq \mathbb{E}_{D^r}\left[\widehat{Y}\right]$ *if $\widehat{X} \subseteq \widehat{Y}$.*

(iii) $\phi \circ \mathbb{E}_{D^r}\left[\widehat{X}\right] = \mathbb{E}_{D^r}\left[\phi \circ \widehat{X}\right]$, *where ϕ is an isometry of M.*

References

1. Bhattacharya, R., Lin, L., Patrangenaru, V.: Fréchet means and nonparametric inference on non-Euclidean geometric spaces. In: A Course in Mathematical Statistics and Large Sample Theory. Springer Texts in Statistics. Springer, New York (2016)
2. Bigot, J.: Fréchet means of curves for signal averaging and application to ECG data analysis. Ann. Appl. Stat. **7**, 2384–2401 (2013)
3. Casella, G., Berger, R.L.: Statistical Inference, 2nd edn. Duxbury, Pacific Grove (2002)
4. Molchanov, I.: Theory of Random Sets. Probability and Its Applications. Springer, London (2005)
5. Murofushi, T., Sugeno, M.: Some quantities represented by the Choquet integral. Fuzzy Sets Syst. **56**, 229–235 (1993)
6. Parthasarathy, K.R.: Probability Measures on Metric Spaces. Academic Press, New York-London (1967)
7. Patrangenaru, V., Ellingson, L.: Nonparametric Statistics on Manifolds and Their Applications to Object Data Analysis. CRC Press, Boca Raton (2015)
8. Shier, D.P.: The monotonicity of power means using entropy. Amer. Statist. **42**, 203–204 (1988)
9. Zong, G., Chen, Z., Shahzad, F.: Comonotonic random sets and its additivity of Choquet integrals. Int. J. Uncertain. Fuzziness Knowl.-Based Syst. **25**, 557–571 (2017)

Empirical Comparison of the Performance of Location Estimates of Fuzzy Number-Valued Data

Beatriz Sinova[1](✉) and Stefan Van Aelst[2]

[1] Departamento de Estadística e I.O. y D.M., Universidad de Oviedo, Oviedo, Spain
`sinovabeatriz@uniovi.es`
[2] Department of Mathematics, KU Leuven, Leuven, Belgium
`stefan.vanaelst@kuleuven.be`

Abstract. Several location measures have already been proposed in the literature in order to summarize the central tendency of a random fuzzy number in a robust way. Among them, fuzzy trimmed means and fuzzy M-estimators of location extend two successful approaches from the real-valued settings. The aim of this work is to present an empirical comparison of different location estimators, including both fuzzy trimmed means and fuzzy M-estimators, to study their differences in finite sample behaviour.

Keywords: Fuzzy number · Location · Simulation · Robustness

1 Introduction

Fuzzy numbers are a useful tool to deal with the imprecision underlying many real-life experiments. For this reason, a methodology to analyze fuzzy number-valued data statistically is of interest and has already provided us with different tools to deal with this kind of data. For example, we could think of regression analysis techniques, clustering, principal components, etc. An important drawback is that a lot of such procedures are based on the use of the Aumann-type mean, which is a generalization of the concept of mean of a random variable. Even when the Aumann-type mean fulfills numerous convenient properties, both from the statistical and probabilistic points of view, it also inherits the lack of robustness of the mean of a random variable. This means that any atypical observation or outlier, or any data changes, may invalidate the conclusions of our study. Unfortunately, it is not uncommon to collect data that include some 'contaminated observations' in real-life experiments. This motivates the search for robust location measures to summarize fuzzy number-valued data sets.

Different robust location alternatives for fuzzy numbers have already been proposed in the literature (see e.g. [1–5]). Among them, the adaptation of trimmed means and M-estimators of location to the fuzzy number-valued settings could be highlighted due to their importance and success for real-valued

© Springer Nature Switzerland AG 2019
S. Destercke et al. (Eds.): SMPS 2018, AISC 832, pp. 191–199, 2019.
https://doi.org/10.1007/978-3-319-97547-4_25

random variables. The main aim of this paper is to empirically compare the behaviour of fuzzy trimmed means and fuzzy M-estimators in presence of outliers, but other location estimates will also be included in the simulations in order to complete the study.

2   Preliminaries

A (bounded) **fuzzy number** is a mapping $\widetilde{U} : \mathbb{R} \to [0,1]$ such that its α-levels

$$\widetilde{U}_\alpha = \begin{cases} \{x \in \mathbb{R} : \widetilde{U}(x) \geq \alpha\} & \text{if } \alpha \in (0,1] \\ \mathrm{cl}\{x \in \mathbb{R} : \widetilde{U}(x) > 0\} & \text{if } \alpha = 0, \end{cases}$$

where cl denotes the closure, are nonempty compact convex sets. Therefore, each fuzzy number $\widetilde{U}$ can be uniquely characterized by means of the infima and suprema of all its α-levels. $\mathcal{F}_c(\mathbb{R})$ will denote the space of fuzzy numbers.

If $\mathcal{X}$ is a random fuzzy number (a fuzzy number-valued mapping associated with a probability space and such that, for each α, the α-level interval-valued mapping is a random interval associated with the probability space), let $\widetilde{\mathbf{x}}_n = (\widetilde{x}_1, \ldots, \widetilde{x}_n)$ be a sample of fuzzy number–valued observations from $\mathcal{X}$. To represent the central tendency of a data set consisting of several fuzzy numbers, the following measures have been proposed.

- The **sample Aumann-type mean** [6] is the fuzzy number $\overline{\widetilde{\mathbf{x}}}_n$ such that for all $\alpha \in [0,1]$ its α-levels are given by

$$(\overline{\widetilde{\mathbf{x}}}_n)_\alpha = \left[\sum_{i=1}^n \inf\, (\widetilde{x}_i)_\alpha/n, \sum_{i=1}^n \sup\, (\widetilde{x}_i)_\alpha/n \right].$$

- The **sample fuzzy trimmed mean** [3] is the fuzzy number $\frac{1}{h} \sum_{j \in \widehat{E}_{\widetilde{\mathbf{x}}_n}} \widetilde{x}_j$, where $\widehat{E}_{\widetilde{\mathbf{x}}_n}$ denotes the corresponding **sample trimming region**, that is,

$$\widehat{E}_{\widetilde{\mathbf{x}}_n} = \arg \min_{\substack{E \subset \{1,\ldots,n\} \\ \#E=h}} \frac{1}{h} \sum_{i \in E} \left(D_\theta \left(\widetilde{x}_i, \frac{1}{h} \sum_{j \in E} \widetilde{x}_j \right) \right)^2$$

$$= \arg \min_{E \in \mathcal{E}} Var(\widetilde{\mathbf{x}}_n | E),$$

with the set $\mathcal{E} = \{E \subset \{1,\ldots,n\} : \#E = h\}$ consisting of all the subsets of h different natural numbers which are up to the sample size, $\theta \in (0, +\infty)$ and D_θ represents the following L^2 metric between fuzzy numbers. Given any $\widetilde{U}, \widetilde{V} \in \mathcal{F}_c(\mathbb{R})$,

$$D_\theta(\widetilde{U}, \widetilde{V}) = \left[\int_{[0,1]} \left(\mathrm{mid}\,\widetilde{U}_\alpha - \mathrm{mid}\,\widetilde{V}_\alpha \right)^2 d\ell(\alpha) \right.$$

$$+ \theta \int_{[0,1]} \left(\operatorname{spr} \widetilde{U}_\alpha - \operatorname{spr} \widetilde{V}_\alpha \right)^2 d\ell(\alpha) \Bigg]^{1/2},$$

where $\operatorname{mid} \widetilde{U}_\alpha = (\inf \widetilde{U}_\alpha + \sup \widetilde{U}_\alpha)/2$ and $\operatorname{spr} \widetilde{U}_\alpha = (\sup \widetilde{U}_\alpha - \inf \widetilde{U}_\alpha)/2$.

- The **sample M-estimator of location** associated with certain loss function ρ [1] is the fuzzy number that minimizes the expression $\frac{1}{n} \sum_{i=1}^{n} \rho(D_\theta(\widetilde{x}_i, \widetilde{U}))$, over $\widetilde{U} \in \mathcal{F}_c(\mathbb{R})$ (if it exists). Concerning the choice of the loss function, Huber's and Hampel's loss functions will be considered along this work. The *Huber loss function*, given by

$$\rho_a^H(x) = \begin{cases} x^2/2 & \text{if } 0 \le x \le a \\ a(x - a/2) & \text{otherwise,} \end{cases}$$

with $a > 0$ a tuning parameter, is a convex function and puts less emphasis on large errors compared to the squared error loss. On the other hand, the *Hampel loss function* corresponds to

$$\rho_{a,b,c}(x) = \begin{cases} x^2/2 & \text{if } 0 \le x < a \\ a(x - a/2) & \text{if } a \le x < b \\ \dfrac{a(x-c)^2}{2(b-c)} + \dfrac{a(b+c-a)}{2} & \text{if } b \le x < c \\ \dfrac{a(b+c-a)}{2} & \text{if } c \le x, \end{cases}$$

where the nonnegative parameters $a < b < c$ allow us to control the degree of suppression of large errors. The smaller their values, the higher this degree. Hampel's family of loss functions is not convex anymore and can better cope with extreme outliers, since observations far from the center ($x \ge c$) all contribute equally to the loss. The following two measures are also particular cases of M-estimators of location.

- The **sample 1-norm median** [5] is the fuzzy number such that for all $\alpha \in [0, 1]$ the corresponding α-level is given by the interval

$$[\operatorname{Me}(\{\inf (\widetilde{x}_i)_\alpha\}_{i=1}^n), \operatorname{Me}(\{\sup (\widetilde{x}_i)_\alpha\}_{i=1}^n)].$$

- The **sample wabl/ldev/rdev-median** [2] is the fuzzy number such that for all $\alpha \in [0, 1]$ the corresponding α-level is given by the interval

$$[\operatorname{Me}(\{\operatorname{wabl} \widetilde{x}_i\}_{i=1}^n) - \operatorname{Me}(\{\operatorname{ldev} (\widetilde{x}_i)_\alpha\}_{i=1}^n),$$

$$\operatorname{Me}(\{\operatorname{wabl} \widetilde{x}_i\}_{i=1}^n) + \operatorname{Me}(\{\operatorname{rdev} (\widetilde{x}_i)_\alpha\}_{i=1}^n)],$$

where wabl, $\operatorname{ldev}_\alpha$ and $\operatorname{rdev}_\alpha$ provide us with an alternative characterization of a fuzzy number. Wabl represents the real number in the interior set $\operatorname{int}(\widetilde{U}_0)$ such that

$$\operatorname{wabl}(\widetilde{U}) = \int_{[0,1]} \operatorname{mid} \widetilde{U}_\alpha \, d\ell(\alpha)$$

with ℓ the Lebesgue measure, and ldev and rdev functions inform of the left and right deviations w.r.t. wabl, respectively

$$\text{ldev}_{\widetilde{U}}(\alpha) = \text{wabl}(\widetilde{U}) - \inf \widetilde{U}_\alpha,$$

$$\text{rdev}_{\widetilde{U}}(\alpha) = \sup \widetilde{U}_\alpha - \text{wabl}(\widetilde{U}).$$

3 Simulation Study

This simulation study aims to empirically compare the different alternatives to summarize the central tendency of fuzzy number-valued data in Sect. 2: fuzzy trimmed means, Huber and Hampel fuzzy M-estimates, 1-norm median and wabl/ldev/rdev-median. In all of them θ is assumed to range in $\{1/3, 1\}$, which are two common choices in the literature (with $\theta = 1/3$ all the points in the α-levels have the same importance for the computation of the D_θ metric, whereas with $\theta = 1$, only the infima and suprema of the α-levels are taken into account). For each of the measures/estimates, the bias, the variance and the mean squared error have been approximated. Different sample sizes ($n = 100, n = 10000$) and different non-contaminated (symmetric and asymmetric) and contaminated distributions have been considered.

Please note that only trapezoidal fuzzy numbers have been considered in order to ease the computation, since a sensitivity analysis has shown that the shape of the fuzzy numbers seems to scarcely affect statistical conclusions (see [7] for more details).

The general scheme of the simulation study is as follows:

Step 1. A sample of n trapezoidal fuzzy number-valued data has been simulated from a random fuzzy number $\mathcal{X}$ for each of the different situations in such a way that

- to generate the trapezoidal fuzzy data, we have considered four real-valued random variables as follows: $\mathcal{X} = \text{Tra}(X_1 - X_2 - X_3, X_1 - X_2, X_1 + X_2, X_1 + X_2 + X_4)$, with $X_1 = \text{mid}\,\mathcal{X}_1$, $X_2 = \text{spr}\,\mathcal{X}_1$, $X_3 = \inf\,\mathcal{X}_1 - \inf\,\mathcal{X}_0$ and $X_4 = \sup\,\mathcal{X}_0 - \sup\,\mathcal{X}_1$ or, alternatively, four ordered real-valued statistics $X_{(1)}, X_{(2)}, X_{(3)}$ and $X_{(4)}$ such that $\mathcal{X} = [X_{(1)}, X_{(2)}, X_{(3)}, X_{(4)}]$, i.e., $X_{(1)} = \inf\,\mathcal{X}_0$, $X_{(2)} = \inf\,\mathcal{X}_1$, $X_{(3)} = \sup\,\mathcal{X}_1$ and $X_{(4)} = \sup\,\mathcal{X}_0$;
- each sample is split into a subsample of size $n(1 - c_p)$ (where c_p denotes the proportion of contamination and ranges in $\{0, 0.1, 0.2, 0.4\}$) associated with a non-contaminated distribution and a subsample of size $n \cdot c_p$ associated with a contaminated one, where an additional contamination role is played by C_D (which measures the relative distance between the distribution of the two subsamples and ranges in $\{0, 1, 5, 10, 100\}$);
- 16 situations with different values of c_p and C_D have been considered. For each of these situations two cases have been selected, namely, one in which random variables X_i (or $X_{(i)}$) are independent (CASES 1 and 3) and another one in which they are dependent (CASES 2, 2' and 4).

Step 2. $N = 1000$ replications of *Step 1* have been considered for the situation $c_p = C_D = 0$ in order to approximate the population measures by using a Monte Carlo approach.

Step 3. $N = 1000$ replications of *Step 1* have been considered for all the situations (c_p, C_D) and the approximated estimates, bias, variance and mean squared error have been computed for each location measure.

The choices of the non contaminated and contaminated distributions in each study will be specified now.

Study 1

In the first study, the sample size is $n = 100$, CASE 1 uses

- $X_1 \sim \mathcal{N}(0, 1)$ and $X_2, X_3, X_4 \sim \chi_1^2$ for the non-contaminated subsample,
- $X_1 \sim \mathcal{N}(0, 3) + C_D$, $X_2, X_3, X_4 \sim \chi_4^2 + C_D$ for the contaminated subsample,

whereas CASE 2 uses

- $X_1 \sim \mathcal{N}(0, 1)$ and $X_2, X_3, X_4 \sim 1/(X_1^2 + 1)^2 + 0.1 \cdot \chi_1^2$ for the non-contaminated subsample,
- $X_1 \sim \mathcal{N}(0, 3) + C_D$ and $X_2, X_3, X_4 \sim 1/(X_1^2 + 1)^2 + 0.1 \cdot \chi_1^2 + C_D$ for the contaminated subsample.

and CASE 2' uses

- $X_1 \sim \mathcal{N}(0, 1)$ and $X_2, X_3, X_4 \sim 1/(X_1^2 + 1)^2 + \sqrt{\chi_1^2}$ for the non-contaminated subsample,
- $X_1 \sim \mathcal{N}(0, 3) + C_D$ and $X_2, X_3, X_4 \sim 1/(X_1^2 + 1)^2 + \sqrt{\chi_1^2} + C_D$ for the contaminated subsample.

Study 2

In the second study, the sample size is $n = 10000$ and in CASES 1, 2 and 2' the distributions for X_1, X_2, X_3 and X_4 in the no-contaminated and the contaminated samples coincide with those for Study 1.

Study 3

In the third study, the sample size is $n = 100$, CASE 3 uses

- $X_{(1)}, X_{(2)}, X_{(3)}, X_{(4)} \sim \text{Beta}(5, 1)$ (they are simply chosen at random and ordered) for the non-contaminated subsample,
- $X_{(1)}, X_{(2)}, X_{(3)}, X_{(4)} \sim \text{Beta}(1, C_D + 1)$ for the contaminated subsample,

whereas CASE 4 uses

- $X_1 \sim \text{Beta}(5, 1)$, $X_2 \sim \text{Uniform}[0, \min\{X_1, 1 - X_1\}]$, $X_3 \sim \text{Uniform}[0, X_1 - X_2]$ and $X_4 \sim \text{Uniform}[0, 1 - X_1 - X_2]$ for the non-contaminated subsample,
- $X_1 \sim \text{Beta}(1, C_D + 1)$, $X_2 \sim \min\{X_1, 1 - X_1\} \cdot \text{Beta}(1, C_D + 1)$, $X_3 \sim (X_1 - X_2) \cdot \text{Beta}(1, C_D + 1)$ and $X_4 \sim (1 - X_1 - X_2) \cdot \text{Beta}(1, C_D + 1)$ for the contaminated subsample.

Study 4
In the fourth study, the sample size is $n = 10000$ and in CASES 3 and 4 the distributions for $X_{(1)}$, $X_{(2)}$, $X_{(3)}$, $X_{(4)}$, X_1, X_2, X_3 and X_4 in the non-contaminated and contaminated samples coincide with those for Study 3.

4   Results

For the bias, variance and mean square error, the conclusions for the different studies are summarized in Tables 1, 2, 3 and 4. The row called Dispersion indicates how variable the choice of the best location measure w.r.t. bias, variance of MSE is ("none" means that the corresponding estimator is the best in all the considered situations; "low", in most of the situations; and "high" if the choice of the best estimator highly depends on the values of the parameters c_p and C_D). For more details about the results, visit http://bellman.ciencias.uniovi.es/SMIRE/Fuzsimul.html.

Table 1. Summary of the main conclusions from Study 1: the best performing (if any) location measures/estimates are indicated for each of the situations

STUDY 1		CASE 1	CASE 2	CASE 2'
Bias	$c_p \leq 0.2$	Hampel	1-norm median	Hampel ($\theta = 1$)
	$c_p = 0.4$	Trimmed	1-norm median	Trimmed ($\theta = 1$)
	Dispersion	None	Low	Low
Variance	$c_p = 0$	1-norm median	1-norm median	1-norm median Mean
	$c_p \leq 0.2$	1-norm median Hampel ($\theta = 1$)	1-norm median	1-norm median
	$c_p = 0.4$	Wabl median 1-norm median	1-norm median	1-norm median
	Dispersion	Low	Low	High
MSE	$c_p = 0$	Huber ($\theta = 1$) 1-norm median	1-norm median	1-norm median Mean
	$c_p \leq 0.2$	Hampel	1-norm median	1-norm median Hampel ($\theta = 1$)
	$c_p = 0.4$	Trimmed	1-norm median	Trimmed ($\theta = 1$)
	Dispersion	Low	Low	High

Table 2. Summary of the main conclusions from Study 2: the best performing (if any) location measures/estimates are indicated for each of the situations

STUDY 2		CASE 1	CASE 2	CASE 2'
Bias	$c_p \leq 0.2$	Hampel	1-norm median	Hampel ($\theta = 1$)
	$c_p = 0.4$	Trimmed	1-norm median	Trimmed ($\theta = 1$)
	Dispersion	None	Medium	Low
Variance	$c_p = 0$	Huber ($\theta = 1$)	1-norm median	1-norm median
	$c_p \leq 0.2$	Hampel	1-norm median	1-norm median
			Hampel ($\theta = 1$)	trimmed
			trimmed ($\theta = 1/3$)	Hampel ($\theta = 1$)
	$c_p = 0.4$	Trimmed	Trimmed	Trimmed
	Dispersion	Medium	Low	Medium
MSE	$c_p = 0$	Huber ($\theta = 1$)	1-norm median	1-norm median
	$c_p \leq 0.2$	Hampel	1-norm median	Hampel ($\theta = 1$)
	$c_p = 0.4$	Trimmed	1-norm median	Trimmed ($\theta = 1$)
	Dispersion	None	Low	Low

Table 3. Summary of the main conclusions from Study 3: the best performing (if any) location measures/estimates are indicated for each of the situations

STUDY 3		CASE 3	CASE 4
Bias	$c_p \leq 0.2$	Hampel ($\theta = 1/3$)	Hampel ($\theta = 1/3$)
		Trimmed ($\theta = 1/3$)	
	$c_p = 0.4$	Trimmed ($\theta = 1/3$)	Trimmed ($\theta = 1/3$)
	Dispersion	None	Low
Variance	$c_p = 0$	Mean	1-norm median
		Huber ($\theta = 1/3$)	
	$c_p \leq 0.2$	Trimmed ($\theta = 1$)	1-norm median
		Wabl median	Huber ($\theta = 1/3$)
	$c_p = 0.4$	Trimmed ($\theta = 1$)	Wabl median
			Trimmed ($\theta = 1/3$)
	Dispersion	High	High
MSE	$c_p = 0$	Mean	Mean
		Huber ($\theta = 1/3$)	1-norm median
	$c_p \leq 0.2$	Trimmed ($\theta = 1/3$)	Huber ($\theta = 1/3$)
		Hampel ($\theta = 1/3$)	Hampel ($\theta = 1/3$)
			Wabl median
	$c_p = 0.4$	Trimmed ($\theta = 1/3$)	Trimmed ($\theta = 1/3$)
			Wabl median
	Dispersion	Medium	High

Table 4. Summary of the main conclusions from Study 4: the best performing (if any) location measures/estimates are indicated for each of the situations

STUDY 4		CASE 3	CASE 4
Bias	$c_p \le 0.2$	Hampel ($\theta = 1/3$) Trimmed ($\theta = 1/3$)	Hampel ($\theta = 1/3$)
	$c_p = 0.4$	Trimmed ($\theta = 1/3$)	Trimmed ($\theta = 1/3$)
	Dispersion	None	Low
Variance	$c_p = 0$	1-norm median	Wabl median
	$c_p \le 0.2$	Trimmed ($\theta = 1$) Hampel ($\theta = 1/3$) 1-norm median	Hampel trimmed ($\theta = 1$) 1-norm median
	$c_p = 0.4$	Trimmed ($\theta = 1$)	Trimmed ($\theta = 1/3$) 1-norm median
	Dispersion	Medium	High
MSE	$c_p = 0$	Wabl median	Mean Wabl median
	$c_p \le 0.2$	Hampel ($\theta = 1/3$) Trimmed ($\theta = 1/3$)	Hampel
	$c_p = 0.4$	Trimmed ($\theta = 1/3$)	Trimmed ($\theta = 1/3$)
	Dispersion	Low	Medium

5   Conclusions

On the basis of the conclusions gathered in Tables 1, 2, 3 and 4, one can conclude that there is no uniformly most appropriate location estimate. Actually, the results seem to indicate that the results depend more on the distributions considered for the non-contaminated and contaminated distributions, or the involved case, than on the sample size. A rather general assertion is that the 1-norm median is the best choice in many cases of Study 1 and Study 2 in terms of any of the considered measures (bias, variance or mean square error), above all in Case 2. In the other cases and studies, the best estimate is not as clear as in Case 2. The Huber and Hampel M-estimators generally behave well for small contamination level while the trimmed means behave well when the proportion of contamination is increased. In Cases 3 and 4, with asymmetric non-contaminated distribution and fuzzy numbers having 0-levels contained in the interval $[0, 1]$, the distinction between the advantages of using these estimates in situations of small or big amounts of contamination is not as evident.

Acknowledgements. This research has been partially supported by the Spanish Ministry of Economy, Industry and Competitiveness Grant MTM2015-63971-P. Its support is gratefully acknowledged.

References

1. Sinova, B., Gil, M.Á., Van Aelst, S.: M-estimates of location for the robust central tendency of fuzzy data. IEEE Trans. Fuzzy Syst. **24**(4), 945–956 (2016)
2. Sinova, B., Pérez-Fernández, S., Montenegro, M.: The Wabl/Ldev/Rdev median of a random fuzzy number and statistical properties. In: Grzegorzewski, P., Gagolewski, M., Hryniewicz, O., Gil, M.Á. (eds.) Strengthening Links Between Data Analysis and Soft Computing. Advances in Intelligent Systems and Computing, vol. 315, pp. 143–150. Springer, Cham (2015)
3. Colubi, A., González-Rodríguez, G.: Fuzziness in data analysis: towards accuracy and robustness. Fuzzy Sets Syst. **281**, 260–271 (2015)
4. Sinova, B., de la Rosa de Sáa, S., Gil, M.Á.: A generalized L1-type metric between fuzzy numbers for an approach to central tendency of fuzzy data. Inf. Sci. **242**, 22–34 (2013)
5. Sinova, B., Gil, M.Á., Colubi, A., Van Aelst, S.: The median of a random fuzzy number. The 1-norm distance approach. Fuzzy Sets Syst. **200**, 99–115 (2012)
6. Puri, M.L., Ralescu, D.A.: Fuzzy random variables. J. Math. Anal. Appl. **114**, 409–422 (1986)
7. Lubiano, M.A., Salas, A., Gil, M.Á.: A hypothesis testing-based discussion on the sensitivity of means of fuzzy data with respect to data shape. Fuzzy Sets Syst. **328**, 54–69 (2017)

Continuity of the Shafer-Vovk-Ville Operator

Natan T'Joens[✉], Gert de Cooman, and Jasper De Bock

ELIS, SYSTeMS, Ghent University, Ghent, Belgium
{natan.tjoens,gert.decooman,jasper.debock}@ugent.be

Abstract. Kolmogorov's axiomatic framework is the best-known app-roach to describing probabilities and, due to its use of the Lebesgue integral, leads to remarkably strong continuity properties. However, it relies on the specification of a probability measure on all measurable events. The game-theoretic framework proposed by Shafer and Vovk does without this restriction. They define global upper expectation operators using local betting options. We study the continuity properties of these more general operators. We prove that they are continuous with respect to upward convergence and show that this is not the case for downward convergence. We also prove a version of Fatou's Lemma in this more general context. Finally, we prove their continuity with respect to point-wise limits of two-sided cuts.

1   Introduction

The most common approach to probability theory is the measure-theoretic framework that originates in Kolmogorov's work [4]. Its popularity is largely due to its interpretational neutrality and the elegant mathematical properties resulting from the use of measure theory. However, this framework requires the definition of a probability measure on all measurable events. Although this is often overlooked, it presents a major drawback, because the actual specification of these probabilities is far from trivial in many practical applications. Hence, the mathematical results are elegant, but the underlying assumptions are very strong.

For dealing with stochastic processes, a more general and intuitive approach was proposed by Shafer and Vovk [5]. Their so-called game-theoretic framework is based on the idea of a supermartingale: a specific way to gamble on the successive outcomes of the process.

In our present contribution, we study the continuity properties of the upper (and therefore also lower) expectation operators that appear in this framework. Our main results are that they are continuous with respect to upward, but not downward, convergence of uniformly bounded below sequences, and continuous with respect to particular limits of two-sided cuts. From our upward convergence result, we also derive a generalised version of Fatou's Lemma.

Due to length constraints, proofs and intermediate results are relegated to the appendix of an extended online version of this paper, available on ArXiv [8].

© Springer Nature Switzerland AG 2019
S. Destercke et al. (Eds.): SMPS 2018, AISC 832, pp. 200–207, 2019.
https://doi.org/10.1007/978-3-319-97547-4_26

2   Preliminaries

We denote the set of all natural numbers, without 0, by $\mathbb{N}$, and let $\mathbb{N}_0 := \mathbb{N} \cup \{0\}$. The set of extended real numbers is denoted by $\overline{\mathbb{R}} := \mathbb{R} \cup \{+\infty, -\infty\}$. The set of positive real numbers is denoted by $\mathbb{R}_{>0}$ and the set of non-negative real numbers by $\mathbb{R}_{\geq 0}$.

We consider sequences of uncertain states $X_1, X_2, ..., X_n, ...$ where the state X_k at each discrete time $k \in \mathbb{N}$ takes a value in some non-empty finite set $\mathscr{X}$, called the *state space*. We call any $x_{1:n} := (x_1, ..., x_n) \in \mathscr{X}_{1:n} := \mathscr{X}^n$, for $n \in \mathbb{N}_0$, a *situation* and we denote the set of all situations by $\mathscr{X}^* := \cup_{n \in \mathbb{N}_0} \mathscr{X}_{1:n}$. So any finite string of possible values for a sequence of consecutive states is called a situation. In particular, the unique empty string $x_{1:0}$, denoted by $\square$, is called the *initial situation*: $\mathscr{X}_{1:0} := \{\square\}$.

An infinite sequence of state values ω is called a *path* and the set of all paths is called the *sample space* $\Omega := \mathscr{X}^{\mathbb{N}}$. For any path $\omega \in \Omega$, the initial sequence that consists of its first n state values is a situation in $\mathscr{X}_{1:n}$ that is denoted by ω^n. The n-th state is denoted by $\omega_n \in \mathscr{X}$.

3   Game-Theoretic Upper Expectations

In order to deal with stochastic processes mathematically, we use variables. A *global variable*, or simply *variable*, f is a map on the set Ω of all paths. An *(extended) real variable* f associates an (extended) real number $f(\omega)$ with any path ω. The set of all extended real variables is denoted by $\overline{\mathbb{V}}$. For any natural $k \leq \ell$, we use $X_{k:\ell}$ to denote the variable that, for every path ω, returns the tuple $X_{k:\ell}(\omega) := (\omega_k, ..., \omega_\ell)$. As such, the state $X_k = X_{k:k}$ at any discrete time k can also be regarded as a variable.

A collection of paths $A \subseteq \Omega$ is called an *event*. The *indicator* $\mathbb{I}_A$ of an event A is defined as the variable that assumes the value 1 on A and 0 elsewhere. With any situation $x_{1:n}$, we associate the *cylinder event* $\Gamma(x_{1:n}) := \{\omega \in \Omega : \omega^n = x_{1:n}\}$: the set of all paths $\omega \in \Omega$ that go through the situation $x_{1:n}$. For a given $n \in \mathbb{N}_0$, we call a variable f *n-measurable* if it is constant on the cylinder events $\Gamma(x_{1:n})$ for all $x_{1:n} \in \mathscr{X}_{1:n}$, that is, if we can write $f = \tilde{f} \circ X_{1:n} = \tilde{f}(X_{1:n})$ for some map $\tilde{f}$ on $\mathscr{X}^n$. We will then use the notation $f(x_{1:n})$ for its constant value $f(\omega)$ on all paths $\omega \in \Gamma(x_{1:n})$.

Variables are inherently uncertain objects and, therefore, we need a way to model this uncertainty. We will do this by means of upper and lower expectations, which requires the introduction of gambles.

For any non-empty set $\mathscr{Y}$, we define a *gamble* f on $\mathscr{Y}$ as a bounded real map on $\mathscr{Y}$. It is then typically interpreted as an uncertain reward $f(y)$ when the outcome of some 'experiment', assuming values in $\mathscr{Y}$, is $y \in \mathscr{Y}$. The set of all gambles on $\mathscr{Y}$ is denoted by $\mathbb{G}(\mathscr{Y})$. In particular, a gamble on Ω is a bounded real variable. When $\mathscr{Y} = \mathscr{X}$, we call the gamble f a *local gamble*.

A coherent upper expectation $\overline{\mathbb{E}}$ on the set $\mathbb{G}(\mathscr{Y})$ is defined as a real functional on $\mathbb{G}(\mathscr{Y})$ that satisfies the following *coherence axioms* [9, 2.6.1]:

E1. $\overline{\mathrm{E}}(f) \leq \sup f$ for all $f \in \mathbb{G}(\mathscr{Y})$;

E2. $\overline{\mathrm{E}}(f + g) \leq \overline{\mathrm{E}}(f) + \overline{\mathrm{E}}(g)$ for all $f, g \in \mathbb{G}(\mathscr{Y})$;

E3. $\overline{\mathrm{E}}(\lambda f) = \lambda \overline{\mathrm{E}}(f)$ for all $f \in \mathbb{G}(\mathscr{Y})$ and real $\lambda \geq 0$.

$\overline{\mathrm{E}}(f)$ can be interpreted as some subject's minimum selling price for the gamble $f \in \mathbb{G}(\mathscr{Y})$ on $\mathscr{Y}$. Alternatively, one can also consider the conjugate lower expectation, defined by $\underline{\mathrm{E}}(f) := -\overline{\mathrm{E}}(-f)$ for all $f \in \mathbb{G}(\mathscr{Y})$. It clearly suffices to focus on only one of the two functionals. We will work with upper expectations.

In an *imprecise probability tree* we attach to each situation $x_{1:n} \in \mathscr{X}^*$ a *local* probability model characterised by a coherent upper expectation $\overline{\mathrm{Q}}(\cdot|x_{1:n})$ on the set $\mathbb{G}(\mathscr{X})$ of all *local* gambles on the next state X_{n+1}. These local probability models $\overline{\mathrm{Q}}(\cdot|x_{1:n})$ are usually known, as, in most practical cases, they can be elicited fairly easily from a subject or learned from data. They express a subject's beliefs or knowledge about the next possible state. However, gathering information or eliciting beliefs about a variable that depends on multiple states or even entire paths is not that straightforward. Therefore, the question arises how we can extend the local probability models (on single states) towards global probability models (on entire paths).

To answer this question, we first need to introduce the concepts of a process and a gamble process. A *process* $\mathscr{L}$ is a map defined on $\mathscr{X}^*$. A real process associates a real number $\mathscr{L}(s) \in \mathbb{R}$ with any situation $s \in \mathscr{X}^*$. A real process is called positive (non-negative) if it is positive (non-negative) in every situation. With any real process $\mathscr{L}$ we associate a sequence of n-measurable gambles $\{\mathscr{L}_n\}_{n\in\mathbb{N}_0}$: for all $n \in \mathbb{N}_0$, we let $\mathscr{L}_n(\omega) := \mathscr{L}(\omega^n)$ for all $\omega \in \Omega$ or, equivalently, $\mathscr{L}_n := \mathscr{L} \circ X_{1:n} = \mathscr{L}(X_{1:n})$. A *gamble process* $\mathscr{D}$ is a process that associates with any situation $x_{1:n} \in \mathscr{X}^*$ a local gamble $\mathscr{D}(x_{1:n}) \in \mathbb{G}(\mathscr{X})$. With any real process $\mathscr{L}$, we can associate a gamble process $\Delta\mathscr{L}$, called its *process difference*. For any situation $x_{1:n}$ the corresponding gamble $\Delta\mathscr{L}(x_{1:n}) \in \mathbb{G}(\mathscr{X})$ is defined by

$$\Delta\mathscr{L}(x_{1:n})(x_{n+1}) := \mathscr{L}(x_{1:n+1}) - \mathscr{L}(x_{1:n}) \text{ for all } x_{n+1} \in \mathscr{X}.$$

We will also use the extended real variables $\liminf \mathscr{L} \in \overline{\mathbb{V}}$ and $\limsup \mathscr{L} \in \overline{\mathbb{V}}$, defined by:

$$\liminf \mathscr{L}(\omega) := \liminf_{n\to+\infty} \mathscr{L}_n(\omega) \text{ and } \limsup \mathscr{L}(\omega) := \limsup_{n\to+\infty} \mathscr{L}_n(\omega)$$

for all $\omega \in \Omega$. If $\liminf \mathscr{L} = \limsup \mathscr{L}$, we denote their common value by $\lim \mathscr{L}$.

For a *given* imprecise probability tree, a *supermartingale* $\mathscr{M}$ is a *real* process for which the process difference $\Delta\mathscr{M}$ has a non-positive local upper expectation everywhere: $\overline{\mathrm{Q}}(\Delta\mathscr{M}(x_{1:n})|x_{1:n}) \leq 0$ for all $x_{1:n} \in \mathscr{X}^*$. In other words, a supermartingale is a process that, according to the local probability models, is expected to decrease. The concept originates in the following 'game-theoretic' argument. Suppose that a forecaster sets minimum selling prices for every gamble f on the next state X_{n+1}, i.e. he defines $\overline{\mathrm{Q}}(f|x_{1:n})$. Non-positive minimum selling prices imply that he is willing to give away these gambles. Suppose now that you take him up on his commitments. The gambles available to you are then exactly the ones with $\overline{\mathrm{Q}}(f|x_{1:n}) \leq 0$. Choosing such an available gamble in every situation $x_{1:n} \in \mathscr{X}^*$ essentially defines a supermartingale. In this way, we can interpret a supermartingale as a strategy for gambling against a forecaster.

We define supermartingales here as *real* processes, whereas Shafer and Vovk define them as extended real processes [6]. For any situation $s \in \mathscr{X}^*$, such an extended real process allows the possibility for $\Delta\mathscr{M}(s)$ to be an extended real function on $\mathscr{X}$. However, it is not immediately obvious to us how to give a behavioural meaning to such extended real process differences, and we therefore prefer to define supermartingales as real processes whose differences are gambles.

We denote the set of all supermartingales *for a given imprecise probability tree* by $\overline{\mathbb{M}}$. The set of all bounded below supermartingales is denoted by $\overline{\mathbb{M}}_b$.

We are now ready to introduce the game-theoretic upper expectation.

Definition 1. The upper expectation $\overline{\mathrm{E}}_V(\cdot|\cdot)$ is defined by

$$\overline{\mathrm{E}}_V(f|s) := \inf\left\{\mathscr{M}(s) \colon \mathscr{M} \in \overline{\mathbb{M}}_b \text{ and } (\forall \omega \in \Gamma(s)) \liminf \mathscr{M}(\omega) \geq f(\omega)\right\}, \quad (1)$$

for all extended real variables $f \in \overline{\mathbb{V}}$ and all $s \in \mathscr{X}^*$.

This definition can be interpreted in the following way: the upper expectation of a variable f when in a situation s, is the infimum starting capital in the situation s such that, by using the available gambles from s onwards, we are able to end up with a capital that dominates f, *no matter the path through s taken by the process*. Importantly, these upper expectations for global variables are defined in terms of supermartingales, and therefore, derived directly from the local models. Moreover, due to [5, Proposition 8.8], for every situation s, the restriction of $\overline{\mathrm{E}}_V(\cdot|s)$ to $\mathbb{G}(\Omega)$ satisfies the coherence axioms E1–E3.

Observe that in defining these global upper expectations, we consider supermartingales that are bounded below, because as is shown in [2, Example 1], for extended real variables, the use of unbounded supermartingales leads to undesirable results, whereas Definition 1 does not.

In the remainder of this contribution, we restrict our attention to upper expectations conditional on the initial situation $\square$ and use the notation $\overline{\mathrm{E}}_V(f) := \overline{\mathrm{E}}_V(f|\square)$. This facilitates the reading and makes the paper conceptually easier. That being said, we stress that all our arguments are easily extendible to upper expectations conditional on a general situation $s \in \mathscr{X}^*$.

4   Continuity with Respect to Upward Convergence

The relevance of continuity properties for (upper) expectation functionals is evident. Not only do they provide the mathematical theory with elegance, they also enhance its practical scope. The continuity of the Lebesgue integral, for instance, is one of the reasons why it is the integral of choice for computing expected values associated with a probability measure. Continuity properties provide constructive ways to calculate expectations that otherwise would be difficult or even impossible to calculate numerically. For example, calculating the upper expectation $\overline{\mathrm{E}}_V(f)$ of an extended real variable f directly is typically practically impossible if it depends on an infinite number of states. However, if we can find a sequence of simpler functions $\{f_n\}_{n \in \mathbb{N}_0}$ that converges in some way to f, such that the upper expectation $\overline{\mathrm{E}}_V$ is continuous with respect to this convergence, then we can easily approximate $\overline{\mathrm{E}}_V(f)$ by $\overline{\mathrm{E}}_V(f_n)$, provided n is large enough. If we can find a sequence for which moreover the individual $\overline{\mathrm{E}}_V(f_n)$ can be calculated directly, we obtain a practical method for calculating $\overline{\mathrm{E}}_V(f)$. Unfortunately, it appears little is known at present about the continuity properties of the functional $\overline{\mathrm{E}}_V$; we aim to remedy this situation here.

It is well-known that every coherent upper expectation $\overline{\mathrm{E}}$ is continuous with respect to uniform convergence [3, p. 63]: if a sequence of gambles $\{f_n\}_{n \in \mathbb{N}_0}$ converges uniformly to a gamble f, meaning that $\lim_{n \to +\infty} \sup\{|f - f_n|\} = 0$, then $\lim_{n \to +\infty} \overline{\mathrm{E}}(f_n) = \overline{\mathrm{E}}(f)$. Hence, since the restriction of $\overline{\mathrm{E}}_V$ to $\mathbb{G}(\Omega)$ is a coherent upper expectation, it is continuous with respect to the uniform convergence of gambles on Ω. This type of continuity is however fairly weak, because the condition of uniform convergence is a very strong one. Moreover, continuity with respect to pointwise convergence is not directly implied by mere coherence [3, p. 63]. The following example demonstrates that, also for the upper expectation operator $\overline{\mathrm{E}}_V$ we are focussing on here, continuity with respect to pointwise convergence, and downward convergence in particular, may fail.

Example 1. Consider, in each situation $x_{1:n} \in \mathscr{X}^*$, a completely vacuous model: $\overline{\mathrm{Q}}(h|x_{1:n}) = \max h$ for all local gambles $h \in \mathbb{G}(\mathscr{X})$ on the next state. Then it can be checked easily that $\overline{\mathrm{E}}_V(f) = \sup f$ for all $f \in \overline{\mathbb{V}}$. Now let $\mathscr{X} := \{0, 1\}$, and consider the decreasing sequence of events A_n, defined by $A_n := \{\omega \in \Omega : \omega_i = 1 \text{ for all } 1 \leq i \leq n\} \setminus \{(1, 1, 1, \dots)\}$. Then $\lim_{n \to +\infty} \mathbb{I}_{A_n} = 0$ pointwise. However, as $\overline{\mathrm{E}}_V(\mathbb{I}_{A_n}) = 1$ for all $n \in \mathbb{N}_0$, we have that $\lim_{n \to +\infty} \overline{\mathrm{E}}_V(\mathbb{I}_{A_n}) = 1$, whereas $\overline{\mathrm{E}}_V(\lim_{n \to +\infty} \mathbb{I}_{A_n}) = \overline{\mathrm{E}}_V(0) = 0$, so $\overline{\mathrm{E}}_V$ is not continuous with respect to downward pointwise convergence of gambles. $\Diamond$

This leads us to the conclusion that, in general, $\overline{\mathrm{E}}_V$ is not continuous with respect to downward—and therefore also pointwise—convergence. Nevertheless, using a version of Lévy's zero-one law [8], we can show that $\overline{\mathrm{E}}_V$ is continuous with respect to upward convergence of extended real variables that are uniformly bounded below, provided that the upper expectation of the limit variable f is finite.

Theorem 1 (Upward Convergence Theorem). *Consider any non-decreasing sequence of extended real variables $\{f_n\}_{n \in \mathbb{N}_0}$ that is uniformly bounded below—i.e. there is an $M \in \mathbb{R}$ such that $f_n \geq M$ for all $n \in \mathbb{N}_0$—and any extended real variable $f \in \overline{\mathbb{V}}$ such that $\lim_{n \to +\infty} f_n = f$ pointwise. If moreover $\overline{\mathrm{E}}_V(f) < +\infty$, then*

$$\overline{\mathrm{E}}_V(f) = \lim_{n \to +\infty} \overline{\mathrm{E}}_V(f_n).$$

The initial idea behind the proof is due to Shafer and Vovk, who proved continuity with respect to non-decreasing sequences of indicator gambles [1, Theorem 6.6]. We have adapted it here to our working with real supermartingales and moreover generalised it to extended real variables.

The following example illustrates the practical relevance of this theorem.

Example 2. In queuing theory or failure estimation, we are often interested in the time until some event happens and, in particular, in the lower and upper expectation of this time. As we will illustrate here, Theorem 1 provides a method to approximate such upper expectations. The lower expectations can also be approximated, using Theorem 3 further on; see Example 3.

Consider the simple case where $\mathscr{X} := \{0,1\}$. Suppose we are interested in the expected time until the first '1' appears. In other words, we are interested in the variable f that returns the number of initial successive '0's in a path:

$$f(\omega) := \inf \left\{ k \in \mathbb{N} \colon \omega_k = 1 \right\} \text{ for all } \omega \in \Omega,$$

where for $\omega = (0,0,0,...)$, $f(\omega) = \inf \emptyset := +\infty$. It is typically infeasible to calculate the upper expectation of this variable directly because it depends on entire paths. We can remedy this by considering instead, for every $n \in \mathbb{N}_0$, the gamble f_n, defined by

$$f_n(\omega) := \min \left\{ f(\omega), n \right\} \text{ for all } \omega \in \Omega.$$

For every $n \in \mathbb{N}_0$, f_n is clearly n-measurable: it only depends on the value of the first n states. Furthermore, $\{f_n\}_{n \in \mathbb{N}_0}$ is bounded below by zero, non-decreasing and converges pointwise to f. Provided that $\overline{\mathrm{E}}_\mathrm{V}(f) < +\infty$, Theorem 1 therefore implies that $\overline{\mathrm{E}}_\mathrm{V}(f) = \lim_{n \to +\infty} \overline{\mathrm{E}}_\mathrm{V}(f_n)$. This allows us to approximate $\overline{\mathrm{E}}_\mathrm{V}(f)$ by $\overline{\mathrm{E}}_\mathrm{V}(f_n)$, for n sufficiently large. Since the n-measurability of f_n will typically make the computation of $\overline{\mathrm{E}}_\mathrm{V}(f_n)$ feasible, we obtain a practical method for computing $\overline{\mathrm{E}}_\mathrm{V}(f)$. $\qquad\qquad\qquad\qquad\qquad\qquad\qquad\qquad\qquad\qquad\quad \Diamond$

As a direct consequence of our Upward Convergence Theorem, we also obtain the following inequality.

Theorem 2 (Fatou's Lemma). *Consider a sequence of extended real variables $\{f_n\}_{n \in \mathbb{N}_0}$ that is uniformly bounded below and let $f := \liminf_{n \to +\infty} f_n$. If $\overline{\mathrm{E}}_\mathrm{V}(f) < +\infty$, then*

$$\overline{\mathrm{E}}_\mathrm{V}(f) \leq \liminf_{n \to +\infty} \overline{\mathrm{E}}_\mathrm{V}(f_n).$$

This result is similar to Fatou's Lemma in measure theory; hence its name. It provides an upper bound on the upper expectation of an extended real variable f, in the form of a limit inferior of the upper expectations of any sequence of extended real variables $\{f_n\}_{n \in \mathbb{N}_0}$ that is uniformly bounded below and whose limit inferior $\liminf_{n \to +\infty} f_n$ is equal to f. Since this last condition is fairly weak, Theorem 2 has wide applicability. In general, the inequality in the statement cannot be reversed, because we do not generally have continuity with respect to pointwise convergence.

5 Continuity with Respect to Limits of Cuts

Historically, the framework of imprecise probabilities as described by Walley [9] has only considered gambles rather than unbounded or even extended real variables. One important reason for this is that they allow us to use less involved mathematics. Moreover, when considering unbounded or extended variables in practice, we are typically obliged to work with approximating gambles rather than the original variables. Considering these arguments, restricting the functional $\overline{E}_V$ to gambles would be very tempting, indeed. However, most practically relevant variables in the context of stochastic processes are in fact unbounded and even extended real-valued; consider for instance hitting or stopping times, as in Examples 2 and 3.

In Theorem 1 we have already shown how to approximate upper expectations for bounded below extended variables. The following theorem allows us to approximate upper expectations for general extended real variables by using sequences of non-increasing lower cuts.

Theorem 3. *Consider an extended real variable $f \in \overline{V}$ and, for every $A \in \mathbb{R}$, the variable f_A defined by $f_A(\omega) := \max\{f(\omega), A\}$ for all $\omega \in \Omega$. Then*

$$\lim_{A \to -\infty} \overline{E}_V(f_A) = \overline{E}_V(f).$$

Combining Theorems 1 and 3, we end up with the following result that allows us to move from upper expectations of gambles to upper expectations of general variables. It also fits within the framework of Troffaes and De Cooman [3, Part 2], which provides a general approach to extending coherent lower and upper expectations from gambles to real variables.

Theorem 4 (Continuity with respect to cuts). *Consider any extended real variable $f \in \overline{V}$ and, for any $A, B \in \mathbb{R}$ such that $B \geq A$, the gamble $f_{(A,B)}$, defined by*

$$f_{(A,B)}(\omega) := \begin{cases} B & \text{if } f(\omega) > B; \\ f(\omega) & \text{if } B \geq f(\omega) \geq A; \quad \text{for all } \omega \in \Omega. \\ A & \text{if } f(\omega) < A, \end{cases}$$

If $\overline{E}_V(f) < +\infty$, then

$$\lim_{A \to -\infty} \lim_{B \to +\infty} \overline{E}_V(f_{(A,B)}) = \overline{E}_V(f).$$

Example 3. Consider the same state space $\mathscr{X}$ and the same variables f and f_n as in Example 2. We have already shown there how to approximate the upper expectation $\overline{E}_V(f)$ of f by $\overline{E}_V(f_n)$. Now, we want to approximate the lower expectation $\underline{E}_V$—defined by $\underline{E}_V(g) := -\overline{E}_V(-g)$ for all $g \in \overline{V}$— of f. As $\{f_n\}_{n\in\mathbb{N}}$ is an increasing sequence of upper cuts of f, $\{-f_n\}_{n\in\mathbb{N}}$ is a decreasing sequence of lower cuts of $-f$. Hence, it follows from Theorem 3 that $\lim_{n \to +\infty} \overline{E}_V(-f_n) = \overline{E}_V(-f)$, and therefore, using conjugacy, $\lim_{n \to +\infty} -\underline{E}_V(f_n) = -\underline{E}_V(f)$, or, equivalently, $\lim_{n \to +\infty} \underline{E}_V(f_n) = \underline{E}_V(f)$. Hence, in the same way as was described in Example 2, we now also have a constructive method for approximating the lower expectation of the variable f. $\Diamond$

6   Conclusion

Among the continuity properties derived in this paper, the continuity with respect to cuts is the more remarkable, as it allows us to limit ourselves, for the larger part, to the study of $\overline{E}_V$ on gambles rather than the study of $\overline{E}_V$ on extended real variables. Although the functional $\overline{E}_V$ is not continuous with respect to general downward convergence, it is thus continuous for a particular way of downward convergence: sequences of non-increasing lower cuts. These results hold provided that the upper expectation $\overline{E}_V(f)$ of the limit variable f is finite. The case where $\overline{E}_V(f) = +\infty$ is largely left unexplored.

There is also an interesting connection between the game-theoretic functional $\overline{E}_V$ and the measure-theoretic Lebesgue integral when all local models are assumed precise. It was already pointed out by Shafer and Vovk [5, Chap. 8] that for indicator gambles $\mathbb{I}_A$ of events A in the σ-algebra created by the cylinder events, $\overline{E}_V(\mathbb{I}_A)$ is equal to the Lebesgue integral of $\mathbb{I}_A$ when the global measure is defined according to the Ionescu-Tulcea Theorem [7, p. 249]. Using our results, we aim to generalise the connection between both operators and study the extend to which they are equal. We leave this as future work.

Another topic of further research is the continuity of $\overline{E}_V$ with respect to the pointwise convergence of n-measurable gambles. We suspect that, in order to establish the behaviour of $\overline{E}_V$ with respect to this particular type of convergence, it will pay to investigate the potentially strong link with the concept of natural extension [9] and, as a consequence, the special status of $\overline{E}_V$ with respect to other extending functionals.

References

1. Augustin, T., Coolen, F., de Cooman, G., Troffaes, M.: Introduction to Imprecise Probabilities. Wiley, Chichester (2014)
2. De Cooman, G., De Bock, J., Lopatatzidis, S.: Imprecise stochastic processes in discrete time: global models, imprecise Markov chains, and Ergodic theorems. Int. J. Approx. Reason. **76**(C), 18–46 (2016)
3. De Cooman, G., Troffaes, M.: Lower Previsions. Wiley, Hoboken (2014)
4. Kolmogorov, A.: Grundbegriffe der Wahrscheinlichkeitsrechnung. Ergebnisse der Mathematik und ihrer Grenzgebiete. Springer, Heidelberg (1933)
5. Shafer, G., Vovk, V.: Probability and Finance. It's Only a Game!. Wiley, Hoboken (2005)
6. Shafer, G., Vovk, V., Takemura, A.: Lévy's zero-one law in game-theoretic probability (2011). arXiv:0905.0254 [math.PR]
7. Shiryaev, A., Wilson, S.: Probability. Graduate Texts in Mathematics. Springer, New York (1995)
8. T'Joens, N., de Cooman, G., De Bock, J.: Continuity of the Shafer-Vovk-Ville operator (2018). arXiv:1804.01980 [math.PR]
9. Walley, P.: Statistical Reasoning with Imprecise Probabilities. Chapman and Hall, London (1991)

Choquet Theorem for Random Sets in Polish Spaces and Beyond

Pedro Terán[⊠]

Departamento de Estadística e I.O. y D.M., Universidad de Oviedo, Gijón, Spain
teranpedro@uniovi.es

Abstract. A fundamental long-standing problem in the theory of random sets is concerned with the possible characterization of the distributions of random closed sets in Polish spaces via capacities. Such a characterization is known in the locally compact case (the Choquet theorem) in two equivalent forms: using the compact sets and the open sets as test sets. The general case has remained elusive. We solve the problem in the affirmative using open test sets.

1 The Problem

The Choquet theorem is a central result in the theory of random sets, allowing one to 'pack' all the information from a probability distribution (acting on sets of sets) in a simpler function, a capacity (acting on sets of points). That is similar to the way the cumulative distribution function contains the distributional information of a random variable. And, like in that case, the harder part is to find the essential properties characterizing those functions which can actually be 'unpacked' to recover a whole distribution.

The *hitting functional* of a random closed set X is given by

$$T_X(A) = P(X \cap A \neq \emptyset).$$

The random set is reconstructed from the information whether it hits (intersects) the sets A in a family of *test sets*. The standard presentation of the Choquet theorem assumes that the carrier space is a locally compact, second countable, Hausdorff space. Those spaces contain $\mathbb{R}^n$ and enjoy a number of its nice topological properties, e.g. they admit a separable complete metric (i.e. they are *Polish spaces*) and are σ-compact (in fact, hemicompact).

The Choquet theorem in locally compact, second countable, Hausdorff spaces was established in 1972 by Matheron [6,7] who provided a proof based on traditional measure extension tools after Choquet's pioneering work [3, Theorem 51.1] which was not explicitly concerned with the problem of characterizing the

---

Dedicated to the memory of my mother Ángeles Vicenta Agraz Viván, who passed away on March 21st, 2017. Research in this paper was partially funded by Asturias's *Consejería de Economía y Empleo* (FC-15-GRUPIN14-101) and by Spain's *Ministerio de Economía y Competitividad* (MTM2015–63971–P).

S. Destercke et al. (Eds.): SMPS 2018, AISC 832, pp. 208–215, 2019.
https://doi.org/10.1007/978-3-319-97547-4_27

distributions of random sets. In 1989, Norberg [12], by entirely different order-theoretical methods, extended it to locally compact, second countable, sober spaces. In 2014, a fourth method allowed us to give a Choquet theorem in locally compact, σ-compact, Hausdorff spaces [17].

But local compactness is a problematic requirement in probability theory. A rather more natural setting, as already established in the 1960s in books like Parthasarathy's [13] and Billingsley's [2], is that of general Polish spaces, to the point that a measurable space whose σ-algebra is isomorphic to the Borel σ-algebra of a Polish space is nowadays known as a *standard measurable space*.

Reflections on the need for a Choquet theorem in Polish spaces date back at least to Goodman *et al.* [5, Chap. 3, p. 93], who wrote

In this Chapter, we consider ... locally compact, Hausdorff and separable spaces. The reason is this. The foundations of random closed sets are based on Choquet theorem on such spaces. It should be noted that the natural domain of probability theory is Polish spaces ... These spaces might not be locally compact, for example, infinite dimensional Banach Spaces. Also ... in other applications, such as optimal control of distributed systems ... it is necessary to consider infinite dimensional topological spaces.

Since, it has explicitly been stated as an open problem in [8, Open Problem 2.29, p. 41], [10, Remark, p. 128], and [9, Open Problem 1.3.24, p. 69].

The theory of stochastic processes involves state or path spaces which are not locally compact. The advent of ever more complex forms of data, such as fuzzy and functional data, also draws attention to carrier spaces which fail to be locally compact or even Polish spaces. Confidence regions, depth functions, and statistics defined as solutions of optimization problems (like M-estimators) all lead naturally to random sets in those spaces.

In this communication, we will solve the problem in the affirmative by extending the Choquet theorem to metrizable Lusin spaces. That generality is sufficient to solve the open problem as stated as well as to additionally cover some examples of state, path and fuzzy set spaces which are not actually Polish but metrizable and Lusin.

2 Preliminaries

As mentioned in Sect. 1, a *Polish space* is a separable space whose topology is compatible with a complete metric. A topological space is called a *Lusin space* if it is the image of a Polish space by a continuous bijective mapping. In other words, its topology is weaker than some Polish topology in the same space (for example, the norm topology of a separable Banach space is Polish while its weak topology is Lusin).

Let $\mathbb{E}$ be a topological space. We will denote by $\mathcal{P}(\mathbb{E})$ the class of parts of $\mathbb{E}$, by $\mathcal{B}(\mathbb{E})$ its Borel σ-algebra (the σ-algebra generated by the open sets), by $\mathcal{F}(\mathbb{E})$ its non-empty closed sets, by $\mathcal{F}'(\mathbb{E})$ its closed sets, by $\mathcal{K}(\mathbb{E})$ its non-empty compact sets, by $\mathcal{K}'(\mathbb{E})$ its compact sets and by $\mathcal{G}(\mathbb{E})$ its open sets.

A subset of $\mathbb{E}$ is called *universally measurable* if it is in the completion of the Borel σ-algebra for every probability measure on $\mathcal{B}(\mathbb{E})$. This defines a larger σ-algebra $\mathcal{B}_u(\mathbb{E})$ called the *universal completion* of $\mathcal{B}(\mathbb{E})$.

A *capacity* in $\mathbb{E}$ is a set function $c : \mathcal{L} \to [0,1]$ on a lattice of sets $\mathcal{L} \subset \mathcal{P}(\mathbb{E})$ such that $\emptyset, \mathbb{E} \in \mathcal{L}$, $c(\emptyset) = 0$ and $c(\mathbb{E}) = 1$ hold, and moreover $c(A) \leq c(B)$ whenever $A \subset B$. It will be called:

- inner continuous, if $c(A_n) \to c(A)$ whenever $A_n \nearrow A$;
- outer continuous, if $c(A_n) \to c(A)$ whenever $A_n \searrow A$;
- completely alternating, if

$$c(\bigcap_{i=1}^{n} A_i) \leq \sum_{I \subset \{1,\ldots,n\}, I \neq \emptyset} (-1)^{|I|+1} c(\bigcup_{i \in I} A_i)$$

for any $n \in \mathbb{N}$;
- completely monotone, if

$$c(\bigcup_{i=1}^{n} A_i) \geq \sum_{I \subset \{1,\ldots,n\}, I \neq \emptyset} (-1)^{|I|+1} c(\bigcap_{i \in I} A_i)$$

for any $n \in \mathbb{N}$.

Often the definitions of complete alternation and monotony are expressed equivalently in terms of successive differences, see [8,10].

A *random closed set* on a probability space $(\Omega, \mathcal{A}, P)$ is a mapping $X : \Omega \to \mathcal{F}(\mathbb{E})$ such that $\{X \cap G \neq \emptyset\} \in \mathcal{A}$ for all $G \in \mathcal{G}(\mathbb{E})$. Equivalently, X is measurable when $\mathcal{F}(\mathbb{E})$ is endowed with the *Effros σ-algebra* $\mathcal{E}(\mathcal{F}(\mathbb{E}))$ generated by all subsets of the form $\{A \in \mathcal{F}(\mathbb{E}) \mid A \cap G \neq \emptyset\}$ where G ranges over $\mathcal{G}(\mathbb{E})$. The distribution P_X of X is the induced probability measure $P_X : \mathcal{E}(\mathcal{F}(\mathbb{E})) \to [0,1]$ given by $P_X(A) = P(X \in A)$.

If X satisfies the generally stronger requirement that $\{X \cap B \neq \emptyset\} \in \mathcal{A}$ for all $B \in \mathcal{B}(\mathbb{E})$, then it is called *strongly measurable*. In that case, also the sets $\{X \subset B\}$ are measurable, since they can easily be obtained using complementation as $\{X \subset B\} = \{X \cap B^c \neq \emptyset\}^c$.

3   Support Results

We collect here the theorems which will be used in the proof of the main result. The following is well known, see e.g. [16, Lemma 9.1.4].

Lemma 1. *Let $\mathbb{E}$ be a separable metrizable space. Then there exists a totally bounded metric which is compatible with the topology of $\mathbb{E}$.*

We will use an old observation of Shafer [15, p. 829].

Lemma 2. *Let $\mathcal{L}_1, \mathcal{L}_2$ be lattices of subsets of $\mathbb{E}_1, \mathbb{E}_2$ respectively, such that $\emptyset, \mathbb{E}_1 \in \mathcal{L}_1$ and $\emptyset, \mathbb{E}_2 \in \mathcal{L}_2$. If $c : \mathcal{L}_1 \to [0, 1]$ is a completely monotone capacity and $\varphi : \mathcal{L}_1 \to \mathcal{L}_2$ is an $\cap$-homomorphism, i.e. $\varphi(\emptyset) = \emptyset$, $\varphi(\mathbb{E}_1) = \mathbb{E}_2$ and $\varphi(A \cap B) = \varphi(A) \cap \varphi(B)$ for all $A, B \in \mathcal{L}_1$, then $c \circ \varphi : \mathcal{L}_2 \to [0, 1]$ is a completely monotone capacity.*

We also need an adaptation to containment functionals of the Choquet theorem.

Lemma 3. *Let $\mathbb{E}$ be a locally compact, second countable, Hausdorff space. Then, the formula*

$$P(X \subset F) = C(F) \quad \forall F \in \mathcal{F}'(\mathbb{E})$$

defines a bijection between the distributions P_X of random closed sets in $\mathbb{E}$ and the outer continuous, completely monotone capacities T on $\mathcal{F}'(\mathbb{E})$.

Proof. The Choquet theorem in the form given in, e.g., [10, pp. 122–123] identifies distributions of random closed sets X and inner continuous, completely alternating capacities T on $\mathcal{G}(\mathbb{E})$ by

$$P(X \cap G \neq \emptyset) = T(G).$$

It suffices to consider the dual capacity given by $C(A) = 1 - T(A^c)$ since

$$P(X \subset F) = 1 - P(X \cap F^c \neq \emptyset) = 1 - T(F^c).$$

This transformation maps inner continuous, completely alternating capacities to their dual capacities which are outer continuous and completely monotone instead. □

One can also retrieve Lemma 3 as a particular case of [17, Theorem 3.2].

Recall the *myopic topology* of $\mathcal{K}(\mathbb{E})$ is generated by the sets $\{A \in \mathcal{K}(\mathbb{E}) \mid A \cap G \neq \emptyset\}$ and $\{A \in \mathcal{K}(\mathbb{E}) \mid A \subset G\}$ for all $G \in \mathcal{G}(\mathbb{E})$.

Lemma 4. *Let $\mathbb{E}$ be a compact metric space. Then $\mathcal{E}(\mathcal{F}(\mathbb{E}))$ is the Borel σ-algebra generated by the myopic topology.*

Proof. Combine [8, Theorem 2.7.(iii), p. 29] with [8, Theorem C.5.(iii), p. 403]. □

Our interest in the myopic topology is due to the following theorem.

Lemma 5. *Let $\mathbb{E}$ be a Polish space. Let P be a probability measure on $(\mathcal{K}(\mathbb{E}), \mathcal{B}(\mathcal{K}(\mathbb{E})))$ (with the Borel σ-algebra of the myopic topology). Then, for each $B \in \mathcal{B}(\mathbb{E})$, the set $\{K \in \mathcal{K}(\mathbb{E}) \mid K \subset B\}$ is universally measurable. Also, letting P_u be the natural extension of P to $\mathcal{B}_u(\mathcal{K}(\mathbb{E}))$, the identity*

$$c(B) = P_u(\{K \in \mathcal{K}(\mathbb{E}) \mid K \subset B\})$$

defines an outer continuous, completely monotone capacity $c : \mathcal{B}(\mathbb{E}) \to [0, 1]$ such that

$$c(B) = \sup_{\substack{K \in \mathcal{K}(\mathbb{E}) \\ K \subset B}} c(K) = \inf_{\substack{G \in \mathcal{G}(\mathbb{E}) \\ B \subset G}} c(G). \tag{1}$$

Moreover, the restriction of c to $\mathcal{G}(\mathbb{E})$ is inner continuous.

Proof. The result is a combination of material from [14], which in turn relies heavily on [3]. Universal measurability is [14, Lemma 1], complete monotony and outer continuity follow from [14, Theorem 2], and (1) from [14, Theorem 1]. Inner continuity on open sets follows then as observed in [14, Proposition 1]. □

Finally, we will also use the following result of Frolík [4, Proposition 7.11].

Lemma 6. *Let* $\mathbb{E}$ *be a Lusin space. If* $\mathbb{F}$ *is a metric space and* $e : \mathbb{E} \to \mathbb{F}$ *embeds* $\mathbb{E}$ *homeomorphically into* $\mathbb{F}$, *then* $e(\mathbb{E}) \in \mathcal{B}(\mathbb{F})$.

4   Main Result

In this section, we state and prove the Choquet theorem.

Theorem 7. *Let* $\mathbb{E}$ *be a metrizable Lusin space. Then, the formula*

$$P(X \cap G \neq \varnothing) = T(G) \quad \forall G \in \mathcal{G}(\mathbb{E})$$

establishes a bijection between the distributions P_X *of random closed sets in* $\mathbb{E}$ *and the inner continuous, completely alternating capacities* T *on* $\mathcal{G}(\mathbb{E})$.

Proof. Showing that the hitting functional of a random closed set is an inner continuous, completely alternating capacity involves basic properties of probabilities and is standard. The fact that it is a capacity is clear. Inner continuity is a consequence of the identity

$$\{X \cap \bigcup_n G_n \neq \varnothing\} = \bigcup_n \{X \cap G_n \neq \varnothing\}$$

and the continuity of probability measures for monotone sequences. Complete alternation is obtained by rewriting it as the statement that the probability of certain events is non-negative (see e.g. [10, p. 116] for details).

For the converse, let $T : \mathcal{G}(\mathbb{E}) \to [0,1]$ be an inner continuous, completely alternating capacity, and define the dual capacity $C : \mathcal{F}'(\mathbb{E}) \to [0,1]$ by $C(F) = 1 - T(F^c)$. It is outer continuous and completely monotone, instead of inner continuous and completely alternating.

Being the continuous image of a separable space, $\mathbb{E}$ is separable. By Lemma 1, it admits a totally bounded metric. The completion $\overline{\mathbb{E}}$ of $\mathbb{E}$ with that metric, being both totally bounded and complete, is a compact metric space.

For the sake of greater clarity, subsets of $\overline{\mathbb{E}}$ will be written in boldface and the complement of $\mathbf{A} \subset \overline{\mathbb{E}}$ will be denoted by $\overline{\mathbb{E}} \backslash \mathbf{A}$.

The natural embedding $e : \mathbb{E} \to \overline{\mathbb{E}}$ identifies homeomorphically $\mathbb{E}$ with $e(\mathbb{E})$. Let $e^\leftarrow : \mathcal{P}(\overline{\mathbb{E}}) \to \mathcal{P}(\mathbb{E})$ be the pre-image mapping given by

$$e^\leftarrow(\mathbf{A}) = \{x \in \mathbb{E} \mid e(x) \in \mathbf{A}\}.$$

By the continuity of e, we have $e^\leftarrow(\mathbf{F}) \in \mathcal{F}'(\mathbb{E})$ for each $\mathbf{F} \in \mathcal{F}'(\overline{\mathbb{E}})$.

Let $\overline{C} : \mathcal{F}'(\overline{\mathbb{E}}) \rightarrow [0,1]$ be given by $\overline{C}(\mathbf{F}) = C(e^{\leftarrow}(\mathbf{F}))$ for any $\mathbf{F} \in \mathcal{F}'(\overline{\mathbb{E}})$. Clearly $\overline{C} = C \circ e^{\leftarrow}$ is outer continuous, and it is a completely monotone capacity by Lemma 2 since $e^{\leftarrow}$ is an $\cap$-homomorphism. Indeed, $e^{\leftarrow}(\emptyset) = \emptyset$, $e^{\leftarrow}(\mathcal{F}'(\overline{\mathbb{E}})) = \mathcal{F}'(\mathbb{E})$ (because each $F \in \mathcal{F}'(\mathbb{E})$ is $e^{\leftarrow}(cl_{\overline{\mathbb{E}}}F)$) and

$$e^{\leftarrow}(\mathbf{F}_1 \cap \mathbf{F}_2) = \{x \in \mathbb{E} \mid e(x) \in \mathbf{F}_1 \cap \mathbf{F}_2\} = e^{\leftarrow}(\mathbf{F}_1) \cap e^{\leftarrow}(\mathbf{F}_2)$$

for all $\mathbf{F}_1, \mathbf{F}_2 \in \mathcal{F}'(\overline{\mathbb{E}})$.

The proof will proceed now by subsequently defining mappings X''', X'', X' and X, of which X will be the random closed set we need.

By Lemma 3, there is a random closed set X''' in $\overline{\mathbb{E}}$, defined on a measurable space endowed with a probability measure Q, such that

$$Q(X''' \subset \mathbf{F}) = \overline{C}(\mathbf{F}) \quad \forall \mathbf{F} \in \mathcal{F}(\overline{\mathbb{E}}).$$

It induces the distribution $Q_{X'''}$ on the measurable space $(\mathcal{F}(\overline{\mathbb{E}}), \mathcal{E}(\mathcal{F}(\overline{\mathbb{E}})))$.

Endow $\mathcal{F}(\overline{\mathbb{E}})$ with the myopic topology (recall compact sets and closed sets coincide in $\overline{\mathbb{E}}$). By Lemma 4, $\mathcal{E}(\mathcal{F}(\overline{\mathbb{E}})) = \mathcal{B}(\mathcal{F}(\overline{\mathbb{E}}))$ so the Effros σ-algebra admits a universal completion $\mathcal{B}_u(\mathcal{F}(\overline{\mathbb{E}}))$. Let $(Q_{X'''})_u$ be the natural extension of $Q_{X'''}$ to $\mathcal{B}_u(\mathcal{F}(\overline{\mathbb{E}}))$ (note we are extending $Q_{X'''}$ by adding to $\mathcal{B}(\mathcal{F}(\overline{\mathbb{E}}))$ the universally null, thus $Q_{X'''}$-null, sets). Consider the identity mapping

$$X'' = \mathrm{id} : (\mathcal{F}(\overline{\mathbb{E}}), \mathcal{B}_u(\mathcal{F}(\overline{\mathbb{E}})), (Q_{X'''})_u) \rightarrow \mathcal{F}(\overline{\mathbb{E}}),$$

which is obviously a random closed set since $\mathcal{E}(\mathcal{F}(\overline{\mathbb{E}})) \subset \mathcal{B}_u(\mathcal{F}(\overline{\mathbb{E}}))$.

By Lemma 5, $\{X'' \subset \mathbf{B}\}$ is a measurable event for each $\mathbf{B} \in \mathcal{B}(\overline{\mathbb{E}})$, and there exists an outer continuous, inner continuous on $\mathcal{G}(\mathbb{E})$, completely monotone capacity $\hat{C} : \mathcal{B}(\overline{\mathbb{E}}) \rightarrow [0,1]$, such that

$$(Q_{X'''})_u(X'' \subset \mathbf{B}) = \hat{C}(\mathbf{B}) = \sup_{\substack{\mathbf{K} \in \mathcal{K}(\overline{\mathbb{E}}) \\ \mathbf{K} \subset \mathbf{B}}} \hat{C}(\mathbf{K})$$

$$= \sup_{\substack{\mathbf{F} \in \mathcal{F}(\overline{\mathbb{E}}) \\ \mathbf{F} \subset \mathbf{B}}} \hat{C}(\mathbf{F}) = \inf_{\substack{\mathbf{G} \in \mathcal{G}(\overline{\mathbb{E}}) \\ \mathbf{B} \subset \mathbf{G}}} \hat{C}(\mathbf{G})$$

for each $\mathbf{B} \in \mathcal{B}(\overline{\mathbb{E}})$.

We still have, for each $\mathbf{F} \in \mathcal{F}(\overline{\mathbb{E}})$,

$$\hat{C}(\mathbf{F}) = (Q_{X'''})_u(X'' \subset \mathbf{F}) = (Q_{X'''})_u(\{\mathrm{id} \cap \mathbf{F}^c \neq \emptyset\}^c)$$
$$= Q_{X'''}(\{\mathrm{id} \cap \mathbf{F}^c \neq \emptyset\}^c) = Q_{X'''}(\mathrm{id} \subset \mathbf{F}) = Q(X''' \subset \mathbf{F})$$
$$= \overline{C}(\mathbf{F}) = C(e^{\leftarrow}(\mathbf{F})).$$

Now define

$$X' = e^{\leftarrow}|_{\mathcal{F}(\overline{\mathbb{E}})} : (\mathcal{F}(\overline{\mathbb{E}}), \mathcal{B}_u(\mathcal{F}(\overline{\mathbb{E}})), (Q_{X'''})_u) \rightarrow \mathcal{F}'(\mathbb{E}).$$

Let us use Y to prove that X' is strongly measurable and $(Q_{X'''})_u(X' \subset F) = C(F)$ for each $F \in \mathcal{F}(\mathbb{E})$. Let $B \in \mathcal{B}(\mathbb{E})$. Then

$$\{X' \subset B\} = \{\mathbf{F} \in \mathcal{F}(\overline{\mathbb{E}}) \mid e^{\leftarrow}(\mathbf{F}) \subset B\} = \{X'' \subset \overline{\mathbb{E}} \backslash e(B^c)\}$$

(since e is injective, $e^{\leftarrow}(\mathbf{F}) \subset B$ if and only if $\mathbf{F}$ is disjoint from $e(B^c)$ i.e. a subset of $\overline{\mathbb{E}} \backslash e(B^c)$).

Now $e(B^c) \in \mathcal{B}(e(\mathbb{E}))$ because e is a homeomorphism and $B^c \in \mathcal{B}(\mathbb{E})$. By Lemma 6, $e(\mathbb{E}) \in \mathcal{B}(\overline{\mathbb{E}})$. We deduce $e(B^c) \in \mathcal{B}(\overline{\mathbb{E}})$, also $\overline{\mathbb{E}} \backslash e(B^c) \in \mathcal{B}(\overline{\mathbb{E}})$ and thus $\{X'' \subset \overline{\mathbb{E}} \backslash e(B^c)\} \in \mathcal{B}_u(\mathcal{F}(\overline{\mathbb{E}}))$. By the arbitrariness of B, the mapping X' is strongly measurable.

Fix an arbitrary $F \in \mathcal{F}(\mathbb{E})$. Then

$$(Q_{X'''})_u(X' \subset F) = (Q_{X'''})_u(X'' \subset \overline{\mathbb{E}} \backslash e(F^c)) = \sup_{\substack{\mathbf{F} \in \mathcal{F}(\overline{\mathbb{E}}) \\ \mathbf{F} \subset \overline{\mathbb{E}} \backslash e(F^c)}} C(e^{\leftarrow}(\mathbf{F})).$$

Since the quantity in the supremum depends on $\mathbf{F}$ only through $e^{\leftarrow}(\mathbf{F})$, we have

$$\sup_{\substack{\mathbf{F} \in \mathcal{F}(\overline{\mathbb{E}}) \\ \mathbf{F} \subset \overline{\mathbb{E}} \backslash e(F^c)}} C(e^{\leftarrow}(\mathbf{F})) = \sup_{\substack{\mathbf{F} \in \mathcal{F}(\overline{\mathbb{E}}) \\ e(e^{\leftarrow}(\mathbf{F})) \subset \overline{\mathbb{E}} \backslash e(F^c)}} C(e^{\leftarrow}(\mathbf{F})) = \sup_{\substack{\mathbf{F} \in \mathcal{F}(\overline{\mathbb{E}}) \\ e^{\leftarrow}(\mathbf{F}) \subset F}} C(e^{\leftarrow}(\mathbf{F})).$$

Since $\{\mathbf{F} \cap e(\mathbb{E})\}_{\mathbf{F} \in \mathcal{F}(\overline{\mathbb{E}})} = \mathcal{F}(e(\mathbb{E}))$ and e is an homeomorphism onto its image, $\{e^{\leftarrow}(\mathbf{F})\}_{\mathbf{F} \in \mathcal{F}(\overline{\mathbb{E}})} = \mathcal{F}(\mathbb{E})$ whence there is some $\mathbf{F} \in \mathcal{F}(\overline{\mathbb{E}})$ for which $e^{\leftarrow}(\mathbf{F})$ is exactly F (precisely, we can take $\mathbf{F}$ to be the closure in $\overline{\mathbb{E}}$ of $e(F)$). Since C is monotone, the supremum must be attained at that $\mathbf{F}$. Therefore

$$\sup_{\substack{\mathbf{F} \in \mathcal{F}(\overline{\mathbb{E}}) \\ e^{\leftarrow}(\mathbf{F}) \subset F}} C(e^{\leftarrow}(\mathbf{F})) = C(F).$$

Through that chain of identities we have proved $(Q_{X'''})_u(X' \subset F) = C(F)$ for an arbitrary non-empty closed $F \subset \mathbb{E}$, or equivalently

$$(Q_{X'''})_u(X' \cap G \neq \emptyset) = 1 - (Q_{X'''})_u(X' \subset G^c)$$
$$= 1 - C(G^c) = 1 - (1 - T((G^c)^c)) = T(G)$$

for all $G \in \mathcal{G}(\mathbb{E})$.

Unfortunately X' may take on empty values, so we are still not done. Since

$$\{X' \neq \emptyset\} = \{\mathbf{F} \in \mathcal{F}(\overline{\mathbb{E}}) \mid \mathbf{F} \cap e(\mathbb{E}) \neq \emptyset\}$$
$$= \{X'' \cap e(\mathbb{E}) \neq \emptyset\} = \{X'' \subset \overline{\mathbb{E}} \backslash e(\mathbb{E})\}^c,$$

the facts that X'' is strongly measurable and $\overline{\mathbb{E}} \backslash e(\mathbb{E}) \in \mathcal{B}(\overline{\mathbb{E}})$ (remember Lemma 6) imply $\{X' \neq \emptyset\} \in \mathcal{B}_u(\mathcal{F}(\overline{\mathbb{E}}))$. We can therefore take the trace measure space $(\Omega, \mathcal{A}, P)$ with the sample space $\Omega = \{X' \neq \emptyset\}$, the σ-algebra $\mathcal{A} = \{A \cap \Omega \mid A \in \mathcal{B}_u(\mathcal{F}(\overline{\mathbb{E}}))\}$ and the measure $P = (Q_{X'''})_u|_{\mathcal{A}}$. Indeed P is a probability measure since

$$P(\Omega) = (Q_{X'''})_u(X' \neq \emptyset) = (Q_{X'''})_u(X' \cap \mathbb{E} \neq \emptyset) = T(\mathbb{E}) = 1.$$

And still

$$P(X \cap G \neq \emptyset) = (Q_{X'''})_u(X' \cap G \neq \emptyset) = T(G)$$

for all $G \in \mathcal{G}(\mathbb{E})$, whence the proof is complete. $\qquad \square$

5   Discussion

Since every Polish space is metrizable and Lusin, Theorem 7 solves the open problem discussed in Sect. 1.

In [11], Nguyen and Nguyen have presented a 'negative version' of the Choquet theorem in Polish spaces. They present a completely alternating capacity on open sets of the Polish (separable Banach) space ℓ_2 which satisfies a limited variant of inner continuity but does not correspond to any random closed set in ℓ_2.

That is in fact compatible with Theorem 7, since their capacity only satisfies $G_n \nearrow G \Rightarrow T(G_n) \to T(G)$ under the additional assumption that $G_n \to G$ in the Hausdorff pseudometric (Nguyen and Nguyen emphasize that their result is *not* a counterexample to the Choquet theorem in Polish spaces with open test sets).

An example of a metric space which is Lusin but not Polish, relevant in the theory of fuzzy sets, is the levelwise L^p-type metric d_p in the space of fuzzy numbers [1]. Therefore, an M-estimator in that space is an example of a random closed set covered by our version of the Choquet theorem.

References

1. Alonso de la Fuente, M.: Variables aleatorias difusas y teorema de Skorokhod. Trabajo fin de grado, Univ. Oviedo, Oviedo (2017)
2. Billingsley, P.: Convergence of Probability Measures. Wiley, New York (1968)
3. Choquet, G.: Theory of capacities. Ann. Inst. Fourier (Grenoble) **5**, 131–295 (1954)
4. Frolík, Z.: A survey of separable descriptive theory of sets and spaces. Czech. Math. J. **20**, 406–467 (1970)
5. Goodman, I.R., Mahler, R.P.S., Nguyen, H.T.: Mathematics of Data Fusion. Kluwer, Norwell (1997)
6. Matheron, G.: Random sets theory and its applications to stereology. J. Microsc. **95**, 15–23 (1972)
7. Matheron, G.: Random Sets and Integral Geometry. Wiley, New York (1975)
8. Molchanov, I.: Theory of Random Sets. Springer, London (2005)
9. Molchanov, I.: Theory of Random Sets, 2nd edn. Springer, London (2018)
10. Nguyen, H.T.: An Introduction to Random Sets. Chapman & Hall/CRC, Boca Raton (2006)
11. Nguyen, H.T., Nguyen, T.N.: A negative version of Choquet theorem for Polish spaces. East-West J. Math. **1**, 61–71 (1998)
12. Norberg, T.: Existence theorems for measures on continuous posets, with applications to random set theory. Math. Scand. **64**, 15–51 (1989)
13. Parthasarathy, K.R.: Probability Measures on Metric Spaces. Academic Press, New York (1967)
14. Philippe, F., Debs, G., Jaffray, J.Y.: Decision making with monotone lower probabilities of infinite order. Math. Oper. Res. **24**, 767–784 (1999)
15. Shafer, G.: Allocations of probability. Ann. Probab. **7**, 827–839 (1979)
16. Stroock, D.W.: Probability theory: an analytic view, 2nd edn. Cambridge University Press, Cambridge (2011)
17. Terán, P.: Distributions of random closed sets via containment functionals. J. Nonlinear Convex Anal. **15**, 907–917 (2014)

Generalising the Pari-Mutuel Model

Chiara Corsato, Renato Pelessoni, and Paolo Vicig[✉]

University of Trieste (DEAMS), 34100 Trieste, Italy
ccorsato@units.it, {renato.pelessoni,paolo.vicig}@deams.units.it

Abstract. We introduce two models for imprecise probabilities which generalise the Pari-Mutuel Model while retaining its simple structure. Their consistency properties are investigated, as well as their capability of formalising an assessor's different attitudes. It turns out that one model is always coherent, while the other is (occasionally coherent but) generally only 2-coherent, and may elicit a conflicting attitude towards risk.

1 Introduction

The term imprecise probability incorporates a large variety of uncertainty models. While being well suited for assessing imprecise, uncertain or vague beliefs, general models, like coherent lower probabilities, may be less manageable for other purposes, like inference or merely checking their coherence. Some special models are nimbler with respect to these issues. In particular, the Pari-Mutuel Model (PMM) is assessed once a reference precise probability P_0 and a parameter are given, and is guaranteed to be coherent [W, PVZ, MMD]. $P_0(A)$ may also be interpreted as the 'true' probability of event A in the assessor's or agent's mind, but unlike its derived PMM may not correspond, in a behavioural or betting scheme, to the agent's selling or buying prices for A.

Our purpose in this paper is to explore further models that generalise the PMM while retaining its simple features. We also focus on what sort of beliefs they can express. After recalling some preliminary matters in Sect. 2, we lay down the general framework for the new models, i.e., the family of what we term Nearly-Linear (NL) models, in Sect. 3. Then, notable instances of NL models, the Vertical Barrier PMM (VB-PMM) and the Horizontal Barrier PMM (HB-PMM), are investigated in Sects. 4 and 5, respectively. The VB-PMM is coherent (even 2-monotone as a lower probability) and may express, so to say, an agent's greedier attitude than the PMM. The HB-PMM is always at least 2-coherent, but may be coherent subject to certain (restrictive) conditions. Its coherence is characterised in a finite setting: for upper probabilities, it is equivalent to subadditivity. Behaviourally, the HB-PMM elicits an agent's conflicting (and partly irrational) beliefs towards risk. Section 6 contains some comparisons with related models in the literature, while Sect. 7 concludes the paper. Due to space limitations, proofs of the results are omitted.

© Springer Nature Switzerland AG 2019
S. Destercke et al. (Eds.): SMPS 2018, AISC 832, pp. 216–223, 2019.
https://doi.org/10.1007/978-3-319-97547-4_28

2   Preliminaries

Let $\underline{P}$ $(\overline{P})$ be a lower (upper) probability, i.e., a map from a set $\mathcal{D}$ of events into $\mathbb{R}$.

$\underline{P}$ is *coherent* on $\mathcal{D}$ iff, $\forall n \in \mathbb{N}, \forall A_0, \ldots, A_n \in \mathcal{D}, \forall s_0, \ldots, s_n \geq 0$, defining $\underline{G} = \sum_{i=1}^{n} s_i(I_{A_i} - \underline{P}(A_i)) - s_0(I_{A_0} - \underline{P}(A_0))$, it holds that $\max \underline{G} \geq 0$ [W].

$\underline{P}$ is *2-coherent* on $\mathcal{D}$ if either $n = 2$ and $s_0 = 0$ or $n \in \{0,1\}$ in the above definition [W,PV].

In a behavioural interpretation, $\underline{P}(A)$ $(\overline{P}(A))$ is an agent's supremum buying price (infimum selling price) for A or its indicator I_A [W]. In the gain $\underline{G}$ above, $I_{A_i} - \underline{P}(A_i)$ is the agent's elementary gain from exchanging event A_i at the price $\underline{P}(A_i)$; coherence and 2-coherence require $\max \underline{G} \geq 0$, i.e., that a finite linear combination of bets on events in $\mathcal{D}$ with certain constraints on the coefficients does not produce a sure loss.

In this paper, $\mathcal{D}$ will be the set $\mathcal{A}(\mathbb{P})$ of events logically dependent on a given partition $\mathbb{P}$ (the powerset of $\mathbb{P}$). Given $\underline{P}$ and $\overline{P}$ on $\mathcal{A}(\mathbb{P})$, they are *conjugate* if $\underline{P}(A) = 1 - \overline{P}(\neg A), \forall A \in \mathcal{A}(\mathbb{P})$.

$\overline{P}$ is *coherent*, alternatively *2-coherent* on $\mathcal{A}(\mathbb{P})$, if its conjugate $\underline{P}$ is.

It is necessary for coherence of $\overline{P}, \underline{P}$ that [W, Sect. 2.7.4]:

(c1) $\overline{P}(A) + \overline{P}(B) \geq \overline{P}(A \vee B)$ (*subadditivity*),
(c2) if $A \wedge B = \emptyset,$ $\underline{P}(A) + \underline{P}(B) \leq \underline{P}(A \vee B)$ (*superadditivity*).

Definition 1. $\underline{P}_{\text{PMM}}$: $\mathcal{A}(\mathbb{P}) \rightarrow \mathbb{R}$ *is a* Pari-Mutuel lower probability *if* $\underline{P}_{\text{PMM}}(A) = \max\{(1 + \delta)P_0(A) - \delta, 0\}, \forall A \in \mathcal{A}(\mathbb{P})$, *where* P_0 *is a given probability and* $\delta \in \mathbb{R}^+$. *Its conjugate upper probability is* $\overline{P}_{\text{PMM}}(A) = \min\{(1 + \delta)P_0(A), 1\}$. $(\underline{P}_{\text{PMM}}, \overline{P}_{\text{PMM}})$ *constitute a* Pari-Mutuel Model *(PMM)*.

$\underline{P}_{\text{PMM}}$ and $\overline{P}_{\text{PMM}}$ are coherent. $\underline{P}_{\text{PMM}}$ is also *2-monotone*: $\forall A, B \in \mathcal{A}(\mathbb{P})$, $\underline{P}_{\text{PMM}}(A \vee B) + \underline{P}_{\text{PMM}}(A \wedge B) \geq \underline{P}_{\text{PMM}}(A) + \underline{P}_{\text{PMM}}(B)$ (while $\overline{P}_{\text{PMM}}$ is 2-alternating) [W,PVZ].

2-coherence is a weaker consistency requirement than coherence (cf. [PV] for details). On $\mathcal{A}(\mathbb{P})$, a still weaker condition is that μ ($\mu = \underline{P}$ or $\mu = \overline{P}$) is a *capacity*: it is requested only that $\forall A, B \in \mathcal{A}(\mathbb{P}) : A \Rightarrow B$, it is $\mu(A) \leq \mu(B)$ (*monotonicity*), and that $\mu(\emptyset) = 0, \mu(\Omega) = 1$ (*normalisation*).

3   Nearly-Linear Imprecise Probability Models

The Pari-Mutuel Model and the models we shall investigate in the next sections belong to the broader family of *Nearly-Linear Models*, which we define next.

Let for this $\mu : \mathcal{A}(\mathbb{P}) \rightarrow \mathbb{R}$ be either a lower or an upper probability.

Definition 2. $\mu : \mathcal{A}(\mathbb{P}) \rightarrow \mathbb{R}$ *is a* Nearly-Linear (NL) *imprecise probability iff* $\mu(\emptyset) = 0, \mu(\Omega) = 1$ *and, given a probability* P_0 *on* $\mathcal{A}(\mathbb{P})$, $a \in \mathbb{R}$, $b > 0$, $\forall A \in \mathcal{A}(\mathbb{P}) \setminus \{\emptyset, \Omega\}$,

$$\mu(A) \overset{\text{def}}{=} \min\{\max\{bP_0(A) + a, 0\}, 1\} = \max\{\min\{bP_0(A) + a, 1\}, 0\}. \quad (1)$$

Lemma 1. *A NL μ is a capacity.*

If μ is given by Definition 2, we shall say shortly that μ is NL(a, b).

An interesting feature of NL models is that they are *self-conjugate*: if μ is NL(a, b), also its conjugate $\mu^c(A) = 1 - \mu(\neg A)$, $\forall A \in \mathcal{A}(\mathbb{P})$, is NL$(a', b')$:

Proposition 1. *If μ is NL(a, b), then μ^c is NL(a', b'), with*

$$a' = 1 - (a + b), \quad b' = b. \tag{2}$$

Example 1. In the PMM, $\overline{P}_{\text{PMM}}$ is NL$(0, 1 + \delta)$, $\underline{P}_{\text{PMM}}$ is NL$(-\delta, 1 + \delta)$, hence here $a = -\delta < 0$, $b = 1 + \delta > 1$, $a + b = 1$.

A NL model typically gives extreme evaluations to a number of events whose probability P_0 is strictly between 0 and 1. We may keep track of this defining:

Definition 3. *Given μ, NL(a, b), define:*

- $\mathcal{N} = \{A \in \mathcal{A}(\mathbb{P}) : \mu(A) = 0\} = \{A \in \mathcal{A}(\mathbb{P}) : P_0(A) \leq -\frac{a}{b}\} \cup \{\emptyset\}$,
- $\mathcal{U} = \{A \in \mathcal{A}(\mathbb{P}) : \mu(A) = 1\} = \{A \in \mathcal{A}(\mathbb{P}) : P_0(A) \geq \frac{1-a}{b}\} \cup \{\Omega\}$,
- $\mathcal{E} = \mathcal{A}(\mathbb{P}) \setminus (\mathcal{N} \cup \mathcal{U}) = \{A \in \mathcal{A}(\mathbb{P}) \setminus \{\emptyset, \Omega\} : -\frac{a}{b} < P_0(A) < \frac{1-a}{b}\}$.

$\mathcal{N}$ is the set of *null events* according to μ, $\mathcal{U}$ the set of *universal events*.

If a generic NL measure μ includes a known model μ^* as a special case, μ is interpreted as either a lower or upper probability if μ^* is so. In general, we may apply the *maximum consistency principle*: μ is a lower probability if it determines a model with a higher degree of consistency than interpreting μ as an upper probability.

In the next two sections, we analyse the two major NL submodels.[1] They relax the PMM condition $a + b = 1$ (cf. Example 1) to, respectively, $a + b \leq 1$ and $a + b \geq 1$, while both keeping $a \leq 0$.

4   The Vertical Barrier Pari-Mutuel Model

To introduce our first model, let μ be a NL(a, b) measure such that

$$0 < a + b \leq 1, \quad a \leq 0. \tag{3}$$

Then, $\forall A \in \mathcal{A}(\mathbb{P}) \setminus \{\Omega\}$, $bP_0(A) + a \leq a + b \leq 1$, and μ in (1) simplifies to $\mu(A) = \max\{bP_0(A) + a, 0\}$. ($\mu(\emptyset)$ is also computed with this formula.)

Note that, when $a + b \leq 0$, μ reduces to the vacuous lower probability $\underline{P}_V(A) = 0$, $\forall A \in \mathcal{A}(\mathbb{P}) \setminus \{\Omega\}$. Hence the constraint $a + b > 0$ rules out (only) this case.

Recalling Example 1, when $a + b = 1$ and $a = -\delta < 0$, μ is the lower probability of a PMM (Definition 1).[2]

---

[1] It can be shown that a third, less relevant, submodel completes the family of NL models.

[2] When $b = 1$, $a = 0$, μ is a probability. We shall hereafter neglect this subcase.

Putting $a = 0$, $b = \varepsilon < 1$, we obtain the lower probability of the ε-*contamination model* (also termed linear-vacuous mixture in [W]):

$$\underline{P}(A) = \varepsilon P_0(A), \quad A \in \mathcal{A}(\mathbb{P}) \setminus \{\Omega\} \quad (\underline{P}(\Omega) = 1).$$

Clearly, requiring conditions (3) μ is a lower probability. Its conjugate upper probability is easily obtained using (2). Summing up, we define

Definition 4. *A* Vertical Barrier Pari-Mutuel Model (VB-PMM) *is a NL model where $\underline{P}$ and its conjugate $\overline{P}$ are given by:*

$$\underline{P}(A) = \max\{bP_0(A) + a, 0\}, \quad \forall A \in \mathcal{A}(\mathbb{P}) \setminus \{\Omega\} \quad (\underline{P}(\Omega) = 1), \quad (4)$$

$$\overline{P}(A) = \min\{bP_0(A) + c, 1\}, \quad \forall A \in \mathcal{A}(\mathbb{P}) \setminus \{\emptyset\} \quad (\overline{P}(\emptyset) = 0), \quad (5)$$

with a, b satisfying (3) and $c = 1 - (a + b) \geq 0$.

A VB-PMM offers very good consistency properties:

Proposition 2. *In a VB-PMM, $\underline{P}$ and $\overline{P}$ are coherent. Further, $\underline{P}$ is 2-monotone, $\overline{P}$ is 2-alternating.*

To justify the name and significance of a VB-PMM, take its upper probability $\overline{P}$, given by (5). Then:

(i) $\overline{P}(A) \geq P_0(A), \forall A$. Obvious when $\overline{P}(A) = 1$ or $A = \emptyset$; otherwise, use Definition 4 to get $\overline{P}(A) = bP_0(A) + 1 - (a+b) \geq P_0(A)$ iff $(1-b)P_0(A) \leq 1 - (a+b)$. If $1 - b \leq 0$, this inequality holds trivially $(1 - (a + b) \geq 0)$; if $1 - b > 0$, it is equivalent to $P_0(A) \leq \frac{1-b-a}{1-b}$, true because $\frac{1-b-a}{1-b} \geq 1$ for $a \leq 0$;

(ii) $\overline{P}(A) \to c \geq 0$ as $P_0(A) \to 0$ (for $P_0(A)$ low enough, $\overline{P}(A) = bP_0(A) + c \to c$);

(iii) $\overline{P}(A) = 1$ iff $P_0(A) \geq \frac{1-c}{b} = \frac{b+a}{b}$.

Now compare $\overline{P}$ with its special case $c = 0$, i.e., $a + b = 1$ (and $b > 1$), which specialises $\overline{P}$ into $\overline{P}_{\mathrm{PMM}}(A) = \min\{bP_0(A), 1\}$. In the behavioural interpretation, both (a generic) $\overline{P}$ and $\overline{P}_{\mathrm{PMM}}$ imply that the agent is essentially unwilling to sell events whose reference or 'true' probability P_0 is too high, by (iii), and in any case her/his selling price is not less than the 'fair' price P_0, by (i). $\overline{P}$ adds a further barrier regarding low probability events: by (ii), if $c > 0$ the agent is not willing to sell (too) low probability events for less than c, whilst $\overline{P}_{\mathrm{PMM}}$ enforces no such barrier. We may deduce that, *ceteris paribus*, the $\overline{P}$-agent is, loosely speaking, greedier than the $\overline{P}_{\mathrm{PMM}}$-agent. This can be easily justified in real-world situations: if the agent is a bookmaker or an insurer, for instance, $c > 0$ may take account of the agent's fixed costs in managing any bet/contract.

While c measures the agent's advantage at $P_0 = 0$, b determines how it varies with P_0 growing. In fact, the advantage is unchanged, decreasing or increasing according to whether it is, respectively, $b = 1$, $b < 1$, $b > 1$.

These features of the VB-PMM can be visualised in a $(P_0, \overline{P})$ plot, as in Fig. 1, 1): the VB-PMM additional barrier is the dotted segment on the $\overline{P}$-axis. The interpretation of $\underline{P}$, defined by (4), is similar. It is easy to check that:

(i') $\underline{P}(A) \leq P_0(A)$, $\forall A$;
(ii') $\underline{P}(A) \to a + b \leq 1$ as $P_0(A) \to 1$;
(iii') $\underline{P}(A) = 0$ iff $P_0(A) \leq -\frac{a}{b}$.

Now the agent using $\underline{P}$ acts as a buyer, but by (ii') does not want to pay more than $a + b$ for any event, even those whose probability P_0 is very high. If $a + b < 1$, this amounts to requiring that the maximum gain $\underline{G}_{\text{MAX}}$ from buying any A for $\underline{P}(A)$ (achieved when A occurs) is $1 - (a+b) > 0$. By contrast, $\underline{G}_{\text{MAX}} \to 0$ as $P_0(A) \to 1$ if $a + b = 1$, as in the PMM. Thus, $\underline{P}$ in the typical VB-PMM (i.e., such that $a + b < 1$) introduces an additional barrier, of width $1 - (a + b)$, with respect to $\underline{P}_{\text{PMM}}$: the dotted segment in the $P_0 = 1$ line of Fig. 1, 1).

5   The Horizontal Barrier Pari-Mutuel Model

Let now μ be a NL(a, b) measure, with the conditions

(k) $a + b > 1$, $2a + b < 1$.

Note that conditions (k) imply $a < 0$, $b > 1$. It can be shown that

Proposition 3. μ *is a 2-coherent lower probability, whilst it is not a 2-coherent upper probability.*

From Proposition 3, and by the maximum consistency principle stated at the end of Sect. 3, μ is conveniently viewed as a lower probability. We define then:

Definition 5. *A Horizontal Barrier Pari-Mutuel Model (HB-PMM) is a NL model where $\underline{P}$ and its conjugate $\overline{P}$ satisfy $\underline{P}(\emptyset) = \overline{P}(\emptyset) = 0$, $\underline{P}(\Omega) = \overline{P}(\Omega) = 1$, $c = 1 - (a + b) < 0$, a, b are as in $(k)^3$ and, for all $A \in \mathcal{A}(\mathbb{P}) \setminus \{\emptyset, \Omega\}$,*

$$\underline{P}(A) = \min\{\max\{bP_0(A) + a, 0\}, 1\}, \tag{6}$$

$$\overline{P}(A) = \max\{\min\{bP_0(A) + c, 1\}, 0\}. \tag{7}$$

Let us discuss the basic features of this model, referring to $\underline{P}$ given by (6). It is easy to check that (in particular, (jj) and (jjj) follow simply from Definition 3):

(j) $\underline{P}(A) > P_0(A)$ iff $1 > P_0(A) > -\frac{a}{b-1}$;
(jj) $\underline{P}(A) = 0$ iff $P_0(A) \leq -\frac{a}{b}$;
(jjj) $\underline{P}(A) = 1$ iff $P_0(A) \geq \frac{1-a}{b}$.

It follows easily from (k) that conditions (j), (jj), (jjj) are not vacuous, i.e., may be satisfied by some events. As for (j), for instance, $-\frac{a}{b-1} < 1$ iff $-a < b - 1$ (by (k)) iff $a + b > 1$, which is true, and it is always $-\frac{a}{b-1} > 0$.

Conditions (j), (jj), (jjj) point out an interesting feature of the HB-PMM: the beliefs it represents may be conflicting and, partly, irrational. In fact, assuming again that P_0 is the 'true' probability for the events in $\mathcal{A}(\mathbb{P})$, by (j) the agent

---

^{3} $a + b > 1$ in (k) could be relaxed to $a + b \geq 1$, thus including the PMM as a special HB-PMM. We left out this case to focus on the 'proper' HB-PMMs.

is willing to buy some events for less, others for more than their probability P_0. In the extreme situations, by (jj) and (jjj), the agent would not buy events whose probability is too low, whilst would certainly buy a high probability event A at the price of 1, gaining from the transaction at most 0 (if A occurs). Thus the agent underestimates the riskiness of a transaction regarding high probability events, but overestimates the risk with low probability events. She/he may be both risk averse and not. Which attitude prevails? In a sense, the prudential one. To see this, note that by (jj) and (jjj) the HB-PMM sets up two horizontal barriers in the $(P_0, \underline{P})$ plane (cf. Fig. 1, 2)). The lower (prudential) barrier is a segment with measure $-\frac{a}{b}$, the upper barrier (in the imprudent area) a segment measuring $1 - \frac{1-a}{b}$, and is narrower: $-\frac{a}{b} > 1 - \frac{1-a}{b}$ iff $2a + b < 1$, true by (k). Similarly, the boundary probability P_0 between the opposite attitudes is set at $-\frac{a}{b-1}$, larger than $\frac{1}{2}$ (by (k)). In this sense, the prudent behaviour prevails.

1)   2)

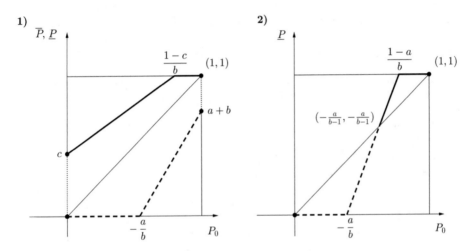

Fig. 1. Plots of $\underline{P}$ or $\overline{P}$ against P_0. (1) A VB-PMM $\overline{P}$ (continuous bold line) and a (non-conjugate) VB-PMM $\underline{P}$ (dashed bold). (2) A HB-PMM $\underline{P}$ (dashed bold line in the prudential part, continuous bold otherwise).

For $\overline{P}$, defined by (7), we can get to specular conclusions. Again the HB-PMM agent is subject to conflicting moods: she/he is unwilling to sell high probability events, but would give away for free low probability events. The lower barrier represents now the imprudent behaviour at its utmost degree, and is narrower than the upper barrier - that emphasising the cautious attitude.

Given this and Proposition 3, one would be tempted to conclude that the HB-PMM can be no more than 2-coherent. While this is true for the 'typical' HB-PMM model, coherence is compatible with some HB-PMM. Even more, there are instances of some HB-PMM $\underline{P}$ (or also $\overline{P}$) which are (0–1 valued) precise probabilities, as in the following example.

Example 2. Let $\mathbb{P} = \{\omega_1, \omega_2, \omega_3\}$, and define $\underline{P}$ by (6), with $a = -0.15$, $b = 1.25$ $(a, b$ comply with $(k))$. The starting probability P_0 takes the following values on $\mathbb{P}$: $P_0(\omega_1) = P_0(\omega_2) = 0.02$, $P_0(\omega_3) = 0.96$. The resulting HB-PMM lower probability $\underline{P}$ is 0–1 valued: $\underline{P}(A) = 0$ if $A \in \{\emptyset, \omega_1, \omega_2, \omega_1 \vee \omega_2\}$, $\underline{P}(A) = 1$ otherwise. Clearly, $\underline{P}$ is a probability on $\mathcal{A}(\mathbb{P})$.

However, these instances are more an exception, rather than the rule. As for $\underline{P}$ being a precise probability, it holds that:

Proposition 4. *If $\underline{P}$ in the HB-PMM is a precise probability, then it is necessarily 0–1 valued. Conversely, if $\underline{P}$ is 0–1 valued, it may be a probability or a lower probability, coherent or only 2-coherent.*

Coherence of $\underline{P}$, or of $\overline{P}$, is subject to rather restrictive conditions. To see this, we suppose in the next two results that $\mathbb{P}$ is finite. We refer to an upper probability $\overline{P}$, because the conditions for coherence are more straightforwardly described in this case. We present first necessary conditions for $\overline{P}$ to be subadditive, which on its turn is necessary for coherence of $\overline{P}$ (Sect. 2, $(c1)$), then state (Proposition 6) that subadditivity alone is also sufficient for coherence of a HB-PMM $\overline{P}$.

Proposition 5. *Let $\overline{P} : \mathcal{A}(\mathbb{P}) \to \mathbb{R}$ be defined by (7), $\mathbb{P}$ finite. Suppose $\overline{P}$ is subadditive. Then (referring to the sets $\mathcal{E}$, $\mathcal{N}$ of Definition 3), for $A \in \mathcal{A}(\mathbb{P})$:*

(a) $A \in \mathcal{E}$ iff $A = \omega^* \vee \bigvee_{h=1}^{k} \omega_{i_h}$, with $\omega^* \in \mathbb{P} \cap \mathcal{E}$, $\omega_{i_h} \in \mathbb{P} \cap \mathcal{N}$, $P_0(\omega_{i_h}) = 0$, $h = 1, \ldots, k$, $k \in \mathbb{N}$;
(b) $A \in \mathcal{N}$ iff $A = \bigvee_{h=1}^{k} \omega_{i_h}$, with $\omega_{i_h} \in \mathbb{P} \cap \mathcal{N}$, $h = 1, \ldots, k$, $k \in \mathbb{N}$;
(c) if $A \in \mathcal{E}$, then $\overline{P}(A) = \overline{P}(\omega^*)$, with $\omega^* \in \mathbb{P} \cap \mathcal{E}$, $\omega^* \Rightarrow A$.

By Proposition 5(a), (c), if $\overline{P}$ is coherent (hence subadditive), its value on any event A in $\mathcal{E}$ is that of the one atom in $\mathbb{P}$, among those implying A, that belongs to $\mathcal{E}$. Put differently, $\overline{P}(A)$ depends on a single atom only, ω^*. It ensues that, on the whole $\mathcal{A}(\mathbb{P})$, $\overline{P}$ may take up at most $n + 2$ distinct values including 0 and 1, if $|\mathbb{P}| = n$. Clearly, these are severe constraints. As for subadditivity, it holds that

Proposition 6. *Let $\overline{P} : \mathcal{A}(\mathbb{P}) \to \mathbb{R}$ be defined by (7), $\mathbb{P}$ finite. Then $\overline{P}$ is coherent iff it is subadditive.*

A corresponding condition for $\underline{P}$ is less immediate, since superadditivity is not the conjugate property of subadditivity. However, superadditivity is necessary for 2-coherence of $\underline{P}$ in the HB-PMM, whatever the cardinality of $\mathbb{P}$:

Proposition 7. *Let $\underline{P} : \mathcal{A}(\mathbb{P}) \to \mathbb{R}$ be defined by (6), with $\mathbb{P}$ arbitrary (finite or not). Then $\underline{P}$ is superadditive (i.e., satisfies $(c2)$ in Sect. 2).*

6   Similar Models

Despite the simplicity of NL models, there are not so many similar or partly overlapping models in the literature, to the best of our knowledge.

In a paper focused on statistical robustness issues, Rieder [R] introduces a specific VB-PMM and proves the 2-monotonicity of $\underline{P}$. His model is a special case of ours, since he requires (using our parametrisation) conditions (3), and the extra condition $a \geq -1$.

Neo-additive capacities, introduced in [CEG], are somewhat similar to NL models, because $\mu(A) = bP_0(A) + a$ there, when $A \in \mathcal{E}$. Yet, the approach is radically different: the sets $\mathcal{N}, \mathcal{E}, \mathcal{U}$ are fixed a priori, and it is required that $A \in \mathcal{N}$ iff $\neg A \in \mathcal{U}$. This condition is unduly restrictive, in our view, for measures that are not precise probabilities. It is usually not met by NL models, not even by the PMM (just think that if $\mu = \underline{P}_{\text{PMM}}$, then $\mathcal{U} = \{\Omega\}$, while generally $\mathcal{N} = \{A \in \mathcal{A}(\mathbb{P}) : P_0(A) \leq -\frac{a}{b}\}$ is larger than $\{\emptyset\}$). Further, μ is only required to be a capacity, while our models ensure at least 2-coherence. Interestingly, neo-additive capacities were introduced to describe both optimistic and pessimistic attitudes towards uncertainty at the same time. This is similar to the agent's waving attitude towards risky contracts expressed by the HB-PMM.

7   Conclusions

In this paper we introduced two models of imprecise probabilities, both generalising the PMM, and studied their basic features and consistency properties. While the VB-PMM is always coherent, the HB-PMM is generally not, but may formalise a conflicting behaviour of the agent towards risk. Further work is needed to complete the analysis of NL models and to explore their connections with other models, for instance probability intervals that were shown to be closely related with the PMM in [MMD]. We also plan to study conditioning with the VB-PMM, its natural extension and relationships with risk measures; this should generalise the analogous work in [PVZ] for the PMM.

References

[CEG] Chateauneuf, A., Eichberger, J., Grant, S.: Choice under uncertainty with the best and worst in mind: neo-additive capacities. J. Econ. Theory **137**, 538–567 (2007)

[MMD] Montes, I., Miranda, E., Destercke, S.: A study of the Pari-Mutuel Model from the point of view of imprecise probabilities. In: Proceedings of the ISIPTA 2017 (2017)

[PV] Pelessoni, R., Vicig, P.: 2-coherent and 2-convex conditional lower previsions. Int. J. Approx. Reason. **77**, 66–86 (2016)

[PVZ] Pelessoni, R., Vicig, P., Zaffalon, M.: Inference and risk measurement with the Pari-Mutuel Model. Int. J. Approx. Reason. **51**, 1145–1158 (2010)

[R] Rieder, H.: Least favourable pairs for special capacities. Annal. Stat. **5**(3), 909–921 (1977)

[W] Walley, P.: Statistical Reasoning with Imprecise Probabilities. Chapman & Hall, London (1991)

A Net Premium Model for Life Insurance Under a Sort of Generalized Uncertain Interest Rates

Dabuxilatu Wang$^{(\boxtimes)}$

Department of Statistics, School of Economics and Statistics, Guangzhou University,
No. 230 Waihuanxi Road, Higher Education Mega Center, Guangzhou 510006,
People's Republic of China
wangdabu@gzhu.edu.cn

Abstract. In this paper, we apply the LR-fuzzy random variables to estimate a discount function that associated with a generalized interest rate as well as fuzzy random variable future lifetimes, and establish some life annuity and endowments models. A novel fuzzy net premium model is obtained. Finally, a statistical simulation is given for illustrating the proposed models.

Keywords: Fuzzy discount function · Fuzzy random variables
Net premium model

1 Introduction

It was well known that the rates of interest were assumed to be constants in the theory of classical life insurance actuarial [6]. However, the interest rates are uncertain in the real life world. The most studies on the uncertainties of the interest rates focused on their randomness. A number of models for random rates of interest have been established. For example, Beekman [5] obtained that the force of interest is a moment of secondary order of the annuities present values under Ornstain-Uhlenbeck process. Schepper [12] obtained that the interest force is the moment generating function of the annuity under the Winner process. Zaks [13] modeled the random rates of interest with Gaussian process and the reflected Winner process. Some studies focused on the subjective factors related to the variation of the interest rates. For example, in recent years, the fuzzy sets approaches have been paid much attention in the modeling of the rates of interest. Andrés-Sánchez [1,3,4] has studied the term structure of interest rates as well as the claim reserves by using fuzzy regression techniques, and some fuzzy term structure of interest rates and fuzzy discount functions were proposed. Thus, it seems quite reasonable that the variation of interest rates are not only heavily influenced by the complex, dynamic and varying economic and social factors, but also recognized as perceived potential fuzzy metric based on the human's perception, subjective judgment on the economic and social politic

© Springer Nature Switzerland AG 2019
S. Destercke et al. (Eds.): SMPS 2018, AISC 832, pp. 224–232, 2019.
https://doi.org/10.1007/978-3-319-97547-4_29

situation. A combination of both stochastic methods and fuzzy sets methods for modeling the behaviors of the interest rates is expected to be developed. Fuzzy random variables [9, 10] have been taken an important role in the modeling. Andrés-Sánchez [2] primarily consider the problem of establishing the fuzzy life annuity through random mortality and fuzzy interest rate structure. Shapiro [10, 11] introduce the concept of fuzzy random variables into the area of insurance and propose to model the future life time as a fuzzy random variable, with which the work of Andrés-Sánchez [2] has been improved.

This article proceeds as follows. In Sect. 1, we introduce some background information on the modeling of the rates of interest. In Sect. 2, some preliminaries about the concepts of fuzzy random variables are introduced. In Sect. 3, a fuzzy random discount function is defined based on considering a generalized interest rate that concerned with an investment returns rate. In Sect. 4, some life insurance models with fuzzy random interest rates as well as fuzzy random variable future lifetimes are established. In Sect. 5, a statistical simulation example is presented for illustrating the proposed models. In Sect. 6, a conclusion of the article is given.

2 Preliminaries

2.1 Fuzzy Numbers

Let $\mathbb{R}$ be the set of all real numbers. A fuzzy set on $\mathbb{R}$ is defined to be a mapping $\tilde{u} : \mathbb{R} \to [0, 1]$ satisfying following conditions: (1) $\tilde{u}_\alpha = \{x|\tilde{u}(x) \geq \alpha\}$ is a closed bounded interval for each $\alpha \in (0, 1]$, i.e. $u_\alpha = [\inf \tilde{u}_\alpha, \sup \tilde{u}_\alpha]$. (2) $\tilde{u}_0 = supp\tilde{u}$ is a closed bounded interval. (3) $\tilde{u}_1 = \{x|\tilde{u}(x) = 1\}$ is nonempty. where $supp\tilde{u} = cl\{x|u(x) > 0\}$, cl denotes the *closure* of a set. Such a fuzzy set is also called a *fuzzy number*. By $\mathcal{F}(\mathbb{R})$ we denote the set of all fuzzy numbers, the arithmetic operation $*$ on $\mathcal{F}(\mathbb{R})$ can be defined by $(\tilde{u} * \tilde{v})(t) = \sup_{\{t_1, t_2 : t_1 * t_2 = t\}} \{\min(\tilde{u}(t_1), \tilde{v}(t_2))\}$, $\tilde{u}, \tilde{v} \in \mathcal{F}(\mathbb{R}), t, t_1, t_2 \in \mathbb{R}, * \in \{+, -, \cdot\}$, where $+, -, \cdot$ denote the addition, subtraction and scalar multiplication among fuzzy numbers, respectively. The following parametric class of fuzzy numbers, the so-called *LR*-fuzzy numbers, are often used in applications:

$$\tilde{u}(x) = \begin{cases} L(\frac{m-x}{l}), & x \leq m \\ R(\frac{x-m}{r}), & x > m \end{cases}$$

Here $L : \mathbb{R}^+ \to [0, 1]$ and $R : \mathbb{R}^+ \to [0, 1]$ are given left- continuous and non-increasing function with $L(0) = R(0) = 1$. L and R are called left and right shape functions, m the central point of $\tilde{u}$ and $l > 0$, $r > 0$ are the left and right spread of $\tilde{u}$. An *LR*-fuzzy number is abbreviated by $\tilde{u} = (m, l, r)_{LR}$. An *LR*-fuzzy number is said to be symmetric if $L(x) = R(x)$ and $l = r$. The triangular fuzzy number is a special case where $L(x) = R(x) = \max\{0, 1 - x\}$. It has been proven that: $(m_1, l_1, r_1)_{LR} + (m_2, l_2, r_2)_{LR} = (m_1 + m_2, l_1 + l_2, r_1 + r_2)_{LR}$,

$$a(m, l, r)_{LR} = \begin{cases} (am, al, ar)_{LR}, & a > 0 \\ (am, -ar, -al)_{RL}, & a < 0 \\ (0, 0, 0), & a = 0 \end{cases}$$

Let $L(\alpha) := \sup\{x \in \mathbb{R}|L(x) \geq \alpha\}, R(\alpha) := \sup\{x \in \mathbb{R}|R(x) \geq \alpha\}$. Then for $\tilde{u} = (m, l, r)_{LR}, \tilde{u}_\alpha = [m - lL(\alpha), m + rR(\alpha)], \alpha \in [0, 1]$.

The distance between $\tilde{u}, \tilde{v} \in F(\mathbb{R})$ defined by the metric δ_2 is given as follows ([7,8]). $\delta_2^2(\tilde{u}, \tilde{v}) := \frac{1}{2} \int_0^1 [(\sup \tilde{u}_\alpha - \sup \tilde{v}_\alpha)^2 + (\inf \tilde{v}_\alpha - \inf \tilde{u}_\alpha)^2] d\alpha$.

2.2   Fuzzy Random Variables

The concept of FRVs is inspired by the attempt to model the randomness and fuzziness that existed in real life phenomenon simultaneously. In practice it can be perceived as a random variable taking values of linguistic data or fuzzy number. Theoretically, one of the definition of FRVs is given by Puri and Ralescu as follows.

Definition 2.1 *(Fuzzy random variables (FRVs) [9]). Let $(\Omega, \mathcal{A}, P)$ be a complete probability space. The mapping $\tilde{X} : \Omega \to F(\mathbb{R})$ is said to be a FRV if $\tilde{X}$ is $\mathcal{A}-\mathcal{B}$ measurable, i.e. for any measurable subset $B \subset \mathbb{R}, \{\omega|\tilde{X}_\alpha(\omega) \cap B \neq \emptyset\} \in \mathcal{A}$, where $\mathcal{B}$ is a σ-algebra on $\mathbb{R}$ induced by $\tilde{X}$ associated with the concerned metric, and $\tilde{X}_\alpha(\omega) := \{x \in \mathbb{R}|\tilde{X}(\omega) \geqslant \alpha\} = [\tilde{X}(\omega)]_\alpha = [\inf(\tilde{X}(\omega))_\alpha, \sup(\tilde{X}(\omega))_\alpha], \alpha \in [0, 1], \omega \in \Omega$.*

An *LR-fuzzy random variable* (*LR*-FRV [8]) on the probability space $(\Omega, \mathcal{A}, P)$ is defined as a measurable mapping $\tilde{X} : \Omega \to \mathcal{F}_{LR}(\mathbb{R}), \tilde{X}(\omega) = (m_x(\omega), l_x(\omega), r_x(\omega))_{LR}, \omega \in \Omega$, in brief we denote $\tilde{X}$ as $\tilde{X} = (m_x, l_x, r_x)_{LR}$, where m_x, l_x, r_x are three real-valued random variables with $P\{l_x \geq 0\} = P\{r_x \geq 0\} = 1$.

Definition 2.2 *(Expectation [9]). The expectation $E\tilde{X}$ of a FRV $\tilde{X}$ is a fuzzy number with the property that $(E\tilde{X})_\alpha = E\tilde{X}_\alpha = [E\inf \tilde{X}_\alpha, E\sup \tilde{X}_\alpha], \alpha \in [0, 1]$.*

Definition 2.3 *(Variance [7,8]). The variance of a FRV $\tilde{X}$ is defined by $Var(\tilde{X}) = \frac{1}{2} \int_0^1 (Var \sup \tilde{X}_\alpha + Var \inf \tilde{X}_\alpha) d\alpha$.*

3   Fuzzy Random Discount Function

In the literature, the random interest rates are modeled with stochastic processes like Gaussian process or Winner process or Possion process $\{\delta t + X(t)\}$, where δ denotes a risk-free interest intensity, and $X(t)$ represents some stochastic process. A present value of a life annuity under the random interest rates simply denoted by $Z = \int_0^{T(x)} e^{-(\delta t + X(t))} dt$, where $T(x)$ is a random variable future lifetime of a life aged x. Here $e^{-(\delta t + X(t))}$ can be viewed as a random discount function.

As that pointed out in the introduction part, the change of an interest rate might be modeled in a way of a combination of the stochastic methods with fuzzy methods. Shapiro [10,11] suggested that the uncertainty with respect to the interest rate is not only due to randomness, but also duo to fuzziness. In fact, we may aware of that an alternation of an interest rate is usually implemented by the management control actions taken by the central bank's decision makers, they expect the finance status is in a normal condition according to the finance rules. Thus, the random nature of the interest rates to a large extent is due to the economic and politic factors, and the fuzzy nature is due to the experts perceive levels on the reality of the economic and politic condition.

In the life insurance issues, random variable future life time $T(x)$ is applied and the actuarial present values or premium for an life aged x for a policy can be carried out. However, human's future life time has many influence factors other than age x. For example, the current health status, the healthy habit, the family history, etc. In a practice of post-retirement financial planning advice, investigators often suggest to describe the future life time with linguistic values like "long future life time", "medium future life time" and "short future life time" etc. Shapiro [10,11] proposed to model the future lifetime as a fuzzy random variable $\tilde{T}(x)$, i.e. a fuzzy number valued random variable defined on the collection of all life aged x.

Though the theory of fuzzy set valued stochastic processes has only been in its infancy, based on which we are abuse of notion to propose a sort of concept of a fuzzy random discount function.

Definition 3.1. *Let $\{\tilde{X}(t)\}$ be a stochastic process consists of FRVs $\tilde{X}(t)$ defined on probability space $(\Omega, \mathcal{A}, P)$, where t is a time variable. The fuzzy random discount function is defined as a stochastic process $\{d(\tilde{X}(t))\}$ consists of FRVs $d(\tilde{X}(t))$ whose α-cuts satisfying*

$$d_\alpha(\tilde{X}(t)) = \left[\min\{e^{-\inf \tilde{X}_\alpha(t)}, e^{-\sup \tilde{X}_\alpha(t)}\}, \max\{e^{-\inf \tilde{X}_\alpha(t)}, e^{-\sup \tilde{X}_\alpha(t)}\} \right],$$

$\alpha \in [0,1]$, *where both $\{\inf \tilde{X}_\alpha(t)\}, \{\sup \tilde{X}_\alpha(t)\}$ are same kinds of stochastic processes, and $e^{-\inf \tilde{X}_\alpha(t)}, e^{-\sup \tilde{X}_\alpha(t)}$ are simple random discount functions.*

In the Definition 3.1, the stochastic process $\{\tilde{X}(t)\}$ can be viewed as a fuzzy random interest rate. In a capital investment case, the investors' expected future return rates from the investment may be viewed as such a stochastic process. The fuzzy random discount function actually is a random discount function taking fuzzy values. The computation of a fuzzy random discount function is subject to the given membership functions of each $\tilde{X}(t)$ in $\{\tilde{X}(t)\}$, the α-cuts of a fuzzy set as well as computational methods of the random discount function.

Now the notion of the present value of a life annuity can be extended to an entirely case where a fuzzy random variable future life time [11] is combined with a fuzzy random discount function.

Definition 3.2. *Let $\tilde{T}(x)$ be a fuzzy random variable future life time [11] of a life aged x, and let $\{d(\tilde{X}(t))\}$ be a fuzzy random discount function defined in Definition 3.1. The present value $\tilde{Z}$ of a life annuity under a combination of fuzzy random variable future life time with fuzzy random discount function is defined as a fuzzy random variable whose α-cuts satisfying $\tilde{Z}_\alpha = \left[\inf \tilde{Z}_\alpha, \sup \tilde{Z}_\alpha \right]$, where*

$$\inf \tilde{Z}_\alpha = \min \left\{ \int_0^{\inf \tilde{T}_\alpha} e^{-\inf \tilde{X}_\alpha(t)} dt, \int_0^{\sup \tilde{T}_\alpha} e^{-\sup \tilde{X}_\alpha(t)} dt \right\},$$

$$\sup \tilde{Z}_\alpha = \max \left\{ \int_0^{\inf \tilde{T}_\alpha} e^{-\inf \tilde{X}_\alpha(t)} dt, \int_0^{\sup \tilde{T}_\alpha} e^{-\sup \tilde{X}_\alpha(t)} dt \right\}.$$

Based on the Definition 3.2, we are able to compute the FRV present value for a life annuity under the given fuzzy random discount function and FRV future life time. In a previous work, the fuzzy number values taken by the fuzzy random variable future life time has been initially considered by Shapiro [11], where the values like short lifetime, medium lifetime and long lifetime represented by some triangular fuzzy numbers are introduced.

In virtue of Definition 3.2, we now suggest to use a generalized triangular fuzzy number, the LR fuzzy numbers as the values taken by the both FRVs $\tilde{T}(x)$ and $\tilde{X}(t)$ aforementioned.

4   Some Life Insurance Models Under FRV Future Lifetime and Fuzzy Random Discount Functions

4.1   An Estimation of the Fuzzy Random Discount Functions

It was well known that the discount function is generated for evaluating a present value of a future uncertain asset, and it has been applied widespread in economic and financial analysis. In virtue of Definition 3.1, the discount function has been extended to a much complicated case. For instance, in a very complicated economic environment, an investment based insurance policy planning ought to consider the impact from the uncertainties of the return rates, and thus the interest rate could be a generalized one, which possesses both stochastic and fuzzy properties. In all of the case, we need to carried out an actual fuzzy random discount function. To facilitate the computation, we assume that the fuzzy random interest rate (generalized interest rate) can be estimated by LR-FRVs. Under the term structure of the interest rates, the tth time interest rate is assumed to be estimated by LR-FRV $(m_i, l_i, r_i)_{LR}(t)$, and the fuzzy random discount function can be estimated by $d((m_i, l_i, r_i)_{LR}(t))$ under Definition 3.1. The FRV future lifetime is assumed to be estimated by LR-FRV $(T, l_T, r_T)_{LR}$.

4.2   The Actuarial Present Value of a Life Annuity

(1) For a continuous life annuity, the actuarial present value is defined by $E\tilde{Z}$, and denoted by $\tilde{a}$, whose α-cuts satisfying $\tilde{a}_\alpha = \left[\inf \tilde{a}_\alpha, \sup \tilde{a}_\alpha \right]$, where

$$\inf \tilde{a}_\alpha = \min \left\{ \int_0^{T-L(\alpha)l_T} Ee^{-(m_i - L(\alpha)l_i)(t)} dt, \int_0^{T+R(\alpha)r_T} Ee^{-(m_i + R(\alpha)r_i)(t)} dt \right\},$$

$$\sup \tilde{a}_\alpha = \max \left\{ \int_0^{T-L(\alpha)l_T} Ee^{-(m_i - L(\alpha)l_i)(t)} dt, \int_0^{T+R(\alpha)r_T} Ee^{-(m_i + R(\alpha)r_i)(t)} dt \right\}.$$

(2) For the n-year temporary life annuity, the actuarial present value is defined as $E\tilde{Z}$, and denoted by $\ddot{\tilde{a}}$, whose α-cuts satisfying $\ddot{\tilde{a}}_\alpha = \left[\inf \ddot{\tilde{a}}_\alpha, \sup \ddot{\tilde{a}}_\alpha\right]$, where

$$\inf \ddot{\tilde{a}}_\alpha = \min\left\{\sum_{t=0}^{n-1} Ee^{-(m_i-L(\alpha)l_i)(t)}{}_{t+L(\alpha)l_K}p_x, \sum_{t=0}^{n-1} Ee^{-(m_i+R(\alpha)r_i)(t)}{}_{t-R(\alpha)r_K}p_x\right\},$$

$$\sup \ddot{\tilde{a}}_\alpha = \max\left\{\sum_{t=0}^{n-1} Ee^{-(m_i-L(\alpha)l_i)(t)}{}_{t+L(\alpha)l_K}p_x, \sum_{t=0}^{n-1} Ee^{-(m_i+R(\alpha)r_i)(t)}{}_{t-R(\alpha)r_K}p_x\right\},$$

where K denotes the integer part of the T, and the FRV integer future lifetim taken as $\tilde{K} = (K, l_K, r_K)_{LR}$.

4.3   The Actuarial Present Value of a Endowments of Duration n

We are abuse of notion to denote the actuarial present value of a endowments of duration n as $\tilde{A}_{x:\overline{n}|}$, whose α-cut satisfying

$$\inf(\tilde{A}_{x:\overline{n}|})_\alpha = \min\left\{\sum_{t=1}^{n} Ee^{-(m_i-L(\alpha)l_i)(t)}P(t \leqslant K - L(\alpha)l_K \leqslant t+1),\right.$$

$$\left.\sum_{t=1}^{n} Ee^{-(m_i+R(\alpha)r_i)(t)}P(t \leqslant K + R(\alpha)r_K \leqslant t+1)\right\},$$

$$\sup(\tilde{A}_{x:\overline{n}|})_\alpha = \max\left\{\sum_{t=1}^{n} Ee^{-(m_i-L(\alpha)l_i)(t)}P(t \leqslant K - L(\alpha)l_K \leqslant t+1),\right.$$

$$\left.\sum_{t=1}^{n} Ee^{-(m_i+R(\alpha)r_i)(t)}P(t \leqslant K + R(\alpha)r_K \leqslant t+1)\right\}.$$

4.4   A Net Premium Model

In our case with respect to an insurance policy, we consider a fuzzy total loss L to the insurer to be the distance between the present value of the benefits and the present value of the premium payments, where the distance is given by δ_2 shown in Sect. 2. On each α level we assume that the equivalence principle $E(L) = 0$ can be applied. Then, we approximately have a net premium $\tilde{P}$ whose α-cut satisfying $\tilde{P}_\alpha = \dfrac{(\tilde{A}_{x:\overline{n}|})_\alpha}{(E\tilde{Z})_\alpha} = \left[\dfrac{\inf(\tilde{A}_{x:\overline{n}|})_\alpha}{\sup \ddot{\tilde{a}}_\alpha}, \dfrac{\sup(\tilde{A}_{x:\overline{n}|})_\alpha}{\inf \ddot{\tilde{a}}_\alpha}\right]$.

5   A Statistical Simulation Example

Consider a life aged 60 who insure a 10 year term endowments for unit. Assume that his or her future lifetime can be modeled by a FRV taken as $(T, l_T, r_T)_{LR}$, where $L(x) = R(x) = \max\{0, 1-x\}$, T is the unknown random variable future lifetime follows approximately a life table CL93, and $l_T \sim U(0,2)$, $r_T \sim U(0,1)$, we approximate $l_T \approx El_T = 1$ and $r_T \approx Er_T = 0.5$. The interest rate is assumed to be a stochastic process $\{(m_i, l_i, r_i)_{LR}(t)\}$, for the simplicity, assume that

Table 1. Discount functions of terms and probabilities for the life annuity ($\alpha \in [0,1]$)

t	$d_\alpha = [\inf d_\alpha, \sup d_\alpha]$	$t-1+\alpha p_{60}$	$t+0.5-0.5\alpha p_{60}$
0	1	1	$0.5-0.5\alpha p_{60}$
1	$\left[\exp\{(0.055\alpha - 0.085)\}, \exp\{(0.01 - 0.04\alpha)\}\right]$	αp_{60}	$1.5-0.5\alpha p_{60}$
2	$\left[\exp\{2(0.055\alpha - 0.085)\}, \exp\{2(0.01 - 0.04\alpha)\}\right]$	$1+\alpha p_{60}$	$2.5-0.5\alpha p_{60}$
3	$\left[\exp\{3(0.055\alpha - 0.085)\}, \exp\{3(0.01 - 0.04\alpha)\}\right]$	$2+\alpha p_{60}$	$3.5-0.5\alpha p_{60}$
4	$\left[\exp\{4(0.055\alpha - 0.085)\}, \exp\{4(0.01 - 0.04\alpha)\}\right]$	$3+\alpha p_{60}$	$4.5-0.5\alpha p_{60}$
5	$\left[\exp\{5(0.055\alpha - 0.085)\}, \exp\{5(0.01 - 0.04\alpha)\}\right]$	$4+\alpha p_{60}$	$5.5-0.5\alpha p_{60}$
6	$\left[\exp\{6(0.055\alpha - 0.085)\}, \exp\{6(0.01 - 0.04\alpha)\}\right]$	$5+\alpha p_{60}$	$6.5-0.5\alpha p_{60}$
7	$\left[\exp\{7(0.055\alpha - 0.085)\}, \exp\{7(0.01 - 0.04\alpha)\}\right]$	$6+\alpha p_{60}$	$7.5-0.5\alpha p_{60}$
8	$\left[\exp\{8(0.055\alpha - 0.085)\}, \exp\{8(0.01 - 0.04\alpha)\}\right]$	$7+\alpha p_{60}$	$8.5-0.5\alpha p_{60}$
9	$\left[\exp\{9(0.055\alpha - 0.085)\}, \exp\{9(0.01 - 0.04\alpha)\}\right]$	$8+\alpha p_{60}$	$9.5-0.5\alpha p_{60}$

Table 2. Discount functions of terms for the endowments ($\alpha \in [0,1]$)

t	$d_\alpha = [\inf d_\alpha, \sup d_\alpha]$
1	$\left[\exp\{(0.055\alpha - 0.085)\}, \exp\{(0.01 - 0.04\alpha)\}\right]$
2	$\left[\exp\{2(0.055\alpha - 0.085)\}, \exp\{2(0.01 - 0.04\alpha)\}\right]$
3	$\left[\exp\{3(0.055\alpha - 0.085)\}, \exp\{3(0.01 - 0.04\alpha)\}\right]$
4	$\left[\exp\{4(0.055\alpha - 0.085)\}, \exp\{4(0.01 - 0.04\alpha)\}\right]$
5	$\left[\exp\{5(0.055\alpha - 0.085)\}, \exp\{5(0.01 - 0.04\alpha)\}\right]$
6	$\left[\exp\{6(0.055\alpha - 0.085)\}, \exp\{6(0.01 - 0.04\alpha)\}\right]$
7	$\left[\exp\{7(0.055\alpha - 0.085)\}, \exp\{7(0.01 - 0.04\alpha)\}\right]$
8	$\left[\exp\{8(0.055\alpha - 0.085)\}, \exp\{8(0.01 - 0.04\alpha)\}\right]$
9	$\left[\exp\{9(0.055\alpha - 0.085)\}, \exp\{9(0.01 - 0.04\alpha)\}\right]$
10	$\left[\exp\{10(0.055\alpha - 0.085)\}, \exp\{10(0.01 - 0.04\alpha)\}\right]$
>10	$\left[\exp\{10(0.055\alpha - 0.085)\}, \exp\{10(0.01 - 0.04\alpha)\}\right]$

$m_i(t) \sim U(0,0.06)$, $l_i(t) \sim U(0.08)$ and $r_i(t) \sim U(0.011)$ and furthermore we approximate $(m_i, l_i, r_i)_{LR}(t) \approx t(Em_i, El_i, Er_i)_{LR} = t(0.03, 0.04, 0.055)_{LR}$. The values of the discount functions of each terms as well as the survive probabilities for the life annuity can be carried out as shown in Table 1. The values of the discount functions of each terms as well as the survive probabilities for the endowments can be carried out as shown in Tables 2 and 3.

For each fixed parameter α, based on the life table the exact results data in Tables 1, 2 and 3 can be carried out, and then the net premium value can be obtained based on the formula of $\tilde{P}_\alpha$.

Table 3. Probabilities for the endowments $(\alpha \in [0, 1])$

t	$P(t+1-\alpha \leqslant K \leqslant t+2-\alpha)$	$P(t-0.5+0.5\alpha \leqslant K \leqslant t+0.5+0.5\alpha)$
1	$P(2-\alpha \leqslant K \leqslant 3-\alpha)$	$P(0.5+0.5\alpha \leqslant K \leqslant 1.5+0.5\alpha)$
2	$P(3-\alpha \leqslant K \leqslant 4-\alpha)$	$P(1.5+0.5\alpha \leqslant K \leqslant 2.5+0.5\alpha)$
3	$P(4-\alpha \leqslant K \leqslant 5-\alpha)$	$P(2.5+0.5\alpha \leqslant K \leqslant 3.5+0.5\alpha)$
4	$P(5-\alpha \leqslant K \leqslant 6-\alpha)$	$P(3.5+0.5\alpha \leqslant K \leqslant 4.5+0.5\alpha)$
5	$P(6-\alpha \leqslant K \leqslant 7-\alpha)$	$P(4.5+0.5\alpha \leqslant K \leqslant 5.5+0.5\alpha)$
6	$P(7-\alpha \leqslant K \leqslant 8-\alpha)$	$P(5.5+0.5\alpha \leqslant K \leqslant 6.5+0.5\alpha)$
7	$P(8-\alpha \leqslant K \leqslant 9-\alpha)$	$P(6.5+0.5\alpha \leqslant K \leqslant 7.5+0.5\alpha)$
8	$P(9-\alpha \leqslant K \leqslant 10-\alpha)$	$P(7.5+0.5\alpha \leqslant K \leqslant 8.5+0.5\alpha)$
9	$P(10-\alpha \leqslant K \leqslant 11-\alpha)$	$P(8.5+0.5\alpha \leqslant K \leqslant 9.5+0.5\alpha)$
10	$P(11-\alpha \leqslant K \leqslant 12-\alpha)$	$P(9.5+0.5\alpha \leqslant K \leqslant 10.5+0.5\alpha)$
>10	$P(11-\alpha \leqslant K \leqslant 12-\alpha)$	$P(9.5+0.5\alpha \leqslant K \leqslant 10.5+0.5\alpha)$

6   Conclusions

We have proposed a new discount function, and have primarily extended the actuarial models of a life annuity, endowments and the net premium to the case of FRV interest rates and future lifetimes. The established models are more suitable to the practical case than the classical ones. The future work is to deepen fuzzy random actuarial theory in a framework of the stochastic fuzzy systems.

Acknowledgment. This research is supported by the NNSF of China under grant number No.11271096. Their financial aid is greatly appreciated.

References

1. Andrés-Sánchez, J., Terceõ, A.: Applications of fuzzy regression in actuarial analysis. J. Risk Insur. **70**, 665–699 (2003)
2. Andrés-Sánchez, J., Gonzalez, L.: Using fuzzy random variable in life annuities pricing. Fuzzy Sets Syst. **188**(1), 27–44 (2012)
3. Andrés-Sánchez, J.: Calculating insurance claim reserves with fuzzy regression. Fuzzy Sets Syst. **157**, 3091–3108 (2006)
4. Andrés-Sánchez, J., Terceõ-Cómez, A.: Estimating a fuzzy term structure of interest rates using fuzzy regression techniques. Eur. J. Oper. Res. **154**, 804–818 (2004)
5. Bekkman, J.A., Fuelling, C.P.: Interest and mortality ranomness in some annuities. Insur.: Math. Econ. **9**(2), 185–196 (1990)
6. Bowers, N.L., Gerber, H.U., Hickman, J.C., Jones, D.A., Nesbitt, C.J.: Actuarial Mathematics, 2nd edn. The Society of Actuares, Schaumburg (1997)
7. Diamond, P., Kloeden, P.: Metric Spaces of Fuzzy Sets. World Scientific, Singapore (1994)
8. Näther, W.: Regression with fuzzy random data. Comput. Stat. Data Anal. **51**, 235–252 (2006)

232   D. Wang

9. Puri, M.L., Ralescu, D.A.: Fuzzy random variables. J. Math. Anal. Appl. **114**, 409–422 (1986)
10. Shapiro, A.F.: Fuzzy random variables. Insur.: Math. Econ. **44**, 307–314 (2009)
11. Shapiro, A.F.: Modeling future life time as a fuzzy random variable. Insur.: Math. Econ. **53**, 864–870 (2013)
12. Schepper, D.F., Vyher, D.F., Gooyerts, M.: Interest randomness in annuities certain. Insur.: Math. Econ. **11**(4), 271–281 (1992)
13. Zaks, A.: Annuities under random rates of interest. Insur. Math. Econ. **28**, 1–11 (2001)

Author Index

© Springer Nature Switzerland AG 2019
S. Destercke et al. (Eds.): SMPS 2018, AISC 832, pp. 233–234, 2019.
https://doi.org/10.1007/978-3-319-97547-4

Printed in the United States
By Bookmasters